Y0-BRL-969

STUDIES IN APPLIED MECHANICS 6

Mechanics of Material Behavior

STUDIES IN APPLIED MECHANICS

1. Mechanics and Strength of Materials (Skalmierski)

2. Nonlinear Differential Equations (Fučík and Kufner)

3. Mathematical Theory of Elastic and Elastico-Plastic Bodies
   An Introduction (Nečas and Hlaváček)

4. Variational, Incremental and Energy Methods
   in Solid Mechanics and Shell Theory (Mason)

5. Mechanics of Structured Media, Parts A and B (Selvadurai, Editor)

6. Mechanics of Material Behavior (Dvorak and Shield, Editors)

7. Mechanics of Granular Materials: New Models and Constitutive
   Relations (Jenkins and Satake, Editors)

STUDIES IN APPLIED MECHANICS 6

Mechanics of Material Behavior

The Daniel C. Drucker Anniversary Volume

Edited by

George J. Dvorak

Department of Civil Engineering, University of Utah, Salt Lake City, Utah, U.S.A.

and

Richard T. Shield

Department of Theoretical and Applied Mechanics, University of Illinois at Urbana-Champaign, Urbana, Illinois, U.S.A.

ELSEVIER

Amsterdam – Oxford – New York – Tokyo 1984

ELSEVIER SCIENCE PUBLISHERS B.V.
Molenwerf 1
P.O. Box 211, 1000 AE Amsterdam, The Netherlands

Distributors for the United States and Canada:

ELSEVIER SCIENCE PUBLISHING COMPANY INC.
52, Vanderbilt Avenue
New York, NY 10017, U.S.A.

ISBN 0-444-42169-6 (Vol. 6)
ISBN 0-444-41758-3 (Series)

Printed in The Netherlands

Preface

Daniel C. Drucker became sixty-five years old on June 3, 1983. This book is a collection of papers by friends and colleagues in grateful and deep appreciation of Dan's outstanding and permanent contributions to the mechanics of material behavior, particularly to the theory of plasticity and its application to the design of engineering structures and components. Most of the papers in this Anniversary Volume were presented at a Symposium on the Mechanics of Material Behavior which was held at the University of Illinois at Urbana-Champaign on June 6 and 7, 1983. Support for the Symposium was provided by the National Science Foundation, the Office of Naval Research and the U.S. Army Research Office.

We thank the contributors for their ready acceptance of the invitation to participate. We regret that limitations on the timing and size of the Volume did not allow all who wished to do so to contribute.

G.J. DVORAK R.T. SHIELD

Contributors

Bernard Budiansky
Harvard University
Cambridge, Massachusetts

W.F. Chen
Purdue University
West Lafayette, Indiana

W.J. Drugan
University of Wisconsin
Madison, Wisconsin

G.J. Dvorak
University of Utah
Salt Lake City, Utah

J. Duffy
Brown University
Providence, Rhode Island

P.D. Griffin
University of Cape Town
Rondebosch, South Africa

W.K. Ho
University of California at Los Angeles
Los Angeles, California

Philip G. Hodge, Jr.
University of Minnesota
Minneapolis, Minnesota

W. Johnson
University Engineering Department
Cambridge, England

L.M. Kachanov
Boston University
Boston, Massachusetts

H. Kolsky
Brown University
Providence, Rhode Island

F.A. Leckie
University of Illinois at Urbana-Champaign
Urbana, Illinois

E.H. Lee
Rensselaer Polytechnic Institute
Troy, New York

T.H. Lin
University of California at Los Angeles
Los Angeles, California

A. Litewka
Technical University
Posnan, Poland

G. Maier
Politecnico di Milano
Milano, Italy

J.B. Martin
University of Cape Town
Rondebosch, South Africa

Robert M. McMeeking
University of Illinois at Urbana-Champaign
Urbana, Illinois

J.M. Mosquera
Brown University
Providence, Rhode Island

S. Murthy
Yale University
New Haven, Connecticut

P.M. Naghdi
University of California at Berkeley
Berkeley, California

A. Nappi
Politecnico di Milano
Milano, Italy

E,T. Onat
Yale University
New Haven, Connecticut

A. Phillips
Yale University
New Haven, Connecticut

J.L. Raphanel
Brown University
Providence, Rhode Island

J.R. Rice
Harvard University
Cambridge, Massachusetts

J. Lyell Sanders Jr.
Harvard University
Cambridge, Massachusetts

A. Sawczuk
Academie Polonaise des Sciences, Warszawa, Poland
Université Aix-Marseille, Marseille, France

R.T. Shield
University of Illinois at Urbana-Champaign
Urbana, Illinois

P.S. Symonds
Brown University
Providence, Rhode Island

S.S. Wang
University of Illinois at Urbana-Champaign
Urbana, Illinois

C.J. Wung
University of Utah
Salt Lake City, Utah

Contents

Daniel C. Drucker

Daniel Charles Drucker was born in New York City on June 3, 1918. His father, Moses Abraham Drucker, was a Civil Engineer and Dan followed his example by studying Civil Engineering at Columbia University, receiving the B.S. degree in 1938. Dan stayed at Columbia for graduate study and in August 1939 he married Ann Bodin, his sweetheart of four years. His Ph.D. research was on three-dimensional photoelastic methods for stress analysis under the supervision of Professor R.D. Mindlin. The research led to his first publication, jointly with Mindlin, in 1940, the year in which he obtained the Ph.D. degree.

The Druckers spent the next three years in Ithaca where Dan was an Instructor in Engineering at Cornell University. Their son Robert David was born in Ithaca in December, 1942. In the summer of 1943 they moved to Chicago so that Dan could assume his next position as Supervisor for Solid Mechanics at the Armour Research Foundation of the Illinois Institute of Technology, a position he held for two years (1943—45). After a year in the Air Force, Drucker returned to the Illinois Institute of Technology as an Assistant Professor of Mechanics. One year later Dan accepted an appointment at Brown University as Associate Professor of Applied Mathematics and of Engineering to begin a long and fruitful association (1947—68). The Druckers' daughter Mady (Miriam) was born in Providence in August 1948.

It was at Brown University that Drucker made the major part of his outstanding contributions to the theory of plasticity and to its application to the design of engineering structures and components. The Brown group in solid mechanics was founded by William Prager and through his fine leadership the group became internationally known for its pioneering work in plasticity. Drucker contributed in full measure and while at Brown published over ninety papers, some in collaboration with colleagues in the Division of Applied Mathematics and in the Division of Engineering. The papers made important contributions to all of the areas involved in the successful development of a realistic and useful theory of plasticity. About fifteen papers dealt with the foundations of plasticity, particularly the theoretical development of plastic stress—strain relations, but the approach was balanced by eleven papers on the experimental evaluation of theoretical models for strain hardening. Drucker introduced the concept of material stability, now known as Drucker's Stability Postulate, which provided a unified approach for the derivation of stress—strain relations for the plastic behavior of metals. He was a co-author of the now

Daniel C. Drucker

classical paper in which the limit load was clearly defined and the theorems of limit analysis were established. The theorems led directly to limit design — a technique to predict the load-carrying capacity of engineering structures and components such as bridges, pressure vessels, machine parts, and so on. Following the publication of this work there was an immediate explosion of applications of the theorems to problems which formerly were beyond the scope of engineering practice. The special merit of limit analysis lies in the fact that engineers can make practical and safe decisions on the design of complex load bearing components on the basis of relatively simple calculations.

Drucker was the first to show how limit analysis could be used in the design of cylindrical shells, and he later applied it effectively to the design of pressure vessels, particularly thin-walled pressure vessels with an A.S. M.E. Standard Torispherical Head. Vessels of this type are widely used and failures, some of them serious, were occurring. With the theorems of limit analysis, calculations were performed which showed that for some typical designs the pressure exceeded the limit value, so that failure was likely and revision of the A.S.M.E. Code was needed. The success of this work played a strong role in the development of the A.S.M.E. Code for Nuclear Pressure Vessels, which bases strength estimates on limit and shakedown concepts. Over 30 of Drucker's papers written at Brown cover important engineering applications of plasticity theory; besides structural mechanics and pressure vessel design, the topics include soil mechanics, metal working and metal cutting.

In his later years at Brown, Drucker became active in a field sometimes known as micromechanics, which attempts to bridge the gap between the material scientist who studies material behavior at the atomic level and the engineer who works with real materials modeled by theories of continuum mechanics. Drucker made significant progress in determining the effect of particle size and shape of hard inclusions on the mechanical properties of certain high strength alloys. Among the phenomena which could be predicted in quantitative terms were dispersion and precipitation hardening of metals. A particularly remarkable prediction was that the flow stress of the ductile matrix of structural metals is determined primarily by microstructural features not visible in the optical microscope; the visible particles previously assumed to be the cause of hardening could not be the agents of this behavior, and particles smaller than a micron must exist to induce the hardening which occurs. Subsequently, refined metallographic observations showed such particles exist.

A related application of limit analysis techniques on the microscale was made by Drucker in the area of composite materials. By calculating upper bounds on limit loads of fibrous and particulate systems, he showed that finite hardening rates may not be attained for certain loading directions in these materials in the absence of matrix hardening. The result is useful in the development of plasticity theories for composite media.

Drucker's first Ph.D. student at Brown was Franz Edelman who received his degree in Applied Mathematics in 1950. Other students in Applied Mathematics were Frederick Stockton (Sc.M. 1949, Ph.D. 1953), William G. Brady (Sc.M. 1953), John Zickel (Ph.D. 1953), John Lyell Sanders, Jr. (Ph.D. 1954), William Bruce Woodward (Sc.M. 1958), and Jack L. Dais (Ph.D. 1967). Students in the Division of Engineering were Vincent William Howard (Sc.M. 1952), Norman C. Small (Ph.D. 1960), Halit Halil Demir (Sc.M. 1962, Ph.D. 1963), Steven Charles Batterman (Ph.D. 1964), David Rubin (Sc.M. 1965, Ph.D. 1967), Andrew Clennel Palmer (Ph.D. 1965), Wai-Fah Chen (Ph.D. 1966), David Christopher McDermott (Sc.M. 1967, Ph.D. 1969), William R. Powell (Sc.M. 1967), Thomas Wesley Butler (Ph.D. 1969), and George J. Dvorak (Ph.D 1969).

In 1968 Drucker assumed his present position as Dean of the College of Engineering at the University of Illinois at Urbana-Champaign. The considerable duties of the position and his other professional commitments leave little free time, but he continues to publish original research. He supervised the research of Luc Palgen who received the Ph.D. degree in Theoretical and Applied Mechanics in 1981.

Drucker's contributions to engineering education other than direct instruction began in earnest when he became Chairman of the Division of Engineering at Brown University in 1953. The Division is presently recognized as having one of the best groups in materials engineering and solid mechanics in the country and this is primarily due to Drucker's leadership and example during the six years as Chairman. He contributed to education in a broader way through his chairmanship of the Physical Sciences Council of Brown during 1961—63. His remarkable abilities as an educator and an administrator made him eminently qualified to assume his present position as Dean of one of the leading Colleges of Engineering in the U.S.A. Since joining the University of Illinois in 1968 he has maintained the College's high standing; his leadership was particularly effective during the period of low engineering enrollments and reduced budgets of the 1970's. His efforts for quality in engineering education have gone beyond the State level and he is recognized as a spokesman on the country's need for highly qualified engineers in an increasingly technological society. Drucker was one of the first to draw attention to the difficulties facing schools of engineering after the mid 1970's in recruiting faculty to educate the engineers to meet industry's demands. His position as a leader in engineering education has been acknowledged by invitations to serve on national committees and to testify before Congress, and by his election to President (1981—1982) of the American Society of Engineering Education.

Drucker participates actively in several professional societies. He has been most active in the American Society for Mechanical Engineers. For twelve years, 1956 to 1968, he was Technical Editor of the A.S.M.E. Journal of Applied Mechanics, the premier journal in the world for theoretical and applied mechanics. He was Chairman of the Executive Committee of the Applied Mechanics Division of A.S.M.E. during 1963–64 and ten years later he served as President of A.S.M.E. He has also served as President of the Society for Experimental Stress Analysis and of the American Academy of Mechanics. Long active in the International Union of Theoretical and Applied Mechanics, he became President of I.U.T.A.M. in 1981.

The respect and admiration Drucker receives from engineers in many fields is indicated by honors and awards from five engineering societies; the von Karman Medal of ASCE (1966), the Marburg Lecture of ASTM (1966), the Murray Lecture of SESA (1967), the Lamme Medal of ASEE (1967), the M.M. Frocht Award of SESA (1971), and Honorary Membership of ASME (1982). Among other honors are an honorary doctorate from Lehigh University (1976), the Egelston Medal of the Columbia University Engineering Alumni Association (1978), and the Gustav Trasenster Medal of the University of Liege (1979). He has been a member of the National Academy of Engineering since 1967, he was the Honorary Chairman of the Canadian Congress of Applied Mechanics held in June 1979, and in 1980 he was elected a Foreign Member of the Polish Academy of Sciences.

This book and the associated Symposium on the Mechanics of Material Behavior held at the University of Illinois at Urbana-Champaign, June 6–7, 1983, are primarily in recognition of Dan Drucker's outstanding technical contributions. His dedication to education and to the advancement of his profession is matched by few. Those who know Dan also admire him for his quiet but strong character, his integrity and his feeling for his fellow men.

R.T. Shield

Technical Publications of Daniel C. Drucker

(With R.D. Mindlin) Stress analysis by three-dimensional photoelastic methods. J. Appl. Phys., 11 (1940) 724–732.

The Photoelastic analysis of transverse bending of plates in the standard transmission polariscope. J. Appl. Mech., 9, Trans. ASME, 64 (1942) A161–A164.

Photoelastic separation of principal stresses by oblique incidence. J. Appl. Mech., 10, Trans. ASME, 65 (1943) A156–A160.

(With C.O. Dohrenwend and Paul Moore) Transverse impact transients. Proc. SESA, 1 (1944) 1–11.

(With H. Tachau) A new design criterion for wire rope. J. Appl. Mech., 12, Trans. ASME, 67 (1945) A33–A38.

(With K.E. Bisshopp) Large deflection of cantilever beams. Quart. Appl. Math., 3 (1945) 272–275.

Stress–strain relations for strain hardening materials — discussion and proposed experiments. Proc. 1st Symp. Applied Mathematics, 1947, Am. Math. Soc., Providence, 1949, pp. 181–187.

(With M.M. Frocht) Equivalence of photoelastic scattering patterns and membrane contours for torsion. Proc. SESA, 5 (1948) 34–41.

A reconsideration of deformation theories of plasticity. Trans. ASME, 71 (1949) 587–592.

Relation of experiments to mathematical theories of plasticity. J. Appl. Mech., 16, Trans. ASME, 71 (1949) A349–A357.

The significance of the criterion for additional plastic deformation of metals. J. Colloid Sci., 4 (1949) 299–311.

An analysis of the mechanics of metal cutting. J. Appl. Phys., 20 (1949) 1013–1021.

A discussion of the theories of plasticity. J. Aero. Sci., 16 (1949) 567.

(With H. Ekstein) A dimensional analysis of metal cutting. J. Appl. Phys., 21 (1950) 104–107.

Some implications of work hardening and ideal plasticity. Quart. Appl. Math., 7 (1950) 411–418.

(With G.N. White, Jr.) Effective stress and effective strain in relation to stress theories of plasticity. J. Appl. Phys., 21 (1950) 1013–1021.

The method of oblique incidence in photoelasticity. Proc. SESA, 8 (1950) 51–66.

Three dimensional photoelasticity. In: M. Hetenyi (Ed.), Handbook of Experimental Stress Analysis, Wiley, New York, 1950, Ch. 17—II, pp. 924—976.

(With F.D. Stockton) Fitting mathematical theories of plasticity to experimental results. J. Colloid Sci. (Rheology Issue) 5 (1950) 239—250.

Stress–Strain relations in the Plastic Range — A Survey of Theory and Experiment. Brown University Report ONR A11-S1, 1950.

Plasticity of metals — mathematical theory and structural applications. Trans. ASCE, 116 (1951) 1059—1082.

(With F. Edelman) Some extensions of elementary plasticity theory. J. Franklin Inst., 251 (1951) 581—605.

(With H.J. Greenberg and W. Prager) The safety factor of an elastic plastic body in plane strain. J. Appl. Mech., 18, Trans. ASME, 73 (1951) A371—A378.

A more fundamental approach to plastic stress—strain relations. Proc. 1st U.S. Nat. Congr. Applied Mechanics, Chicago, June 11—16, 1951, ASME, New York, 1951, pp. 487—491.

(With H.J. Greenberg, E.H. Lee and W. Prager) On plastic rigid solutions and limit design theorems for elastic plastic bodies. Proc. 1st U.S. Nat. Congr. Applied Mechanics, Chicago, June 11–16, 1951, ASME, New York, 1951, pp. 533—538.

(With F.D. Stockton) Instrumentation and fundamental experiments in plasticity. Proc. SESA, 10 (1951) 127—142.

(With H.J. Greenberg and W. Prager) Extended limit design theorems for continuous media. Quart. Appl. Math., 9 (1952) 381—389.

(With W. Prager) Soil mechanics and plastic analysis or limit design. Quart. Appl. Math., 10 (1952) 157—165.

(With E.H. Lee) Notes on the Eighth International Congress on Theoretical and Applied Mechanics. Appl. Mech. Rev., 5 (1952) 497—498.

(With E.T. Onat) Inelastic instability and incremental theories of plasticity. J. Aero. Sci., 20 (1953) 181—186.

Limit analysis of two and three dimensional soil mechanics problems. J. Mech. and Phys. Solids, 1 (1953) 217—226.

(With R.T. Shield) The application of limit analysis to punch indentation problems. J. Appl. Mech., 20, Trans. ASME, 75 (1953) 453—460.

Limit analysis of cylindrical shells under axially-symmetric loading. Proc. 1st Midwest. Conf. Solid Mechanics, Urbana, Illinois, April 24—25, 1953, Eng. Exper. Station, Urbana, 1953, pp. 158—163.

(With W.G. Brady) An experimental investigation and limit analysis of net area in tension. Proc. ASCE Sep. 296 (1953), Trans. ASCE, 120 (1955) 1133—1154.

Coulomb friction, plasticity, and limit loads. J. Appl. Mech., 21, Trans. ASME, 76 (1954) 71—74.

An Evaluation of Current Knowledge of the Mechanics of Brittle Fracture. Serial No. SSC-69, Ship Structure Committee, Amer. Bur. Shipping, New York, 1954.

(With W.B. Woodward) Interpretation of photoelastic transmission patterns for a three-dimensional model. J. Appl. Phys., 25 (1954) 510—512.

On obtaining plane strain or plane stress conditions in plasticity. Proc. 2nd U.S. Nat. Congr. Applied Mechanics, Ann Arbor, June 14—18, 1954, ASME, New York, 1954, pp. 485—488.

(With H.G. Hopkins) Combined concentrated and distributed load on ideally-plastic circular plates. Proc. 2nd U.S. Nat. Congr. Applied Mechanics, Ann Arbor, June 14—18, 1954, ASME, New York, 1954, pp. 517—520.

(With E.T. Onat) On the concept of stability of inelastic systems. J. Aero Sci., 21 (1954) 543—548.

Limit analysis and design. Appl. Mech. Rev., 7 (1954) 421—423.

(With J. D'Agostino, C.K. Liu and C. Mylonas) An analysis of plastic behavior of metals with bonded birefringent plastic. Proc. SESA, 12 (1955) 115-122.

(With J. D'Agostino, C.K. Liu and C. Mylonas) Epoxy adhesives and casting resins and photoelastic plastics. Proc. SESA, 12 (1955) 123—128.

(With R.E. Gibson and D.J. Henkel) Soil mechanics and work-hardening theories of plasticity. Proc. ASCE Sep. No. 798 (1955) Trans. ASCE, 122 (1957) 338—346.

Action of towed and driving wheels in a plastic soil — A preliminary study. Part I, Land Locomotion Bulletin, Report No. 1, 1955, pp. 14—18. Part II, Land Locomotion Bulletin, Report No. 2, 1955 pp. 5—7. Detroit Arsenal, Center Line, Michigan.

Analogous strain rate effects in jet formation and metal cutting. Proc. Shaped Charge Symp., U.S. Army, Aberdeen, Maryland, 1956.

On uniqueness in the theory of plasticity. Quart. of Appl. Math., 14 (1956) 35–42.

Stress–strain relations in the plastic range of metals — experiments and basic concepts. In: Source Book of Rheology, Vol. I, Academic Press, New York, 1956, pp. 97–119.

The effect of shear on the plastic bending of beams. J. Appl. Mech., 23, Trans. ASME, 78 (1956) 509–514.

Variational principles in the mathematical theory of plasticity. Proc. 8th Symp. Applied Mathematics, 1956, McGraw-Hill, New York, 1958, pp. 7–22.

(With R.T. Shield) Design for minimum weight. Proc. 9th Int. Congr. Applied Mechanics, Brussels, 1957, Vol. 5, pp. 212–222.

(With R.T. Shield) Bounds on minimum weight design. Quart. Appl. Math., 15 (1957) 269–281.

Plastic design methods — advantages and limitations. Trans. Soc. Naval Arch. and Marine Eng., 65 (1958) 172–196.

(With R.T. Shield) Limit strength of thin walled pressure vessels with an ASME standard torispherical head. In: R.M. Haythornthwaite (Ed.), Proc. 3rd U.S. Nat. Congr. Applied Mechanics, Providence, June 11–14, 1958, ASME, New York, 1958, pp. 665–672.

(With C. Mylonas and J.D. Brunton) Static brittle fracture initiation at net stress 40% of yield. Welding J. (Res. Suppl.), 37 (1958) 473s–479s.

(With R.T. Shield) Limit analysis of symmetrically loaded thin shells of revolution. J. Appl. Mech., 26, Trans. ASME, 81 (1959) 61–68.

A definition of stable inelastic material. J. Appl. Mech., 26, Trans. ASME, 81 (1959) 101–106.

On minimum weight design and strength of non-homogeneous plastic bodies. In: W. Olszak (Ed.), Proc. 1st Symp. Non-Homogeneity in Elasticity and Plasticity, Warsaw, September 1958, Pergamon Press, Oxford, 1959, pp. 139–146.

Plasticity. In: J.N. Goodier and N.J. Hoff (Eds.), Structural Mechanics, Proc. 1st Symp. Naval Structural Mechanics, Stanford, August 11–14, 1958, Pergamon Press, Oxford, 1960, pp. 407–455.

(With C. Mylonas and G. Lianis) On the exhaustion of ductility of E-steel in tension following compressive pre-strain. Welding J. (Res. Suppl.), 39 (1960) 117s–120s.

Extension of the stability postulate with emphasis on temperature changes. In: E.H. Lee and P.S. Symonds (Eds.), Plasticity, Proc. 2nd Symp. Naval

Structural Mechanics, Providence, April 5—7, 1960, Pergamon Press, Oxford, 1960, pp. 170—184.

(With J.H. Ludley) A reversed bend test to study ductile to brittle transition. Welding J. (Res. Suppl.), 39 (1960) 543s—546s.

Contributor to Dictionary of Applied Mathematics, D. Van Nostrand Company, Princeton, 1961.

(With J.H. Ludley) Size effect in brittle fracture of notched E-steel plates in tension. J. Appl. Mech., 28, Trans ASME, 83 (1961) 137—139.

(With C. Mylonas) Twisting stresses in tape. Exper. Mech., 1 (1961) 23—32.

(With R.T. Shield) Design of thin-walled torispherical and toriconical pressure vessel heads. J. Appl. Mech., 28, Trans. ASME, 83 (1961) 292—297.

On structural concrete and the theorems of limit analysis. IABSE Publications, 21 (1961) 49—59.

On stress—strain relations for soils and load carrying capacity. Proc. 1st Int. Conf. Mechanics of Soil Vehicle Systems, Turin, June 1961, Minerva Tecnica, Bergamo, 1961, pp. 15—23.

On the macroscopic theory of inelastic stress—strain—time—temperature behavior. In: Advances in Materials Research in the NATO Nations, Proc. AGARD Int. Symp. Materials Science, Paris, May 1961, Pergamon Press, Oxford, 1963, pp. 193—221.

(With C.R. Calladine) Nesting surfaces of constant rate of energy dissipation in creep. Quart. Appl. Math., 20 (1962) 79—84.

(With C.R. Calladine) A bound method for creep analysis of structures: direct use of solutions in elasticity and plasticity. J. Mech. Eng. Sci., 4 (1962) 1—11.

Basic concepts in plasticity. In: W. Flugge (Ed.), Handbook of Engineering Mechanics, McGraw-Hill, New York, 1962, pp. 46-3—46-15.

On the role of experiment in the development of theory. Proc. 4th U.S. Nat. Congr. Applied Mechanics, Berkeley, June 18—21, 1962, ASME, New York, 1962, pp. 15—33.

Survey on second-order plasticity. In: M. Reiner and D. Abir (Eds.), Proc. IUTAM Symp. Second-Order Effects in Elasticity, Plasticity and Fluid Dynamics, Haifa, Israel, April 21—29, 1962, Pergamon Press, Oxford, 1964, pp. 416—423.

Stress—strain—time relations and irreversible thermodynamics. In: M. Reiner and D. Abir (Eds.), Proc. IUTAM Symp. Second Order Effects in Elasticity, Plasticity and Fluid Dynamics, Haifa, Israel, April 21—29, 1962, Pergamon Press, Oxford, 1964, pp. 331—351.

A continuum approach to the fracture of metals. In: D.C. Drucker and J.J. Gilman (Eds.), Fracture of Solids, Proc. Int. Conf. Fracture, Seattle, August 20—24, 1962, Wiley-Interscience, New York, 1963, pp. 3—50.

(With H.H. Demir) An experimental study of cylindrical shells under ring loading. In: Progress in Applied Mechanics — The Prager Anniversary Volume, Macmillan, New York, 1963, pp. 205—220.

Some remarks on flow and fracture. In: S. Ostrach and R.H. Scanlan (Eds.), Developments in Mechanics, Vol. 2, Proc. 8th Midwest. Conf. Solid Mechanics, Cleveland, April 1—3, 1963, Pergamon Press, Oxford, 1965, pp. 201—218.

The experimental and analytical significance of limit pressures for cylindrical shells — a discussion. J. Eng. Mech. Div. ASCE, 90 (1964) 295—301.

On the postulate of stability of material in the mechanics of continua. J. Mechanique, 3 (1964) 235—249.

Plasticity, perfect or ideal plastic flow. In: Encyclopaedic Dictionary of Physics, Pergamon Press, Oxford, 1964.

Engineering and continuum aspects of high strength materials. In: V.F. Zackay (Ed.), High Strength Materials, Proc. 2nd Int. Symp. High Strength Materials, June 1964, Wiley, New York, 1965, pp. 795—833.

Concepts of path independence and material stability for soils. In: J. Kravtchenko and P.M. Sirieys (Eds.), Rheologie et Mechanique des Sols, Proc. IUTAM Symp., Grenoble, April 1964, Springer, Berlin, 1966, pp. 23—43.

(With W.N. Findley) An experimental study of plane plastic straining of notched bars. J. Appl. Mech., 32, Trans. ASME, 87 (1965) 493—503.

(With G. Maier) Elastic-plastic continua containing unstable elements obeying normality and convexity relations. Schweiz. Bauzeit., 84 (1966) 447—450.

On time-independent plasticity and metals under combined stress at elevated temperature. In: B. Broberg, J. Hult and F. Niordson (Eds.), Recent Progress in Applied Mechanics, The Folke Odqvist Volume, Almqvist and Wiksell, Stockholm, 1967, pp. 209—222.

The continuum theory of plasticity on the macroscale and the microscale (1966 Marburg Lecture), J. Mat. ASTM, 1 (1966) 873—910.

(With D. Rubin) On stability of viscoplastic systems with thermomechanical coupling. In: D. Abir (Ed.), Contributions to Mechanics, The

Markus Reiner 80th Anniversary Volume, Pergamon Press, Oxford, 1968, pp. 171—179.

(With J.R. Rice) Energy changes in stressed bodies due to void and crack growth. Int. J. of Fracture Mech., 3 (1967) 19—27.

(With T.T. Wu) Continuum plasticity theory in relation to solid solution, dispersion, and precipitation hardening. J. Appl. Mech., 34, Trans. ASME, 89 (1967) 195—199.

(With A.C. Palmer and G. Maier) Normality relations and convexity of yield surfaces for unstable materials or structural elements. J. of Appl. Mech., 34, Trans. ASME, 89 (1967) pp. 464—470.

Introduction to Mechanics of Deformable Solids. McGraw-Hill, New York, 1967.

On the continuum as an assemblage of homogeneous elements or states. In: H. Parkus and L.I. Sedov (Eds.), Irreversible Aspects of Continuum Mechanics and Transfer of Physical Characteristics in Moving Fluids, Proc. IUTAM Symp., Vienna, 1966, Springer, Vienna, 1968, pp. 77—93.

(With W.F. Chen) On the use of simple discontinuous fields to bound limit loads. In: J. Heyman and F.A. Leckie (Eds.), Engineering Plasticity, Cambridge Univ. Press, 1968, pp. 129—145.

Thoughts on the present and future interrelation of theoretical and experimental mechanics (1967 William M. Murray Lecture). Exper. Mech., 8 (1968) 1—10.

Macroscopic fundamentals in brittle fracture. In: H. Liebowitz (Ed.), Treatise on Fracture, Academic Press, 1968, Ch. 8, pp. 473—531.

Closing comments of session II. In: U.S. Lindholm (Ed.), Symp. Mechanical Behavior of Materials Under Dynamic Loads, Springer, New York, 1968, pp. 405—409.

(With J.R. Rice) Plastic deformation in brittle and ductile fracture. J. Eng. Fracture Mech., 1 (1970) 577—602.

(With W.F. Chen) Bearing capacity of concrete blocks or rock. J. Eng. Mech. Div. ASCE, 95 (1969) 955—978.

Metal deformation processing from the viewpoint of solid mechanics. In: Developments in Mechanics, Vol. 6, Proc. 12th Midwest. Mech. Conf., Notre Dame, Indiana, August 16—18, 1971, Univ. Notre Dame, 1971, pp. 3—16.

Tensile strength and extensibility of concrete and mortar idealized as elastic- or plastic-brittle. Arch. Mech. Stosow., 24 (1972) 1073—1081.

(With R.H. Hawley) Brittle fracture of precompressed steel as affected by hydrostatic pressure, temperature and strain concentration. Exper. Mech., 13 (1973) 1—6.

Plasticity theory, strength-differential (SD) phenomenon, and volume expansion in metals and plastics. Metall. Trans., 4 (1973) 667—673.

(With T.W. Butler) Yield strength and microstructural scale: a continuum study of pearlitic versus spheroidized steel. J. Appl. Mech., 40 (1973) 780—784.

(With G. Maier) Effects of geometry change on essential features of inelastic behavior. J. Eng. Mech. Div. ASCE, 99 (1973) 819—834.

(With J. Duffy and E.A. Fox) Remarks on the past, present, and likely future of photoelasticity. In: R.D. Mindlin and Applied Mechanics, Pergamon Press, Oxford, 1974, pp. 1—24.

On plastic analysis of the microstructure of metallic alloys. In: A. Sawchuk (Ed.), Problems of Plasticity, Proc. Symp. Foundations of Plasticity, Warsaw, August 30—September 2, 1972, Noordhoff Int., Leyden, 1974, pp. 25—44.

Typescript of the panel on the occasion of the Symposium on Foundations of Plasticity, Warsaw, August 30—September 2, 1972, Nauka Polska, Rok XXI, 4 (106) (1973) 92—93.

(With E.H. Lee and B. Budiansky) On the influence of variations of material properties on stress wave propagation through elastic slabs. J. Appl. Mech., 42 (1975) 417—422.

Yielding, flow, and failure. In: C.T. Herakovich (Ed.), Inelastic Behavior of Composite Materials, AMD Vol. 13, ASME, New York, 1975, pp. 1—15.

Concluding summary. In: The Photoelastic Effect and Its Applications, Proc. IUTAM Symp., Ottignies, Belgium, 1973, Springer, Berlin, 1975, pp. 635—638.

Elementary results of dimensional analysis for friction and wear in steady state sliding. Wear, 40 (1976) 129—133.

Approximate calculations of spall and cratering in high speed impact. In: K. Kawata and J. Shioiri (Eds.), High Velocity Deformation of Solids, Proc. IUTAM Symp., Tokyo, August 24—27, 1977, Springer, Berlin, 1979, pp. 214—227.

(With B. Budiansky, G.S. Kino and J.R. Rice) Pressure sensitivity of a clad optical fiber. Appl. Optics, 18 (1979) 4085—4088.

Taylor instability of the surface of an elastic-plastic plate. In: S. Nemat-Nasser (Ed.), Mechanics Today, Vol. 5, Pergamon Press, Oxford, 1980, pp. 37—47.

A further look at Raleigh—Taylor and other surface instabilities in solids. Ing. Arch., 49 (1980) 361—376 (special issue in honor of Hans Ziegler).

Some classes of inelastic materials-related problems basic to future technologies. Nucl. Eng. and Design, 57 (1980) 309—322.

(With Luc Palgen) On stress—strain relations suitable for cyclic and other loading. J. Appl. Mech., 48 (1981) 479—485.

Preliminary design on the microscale and macroscale. In: S.S. Wang and W.D. Renton (Eds.), Advances in Aerospace Structures and Materials, ASME, New York, 1981, pp. 1—3.

Anisotropic Plasticity of Plane-Isotropic Sheets

BERNARD BUDIANSKY

Division of Applied Sciences, Harvard University, Cambridge, MA 02138 (U.S.A.)

Abstract

A general equation is written for the plane-stress yield condition of orthotropic sheet material that is isotropic in its plane. A set of constitutive relations for planar-isotropic strain hardening is then formulated on the basis of the usual normality rule for strain rates, together with the assumption of a scalar relation between an effective stress and a work-equivalent effective strain. It is shown how particular laws proposed earlier for sheet-metal plasticity fit into the general scheme. Solutions to the problem of balanced biaxial stretching of a sheet containing a circular hole illustrate a wide range of effects produced by various realizations of the general constitutive relations.

Introduction

Interest in the technology of sheet-metal forming has provoked the formulation of various kinds of constitutive equations of plane-stress plasticity that incorporate effects of anisotropy. Most of these equations, as well as the work in this paper, focus on that special anisotropy in which isotropy in the plane of the sheet is preserved. The earliest work is by Hill [6], in which a one-parameter, quadratic yield function, generalizing the Mises ellipse, was proposed. Later yield-condition generalizations by Bassani [1] and Hill [7] (the latter quoted and used earlier by Parmar and Mellor [9, 10]) involve non-quadratic algebraic functions, with more disposable parameters available. The present paper exhibits a general yield condition in a form that appears attractive for applications, especially when coupled with appropriate in-plane constitutive relations. The special problem of finding the stress concentration at a circular hole in an infinite sheet subjected to balanced biaxial tension, solved earlier [2] for a power-hardening Mises material [2], and [11, 3] for Hill's older (1948) anisotropic relations, will be found also to have an easy exact solution for the general orthotropic, in-plane isotropic, power-law material.

Yield condition and constitutive relations

GENERAL YIELD CONDITION

Planar isotropy permits any plane-stress yield condition to be expressed in terms of $(\sigma_2 + \sigma_1)$ and $(\sigma_2 - \sigma_1)$, where σ_1 and σ_2 are the in-plane principal stresses. Accordingly, the yield condition can be prescribed parametrically in the polar-coordinate form

$$x \equiv \frac{\sigma_2 + \sigma_1}{2\sigma_{bt}} = g(\alpha)\cos\alpha, \qquad y \equiv \frac{\sigma_2 - \sigma_1}{2\sigma_s} = g(\alpha)\sin\alpha \tag{1}$$

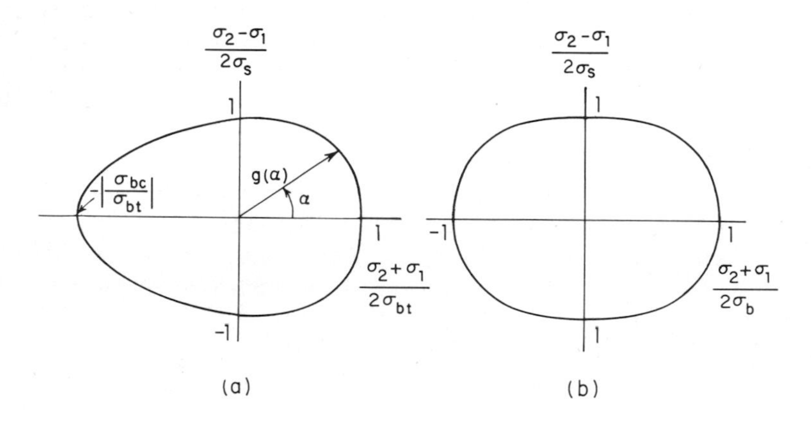

Fig. 1. Normalized yield loci.

Here (see Fig. 1a) $g(\alpha) > 0$ is the length of the radius to a point on the normalized yield locus Γ and α is the associated polar angle. The normalizing parameters σ_s and σ_{bt} are respectively the yield stresses in pure shear and balanced biaxial tension, so that $g(0) = g(\pm\pi/2) = 1$. Isotropy imposes the further requirement that the yield locus be symmetrical about the x-axis, because for a given value of $(\sigma_1 + \sigma_2)$ the yield condition must be unaffected when σ_1 and σ_2 are interchanged. However, isotropy does not imply symmetry about the y-axis. (Thus the compressive yield stress σ_{bc} in biaxial compression need not equal σ_{bt}.) This extra symmetry has, however, usually been assumed, and would be implied by the presumption that the yield condition is unchanged by sign reversal of the stress tensor. Under these circumstances, we have $g(\alpha) = g(\pi - \alpha)$, as in Fig. 1b. For simplicity, this additional restriction will be adopted in the rest of this paper, and we will write $\sigma_b = \sigma_{bc} = \sigma_{bt}$. The yield stress σ_u in uniaxial tension is given by

$$\sigma_u = 2\sigma_s g(\alpha_u)\sin\alpha_u = 2\sigma_b g(\alpha_u)\cos\alpha_u \tag{2}$$

where

$$\alpha_u = \tan^{-1}(\sigma_b/\sigma_s) \qquad (0 < \alpha_u < \pi/2) \tag{3}$$

The introduction of the stress ratios

$$X = \sigma_b/\sigma_u \tag{4}$$

$$Y = \sigma_b/\sigma_s \tag{5}$$

as characteristic non-dimensional material parameters permits (1) to be rewritten as

$$\frac{\sigma_2 + \sigma_1}{2\sigma_u} = Xg(\alpha)\cos\alpha, \qquad \frac{\sigma_2 - \sigma_1}{2\sigma_u} = \frac{Xg(\alpha)\sin\alpha}{Y} \tag{6}$$

The uniaxial yield stress σ_u will be chosen later to play the role of the *effective stress* $\sigma \equiv \sigma_u$ that measures the level of strain hardening. In terms of σ_1 and σ_2, α and σ are given by

$$\alpha = \tan^{-1}\left[\frac{Y(\sigma_2 - \sigma_1)}{\sigma_2 + \sigma_1}\right] \tag{7}$$

$$\sigma = \frac{[(\sigma_2 + \sigma_1)^2 + Y^2(\sigma_2 - \sigma_1)^2]^{1/2}}{2Xg(\alpha)} \tag{8}$$

The appropriate quadrant for α is implied by Eqns. (1) and Fig. 1a. (In Fortran notation, $\alpha = \text{ATAN2}[Y(\sigma_2 - \sigma_1), (\sigma_2 + \sigma_1)]$.)

Explicit parametric representations for σ_1 and σ_2 that follow from Eqns. (2)—(6) are

$$\frac{\sigma_1}{\sigma} = -\frac{Xg(\alpha)\sin(\alpha - \alpha_u)}{\sin\alpha_u}, \qquad \frac{\sigma_2}{\sigma} = \frac{Xg(\alpha)\sin(\alpha + \alpha_u)}{\sin\alpha_u} \tag{9}$$

In addition to the conditions $g(\alpha) = g(-\alpha) = g(\pi - \alpha)$ implied by in-plane isotropy and stress-sign-reversal symmetry, the usual requirement of convexity of the yield locus [4] imposes the restriction

$$g^2 + 2(g')^2 - gg'' \geqslant 0 \tag{10}$$

In (σ_1, σ_2) space, the interior of any convex yield locus must contain the square $|\sigma_2 \pm \sigma_1| = \sigma_u$. Hence $\sigma_b \geqslant \sigma_u/2$, $\sigma_s \geqslant \sigma_u/2$, or

$$X \geqslant \tfrac{1}{2}, \qquad Y \leqslant 2X \tag{11}$$

CONSTITUTIVE RELATIONS

Suppose the stresses (σ_1, σ_2) satisfy the yield condition, and let $(\dot{\epsilon}_1, \dot{\epsilon}_2)$ denote the non-zero strain-rate vector produced by a "loading" stress rate $(\dot{\sigma}_1, \dot{\sigma}_2)$ pointing to the exterior of the yield locus C in (σ_1, σ_2) space. The usual normality rules of plasticity theory require that $(\dot{\epsilon}_1, \dot{\epsilon}_2)$ also point to the outside of C and, if C is smooth, that

$$\sigma_1' \dot{\epsilon}_1 + \sigma_2' \dot{\epsilon}_2 = 0 \tag{12}$$

for the vector (σ_1', σ_2') tangent to C. Equivalently,

$$\tfrac{1}{2}(\sigma_2' + \sigma_1')(\dot{\epsilon}_2 + \dot{\epsilon}_1) + \tfrac{1}{2}(\sigma_2' - \sigma_1')(\dot{\epsilon}_2 - \dot{\epsilon}_1) = 0 \tag{13}$$

or

$$(\dot{\epsilon}_2 + \dot{\epsilon}_1) \frac{\mathrm{d}}{\mathrm{d}\alpha} [\tfrac{1}{2}(\sigma_2 + \sigma_1)] + (\dot{\epsilon}_2 - \dot{\epsilon}_1) \frac{\mathrm{d}}{\mathrm{d}\alpha} [\tfrac{1}{2}(\sigma_2 - \sigma_1)] = 0 \tag{14}$$

where the bracketed quantities are given by (6). It follows that for plastic loading

$$\dot{\epsilon}_2 - \dot{\epsilon}_1 = -\lambda X \frac{\mathrm{d}}{\mathrm{d}\alpha} [g(\alpha) \cos \alpha], \qquad \dot{\epsilon}_2 + \dot{\epsilon}_1 = \lambda (X/Y) \frac{\mathrm{d}}{\mathrm{d}\alpha} [g(\alpha) \sin \alpha] \tag{15}$$

where λ is a positive scalar that may depend on α and σ as well as $(\dot{\sigma}_1, \dot{\sigma}_2)$.
A *work-equivalent* effective plastic strain rate $\dot{\epsilon}$ [7] is defined by

$$\begin{aligned} \sigma\dot{\epsilon} &= \sigma_1 \dot{\epsilon}_1 + \sigma_2 \dot{\epsilon}_2 \\ &= \tfrac{1}{2}(\sigma_2 + \sigma_1)(\dot{\epsilon}_2 + \dot{\epsilon}_1) + \tfrac{1}{2}(\sigma_2 - \sigma_1)(\dot{\epsilon}_2 - \dot{\epsilon}_1) \end{aligned} \tag{16}$$

Substitution from (6) and (15) gives

$$\lambda = \left[\frac{Y}{X^2 g^2}\right] \dot{\epsilon}$$

so that the constitutive equation can be written

$$\frac{\dot{\epsilon}_2 - \dot{\epsilon}_1}{\dot{\epsilon}} = -\frac{Y(g \cos \alpha)'}{X g^2}, \qquad \frac{\dot{\epsilon}_2 + \dot{\epsilon}_1}{\dot{\epsilon}} = \frac{(g \sin \alpha)'}{X g^2} \tag{17}$$

where now $(\ )' \equiv \mathrm{d}/\mathrm{d}\alpha(\ )$.

Continuity of $g'(\alpha)$ has been presumed in reaching (17). If (σ_1, σ_2) is at a corner on the yield locus, the individual strain rates are not determinate, but in this case a consistent rederivation permitting a jump in g' provides the constraint

$$\frac{(\dot{\epsilon}_2 - \dot{\epsilon}_1)}{\dot{\epsilon}} \sin \alpha + Y \frac{(\dot{\epsilon}_2 + \dot{\epsilon}_1)}{\dot{\epsilon}} \cos \alpha = \frac{Y}{X g^2} \tag{18}$$

(This follows formally from (17) by elimination of g'.)

The assumption of isotropic strain hardening implies that the normalized yield locus Γ in Fig. 1a or 1b remains unchanged during plastic deformation. The relations (17) may then be rendered explicit by the additional assumption that integration of $\dot{\epsilon}$ over the loading history provides an effective plastic strain ϵ that is a function only of the final effective stress σ. Since σ is also the current tensile yield stress, this function is provided by the stress–strain relation $\epsilon = f(\sigma)$ found in a tension test. Thus, in (17),

$$\dot{\epsilon} = f'(\sigma)\dot{\sigma} = \left[\frac{1}{E_{\tan}} - \frac{1}{E}\right]\dot{\sigma} \tag{19}$$

where $E_{\tan}(\sigma)$ is the slope of the uniaxial stress–strain curve, and E is the elastic tensile modulus.

For the case of *proportional loading*, in which the stress ratio σ_2/σ_1, and hence α, remain fixed, (17) can be integrated to provide the finite form

$$\frac{\epsilon_2 - \epsilon_1}{\epsilon} = -\frac{Y(g\cos\alpha)'}{Xg^2}, \qquad \frac{\epsilon_2 + \epsilon_1}{\epsilon} = \frac{(g\sin\alpha)'}{Xg^2} \tag{20}$$

With $\epsilon = f(\sigma) = \sigma/E_{\text{sec}}$, where $E_{\text{sec}}(\sigma)$ is the uniaxial secant modulus, (20) may be regarded as the constitutive equations of a deformation theory of plasticity, with a possible range of approximate validity even in the absence of strictly proportional loading.

Equations (17) are equivalent to

$$\frac{\dot{\epsilon}_1}{\dot{\epsilon}} = \frac{[g\sin(\alpha + \alpha_u)]'}{2Xg^2\cos\alpha_u}, \qquad \frac{\dot{\epsilon}_2}{\dot{\epsilon}} = \frac{[g\sin(\alpha - \alpha_u)]'}{2Xg^2\cos\alpha_u} \tag{21}$$

and dropping the dots gives the corresponding deformation theory expressions.

A parameter widely used in the characterization of sheet-metal orthotropy is R, defined on the basis of a uniaxial, in-plane tension test as the ratio of the in-plane transverse plastic strain to the plastic strain in the thickness direction. Invoking the assumption of plastic incompressibility (for the first time in this paper) gives $R = -\epsilon_1/(\epsilon_1 + \epsilon_2)$ for $\alpha = \alpha_u$, corresponding to pure tension in the 2-direction, and from (21) this leads to

$$\frac{g'(\alpha_u)}{g(\alpha_u)} = \frac{Y^2 - 1 - 2R}{2Y(1+R)} \tag{22}$$

This result, the relation $Y = \tan\alpha_u$, and the connection

$$2Xg(\alpha_u) = \sqrt{1 + Y^2} \tag{23}$$

given by (2), may be regarded as constraints set by X, Y, and R on $g(\alpha_u)$ and $g'(\alpha_u)$.

Since negative values of R have not been observed (even though they cannot be ruled out on mathematical or physical grounds) it is reasonable to impose the restriction $R \geqslant 0$. By the normality rule, this implies $d\sigma_1/d\sigma_2(0, \sigma_u) \geqslant 0$ on the yield locus in (σ_1, σ_2) space, and then convexity imposes the limit $\sigma_s \leqslant \sigma_u$. This, with (11), provides the continued inequality $2X \geqslant Y \geqslant X \geqslant 1/2$.

It may be noted that with $X = 1$, and suitable restrictions on $g(\alpha)$, the formulation (6) includes yield loci for incompressible, fully isotropic materials, again indifferent to sign reversal of the stress tensor. The yield condition $F(\sigma_1, \sigma_2) = 0$ of such a material obeys the rules $F(\sigma_1, \sigma_2) = F(\sigma_2 - \sigma_1, \sigma_2) = F(\sigma_1 - \sigma_2, \sigma_1)$, and it follows that $g(\alpha)$ is determined everywhere by its values just in $0 \leqslant \alpha \leqslant \tan^{-1} Y/3$.

SOME SPECIAL CASES

Several previously proposed yield conditions will be exhibited, together with the equivalent choices for X, Y, and $g(\alpha)$ in the present formulation.

(a) Mises

$$3\left(\frac{\sigma_2 - \sigma_1}{\sigma}\right)^2 + \left(\frac{\sigma_2 + \sigma_1}{\sigma}\right)^2 = 4 \tag{24}$$

This is the isotropic, quadratic case, where $g(\alpha) \equiv 1$, $R = 1$, $X = 1$, and $Y = \sqrt{3}$.

(b) Hill [6]

$$(1 + 2R)\left(\frac{\sigma_2 - \sigma_1}{\sigma}\right)^2 + \left(\frac{\sigma_2 + \sigma_1}{\sigma}\right)^2 = 2(1 + R) \tag{25}$$

Again $g(\alpha) \equiv 1$, but now varying R provides a one-parameter family of yield-locus shapes. Consistent with (22) and (23) we have $X = \sqrt{(1 + R)/2}$ and $Y = \sqrt{1 + 2R}$.

(c) Hill [7]

Here Hill suggested several possible algebraic generalizations of (25), but those that have received the most attention (e.g. Parmar and Mellor, [10]) are represented by the two-parameter family obtained by varying R and p in

$$(1 + 2R)\left|\frac{\sigma_2 - \sigma_1}{\sigma}\right|^p + \left|\frac{\sigma_2 + \sigma_1}{\sigma}\right|^p = 2(1 + R) \tag{26}$$

where, for convexity, $p \geqslant 1$. This corresponds to

$$g(\alpha) = [\,|\cos\alpha|^p + |\sin\alpha|^p\,]^{-1/p} \tag{27}$$

$$X = \tfrac{1}{2}[2 + 2R]^{1/p} \tag{28}$$

$$Y = (1 + 2R)^{1/p} \tag{29}$$

(d) Bassani [1]

Still more flexibility was provided by the Bassani shape

$$\left|\frac{\sigma_2 - \sigma_1}{2\sigma_s}\right|^p + \left|\frac{\sigma_2 + \sigma_1}{2\sigma_b}\right|^q = 1 \tag{30}$$

or

$$\left|\frac{Y(\sigma_2 - \sigma_1)}{2X\sigma}\right|^p + \left|\frac{\sigma_2 + \sigma_1}{2X\sigma}\right|^p = 1 \tag{31}$$

where convexity demands $p \geqslant 1$, $q \geqslant 1$, and where X, Y, p, q are linked by

$$\left(\frac{Y}{2X}\right)^p + \left(\frac{1}{2X}\right)^q = 1 \tag{32}$$

This gives a three-parameter family of yield-locus shapes. The function $g(\alpha)$ in Eqn. (6) is given by the appropriate root of

$$|g(\alpha) \sin \alpha|^p + |g(\alpha) \cos \alpha|^q = 1 \tag{33}$$

and the parameters X, R, p, q satisfy

$$p\,[(2X)^q - 1] = q(1 + 2R) \tag{34}$$

Using (32) and (34) to eliminate X and Y from (31) gives the yield-locus equation

$$\frac{q(1 + 2R)}{p}\left|\frac{\sigma_2 - \sigma_1}{\sigma}\right|^p + \left|\frac{\sigma_2 + \sigma_1}{\sigma}\right|^q = \left[1 + \frac{q(1 + 2R)}{p}\right] \tag{35}$$

found by Bassani, which shows more directly how the Hill relation (20) was generalized.

Other possible yield-locus formulations come readily to mind. With enough terms, the series

$$g(\alpha) = \sum a_n \cos 2n\alpha \tag{36}$$

could fit any experimental yield locus, and would be explicit, in contrast to, say, (33). However, enough terms would have to be included to smooth out unphysical variations in $g'(\alpha)$.

Circular hole problem

An infinite sheet containing a circular hole, and loaded at infinity by balanced biaxial tension (Fig. 2), has been studied on the basis of both the Mises and Hill (1948) yield criteria and associated flow rules [2, 3, 11].

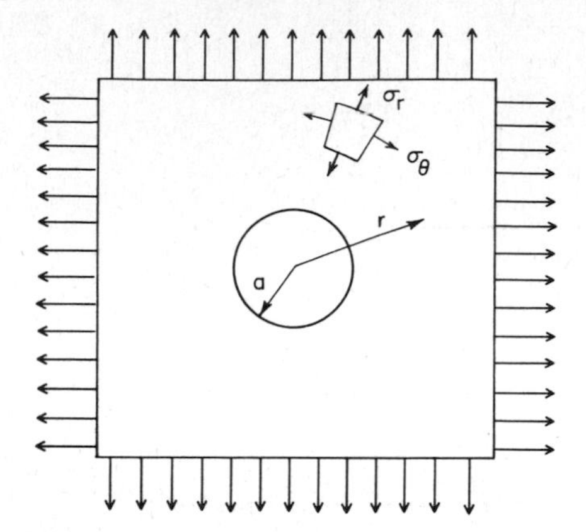

Fig. 2. Hole in sheet under balanced biaxial tension.

Under the assumptions of rigid-plastic, power-law hardening (and with geometry changes neglected) the stress and strain concentrations at the hole were found explicitly. A similarly simple solution will now be shown for the general yield condition and flow laws (9) and (21).

ANALYSIS

By the Ilyushin [8] theorem, the power hardening assumption

$$\epsilon = A\sigma^n \tag{37}$$

permits solution of the problem on the basis of the integrated, deformation-theory constitutive relations for the strains, with results identical to those of flow theory. As in the earlier solutions the equilibrium equation

$$\frac{d\sigma_r}{dr} + \frac{1}{r}(\sigma_r - \sigma_\theta) = 0 \tag{38}$$

and compatibility equation

$$\frac{d\epsilon_\theta}{dr} + \frac{1}{r}(\epsilon_\theta - \epsilon_r) = 0 \tag{39}$$

are combined to give

$$\frac{d\sigma_r}{\sigma_\theta - \sigma_r} = \frac{d\epsilon_\theta}{\epsilon_r - \epsilon_\theta} \tag{40}$$

Here $d\sigma_r$, $d\epsilon_\theta$ refer to spatial variations, with the loading S fixed. With $\sigma_1 \equiv \sigma_r$, $\sigma_2 \equiv \sigma_\theta$, $\epsilon_1 \equiv \epsilon_r$, $\epsilon_2 \equiv \epsilon_\theta$, the use of (6), (9), (20) and the finite, integrated version of (21) then gives

$$-\frac{Y \sin(\alpha - \alpha_u)}{\sin \alpha} \frac{d\sigma}{\sigma} - \frac{[g \sin(\alpha - \alpha_u)]'}{[g \cos \alpha]'} \frac{d\epsilon}{\epsilon}$$

$$= \left\{ \frac{Y[g \sin(\alpha - \alpha_u)]'}{g \sin \alpha} + \frac{g^2([g \sin(\alpha - \alpha_u)]'/g^2)'}{(g \cos \alpha)'} \right\} d\alpha \qquad (41)$$

Now use $d\epsilon/\epsilon = n d\sigma/\sigma$ from (37); note that $\alpha = 0$ at $r = \infty$, $\alpha = \alpha_u$ at $r = a$; and observe that $S = X\sigma(\infty)$, $\sigma_r(a) = \sigma(a)$. Hence

$$\int_{r=\infty}^{r=a} \frac{d\sigma}{\sigma} = \log(\sigma(a)/\sigma(\infty)) = \log(KX)$$

where $K = \sigma_\theta(a)/S$ is the stress-concentration factor. Integration of (41) then produces the result

$$K = \frac{1}{X} \exp \int_0^{\alpha_u} - \frac{Y(g \cos \alpha)'[g \sin(\alpha - \alpha_u)]' + g^3 \sin \alpha([g \sin(\alpha - \alpha_u)]'/g^2)'}{Y(g \cos \alpha)' g \sin(\alpha - \alpha_u) + ng \sin \alpha [g \sin(\alpha - \alpha_u)]'} d\alpha \qquad (42)$$

Let $\epsilon_t(r)$ be the thickness-strain in the sheet. A convenient strain-concentration factor may be defined by

$$C_T = \epsilon_\theta(a)/|\epsilon_t(\infty)| = \epsilon_\theta(a)/2\epsilon_\theta(\infty) \qquad (43)$$

Since, by (20), $2\epsilon_\theta(\infty) = \epsilon(\infty)/X$, and $\epsilon_\theta(a) = \epsilon(a)$, it follows from (37) that

$$C_T = X(KX)^n \qquad (44)$$

EXAMPLES

(a) Hill [6] yield condition

Here $g \equiv 1$, the corresponding integral (42) was evaluated analytically earlier [3, 11], and the result, shown here for convenience, is

$$K = \sqrt{\frac{2}{1+R}} \left[\frac{n+1+2R}{n\sqrt{2+2R}}\right]^{\frac{n+1+2}{n^2+1+2R}} \exp\left[\frac{(n-1)\sqrt{1+2R}}{n^2+1+2R}\right] \tan^{-1}\sqrt{1+2R} \qquad (45)$$

(b) Hill [7] yield condition

With $g(\alpha)$ given by (27), the integrand in (42) can be written explicitly, and numerical integration executed. For $n = 3$, sample results for K are shown in Fig. 3 and for C_T in Fig. 4. The result (45) for the older Hill condition occurs at $p = 2$. For $p = 1$, we have $X = 1 + R$, $Y = 1 + 2R$, and analytic integration of (42) yields

$$K = \left[\frac{1}{1+R}\right]\left[\frac{n}{1+2R}\right]^{\frac{1+2R}{n-1-2R}} \tag{46}$$

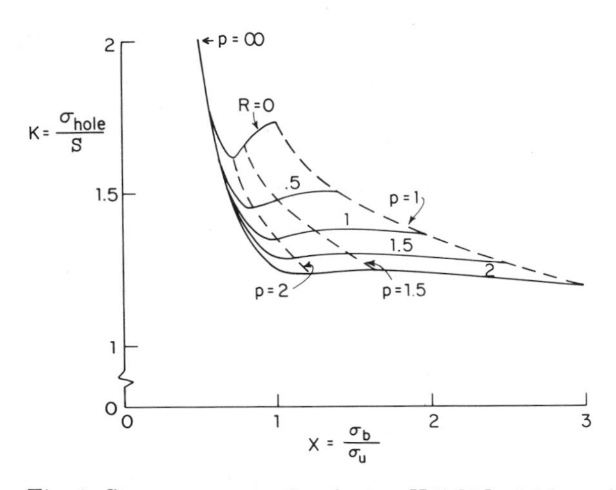

Fig. 3. Stress concentration factor, Hill [7] yield condition; $n = 3$.

A possibly useful observation that emerges from Fig. 4 is that the strain-concentration factor is much more sensitive to X than to R. Thus, direct measurements of R and C_T should establish X with good precision, assuming, of course, that the yield condition is valid.

Using the same new Hill yield condition, Durban and Birman [5] have recently analyzed the hole problem by elastic-plastic deformation theory, taking into account large strains and thickness changes. They reduce the problem to a first order differential equation analogous to (41), which they solve numerically. Parmar and Mellor [10] solved the same large-strain problem in sheets of finite size, neglecting elastic strains, but using flow theory, and a finite-difference numerical procedure. For sufficiently small holes, for plastic strains large compared to the elastic ones but still small compared to unity, and for power-law hardening, these solutions and the present one should agree closely. Some numerical examples shown by Durban and Birman corroborate this.

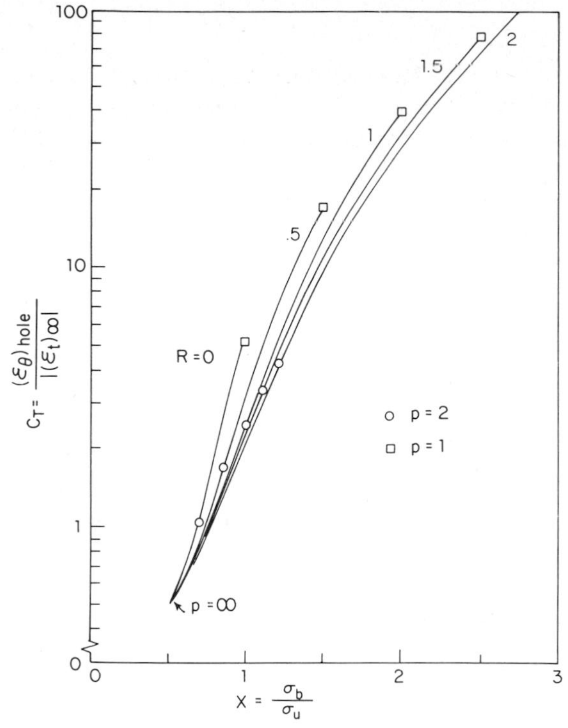

Fig. 4. Strain concentration factor, Hill [7] yield condition; $n = 3$.

(c) Bassani [1] yield condition

To implement the Bassani yield condition in (42), $q(\alpha)$ may be found by numerical solution of (33), $q'(\alpha)$ and $q''(\alpha)$ written in terms of $q(\alpha)$, and the integral evaluated numerically. Figure 5 has been prepared to show the spread in values of C_T that can emerge from those for $X = 1$ in the previously discussed solution for the 1979 Hill criterion, shown by the small circles.

(d) Maximum shear yield conditions

Calculations were also made for an idealized, anisotropic maximum-shear yield condition not contained in the families considered so far. Suppose yield is governed by *two* shearing yield stresses τ_y and $\bar{\tau}_y$, according to the rule that yield occurs whenever one of the conditions

$$\tfrac{1}{2}|\sigma_1 - \sigma_2| = \tau_y, \qquad \tfrac{1}{2}|\sigma_1| = \bar{\tau}_y, \qquad \tfrac{1}{2}|\sigma_2| = \bar{\tau}_y \tag{47}$$

is met. Thus τ_y is the in-plane, maximum-shear yield-stress, while $\bar{\tau}_y$ is the yield stress associated with the other two principal shears. The yield loci in the (σ_1, σ_2) plane for various ranges of $\lambda = \bar{\tau}_y/\tau_y$ are sketched in Fig. 6, together with the corresponding loci in the normalized coordinates

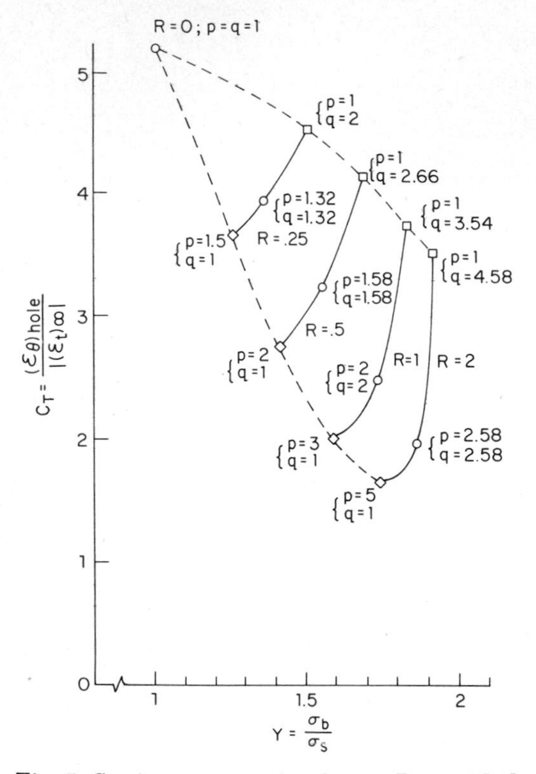

Fig. 5. Strain concentration factor, Bassani [1] yield condition; $x = 1$, $n = 3$.

of the present paper. Values of X, Y, and R are also shown in Fig. 6. (Note the sudden transition of R from one extreme to the other at $\lambda = 1$.) The heavy solid lines show the anticipated range of stress ratios in the hole problem, wherein there is a transition from balanced biaxial tension at infinity to uniaxial tension at the hole.

In range (i) (see Fig. 6) we have

$$g(\alpha) = (\cos\alpha + \sin\alpha)^{-1} \tag{48}$$

and $\alpha_u = \pi/4$. Evaluation of the integral (42) gives

$$K = n^{\left(\frac{1}{n-1}\right)} \tag{49}$$

(Note that this is the same as (46) with $R = 0$, as it should be.) The same result obviously must hold in range (ii), since the identical relevant yield line is operative in (σ_1, σ_2) space. Thus (49) applies for all $\lambda \leqslant 1$.

In range (iii) we have

$$\begin{aligned} g(\alpha) &= \frac{2\lambda}{\sin\alpha + 2\lambda\cos\alpha} \qquad \text{for} \quad 0 < \alpha < \bar{\alpha} \\ &= 1/\sin\alpha \qquad \text{for} \quad \alpha > \bar{\alpha} \end{aligned}$$

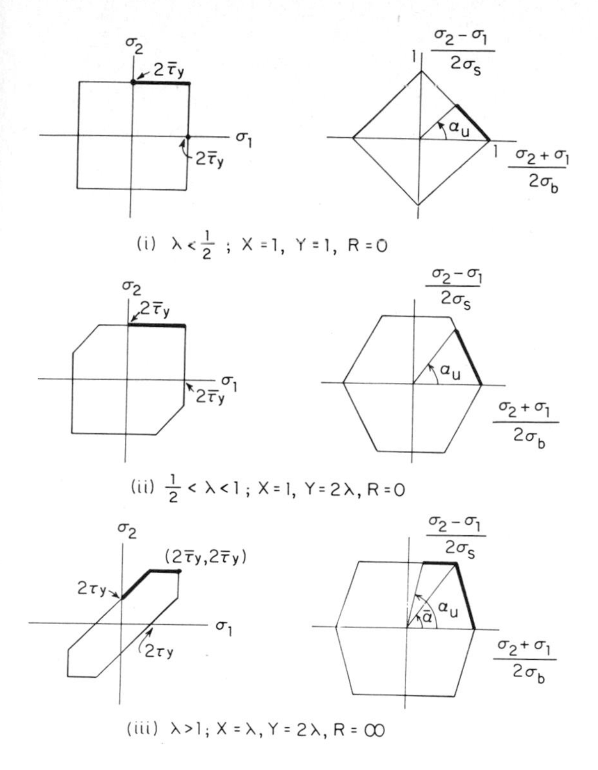

Fig. 6. Maximum-shear yield loci; $\lambda = \bar{\tau}_y/\tau_y$.

where

$$\bar{\alpha} = \tan^{-1} 2\lambda/(2\lambda - 1)$$

is the polar angle of the corner. The integral in (42) produces the three contributions

$$\int_0^{\bar{\alpha}_-} (\ \ ) \mathrm{d}\alpha = \left(\frac{1}{n-1}\right) \log\left(\frac{n+\lambda-1}{\lambda}\right)$$

$$\int_{\alpha_-}^{\bar{\alpha}_+} (\ \ ) \mathrm{d}\alpha = \left(\frac{1}{n+2\lambda-1}\right) \log\left(\frac{n+\lambda-1}{\lambda(n+2\lambda-2)}\right)$$

$$\int_{\alpha_+}^{\alpha_u} (\ \ ) \mathrm{d}\alpha = \log\left(\frac{n+2\lambda-2}{n}\right)$$

where the second piece arises from the jump in $g'(\alpha)$ at $\bar{\alpha}$. With $X = \lambda$, the final result given by (42) is

$$K = \left[\frac{n+2\lambda-2}{\lambda n}\right]\left[\frac{n+\lambda-1}{\lambda}\right]^{\frac{1}{n-1}}\left[\frac{n+\lambda-1}{\lambda(n+2\lambda-2)}\right]^{\frac{1}{n+2\lambda-1}} \tag{50}$$

For $\lambda \to \infty$, this gives $K = 2/n$.

For $\lambda = 1$, the material obeys the standard isotropic Tresca yield condition. A comparison of K for this case (Eqn. (49)) with that for the isotropic Mises material given by Eqn. (45) with $R = 1$, is shown in Fig. 7, in the range $1 \leqslant n < \infty$.

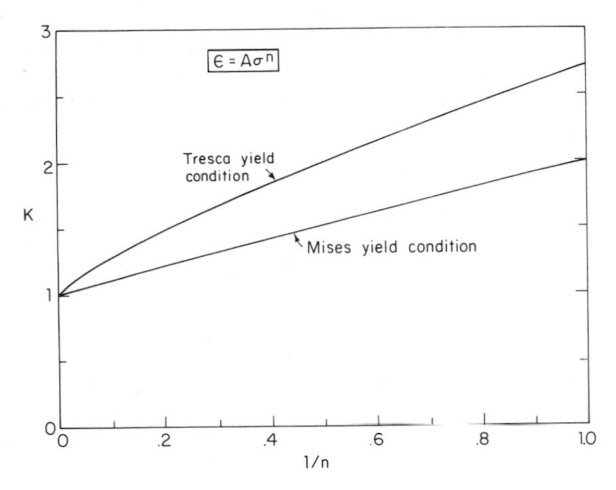

Fig. 7. Comparison of stress concentration factors, isotropic sheets: Tresca vs. Mises.

Concluding remarks

The general constitutive equations proposed may provide a convenient unified basis for the characterization of orthotropic, planar-isotropic thin sheet plasticity. With isotropic power-law hardening, exact solutions are possible for small-strain, axisymmetric stress problems other than the one treated here. For example, there should be no difficulty in considering radial stretching stress at the hole boundary. Further, it should be possible to get a neat *elastic*-plastic solution for a material governed by a pure power law beyond the elastic limit.

Acknowledgment

This work was supported in part by the National Science Foundation under Grant CME78-10756, and by the Division of Applied Sciences, Harvard University.

References

1 Bassani, J.L., 1977. Yield characterization of metals with transversely isotropic plastic properties. Int. J. Mech. Sci., 19: 651—660.
2 Budiansky, B. and Mangasarian, O.L., 1960. Plastic concentration at a circular hole in an infinite sheet subjected to equal biaxial tension. Jour. Appl. Mech., 27: 59—64.
3 Budiansky, B., 1971. An exact solution to an elastic-plastic stress-concentration problem, (in Russian). Prikl. Mat. Mekh., 35: 40—48; in English: Jour. Appl. Math. Mech., 35 (1): 18—25.
4 Drucker, D.C., 1950. A more fundamental approach to plastic stress-strain relations. Proc. 1st U.S. Nat. Cong. Appl. Mech., 1950, pp. 487—492.
5 Durban, D. and Birman, V., 1982. On the elasto-plastic stress concentration at a circular hole in an anisotropic sheet. Acta Mechanica, 43 (1-2): 73—84.
6 Hill, R., 1948. A theory of the yielding and plastic flow of anisotropic metals. Proc. R. Soc., London, A 193, pp. 281—297.
7 Hill, R., 1979. Theoretical plasticity of textured aggregates. Math. Proc. Camb. Phil. Soc., 55: 179—191.
8 Ilyushin, A.A., 1946. The theory of small elastic-plastic deformations. Prikl. Mat. Mekh., 10 (3): 347—356.
9 Parmar, A. and Mellor, P.B., 1978a. Predictions of limit strains in sheet metal using a more general yield condition. Int. J. Mech. Sci., 20: 385—391.
10 Parmar, A. and Mellor, P.B., 1978b. Plastic expansion of a circular hole in sheet metal subjected to biaxial tensile stress. Int. J. Mech. Sci., 20: 707—720.
11 Yang, W.H., 1969. Axisymmetric plane stress problems in anisotropic plasticity. Jour. Appl. Mech., 36: 7—14.

Soil Mechanics, Plasticity and Landslides

W.F. CHEN

School of Civil Engineering, Purdue University, West Lafayette, IN 47907 (U.S.A.)

Abstract

After a brief historical introduction, the impact of D.C. Drucker's ideas on the modern development of the theory of soil plasticity is discussed. Recent achievements in plasticity as applied to soil mechanics are summarized with emphasis on applications to earthquake-induced landslide problems.

To this end, a detailed description of the following three subjects is given: (1) stress—strain relations for soil; (2) limit analysis for seismic stability of slopes; and (3) finite element analysis for progressive failure behavior of slopes under earthquake loading. In this way, some of the interrelationships between the limit analysis of perfect plasticity and the finite element analysis of work-hardening plasticity are demonstrated, and their power and their relative merits and limitations for practical applications are evaluated within the context of their use in the seismic analyses of soil mass involving slope failures and landslides.

1. Introduction

In the 1950s major advances were made in the theory of metal plasticity by the development of (i) the fundamental theorems of limit analysis; (ii) Drucker's postulate or definition of stability of material; and (iii) the concept of normality condition or associated flow rule. The theory of limit analysis of perfect plasticity leads to practical methods that are needed to estimate the load-carrying capacity of structures in a more direct manner. The concept of a stable material provides a unified treatment and broad point of view of the stress—strain relations of plastic solids. The normality condition provides the necessary connection between the yield criterion or loading function and the plastic stress—strain relations. All these have led to a rigorous basis for the theory of classical plasticity, and laid down the foundations for subsequent notable developments.

The initial applications of the classical theory of plasticity were almost exclusively concerned with perfectly plastic metallic solids such as mild steel which behaves approximately like a perfectly plastic material [62]. For these materials, the angle of internal friction ϕ is zero, no plastic volume change occurs and the only material property is the shear strength k or cohesion c in the terminology of soil mechanics. Numerical calculations were restricted to the method of characteristics based on the theory of the plane slip-line field analysis to derive the stress and velocity distribution in the plastic region [36]. Since the plane slip-line field analysis is rarely applicable to structures, exact and approximate calculations of the plastic collapse load were made exclusively by the methods of limit analysis [29].

The development of the modern theory of soil plasticity, as the new field was called, was strongly influenced by the well-established theory of metal plasticity. Soil mechanics specialists have been preoccupied with extending these concepts to answer the complex problems of soil behavior. Tresca's yield condition, used widely in metal plasticity, can be regarded as a special case of the condition of Coulomb on which the important concept of the limiting equilibrium of a soil medium had been firmly established in soil mechanics [80].

It is a relatively straightforward matter to extend the method of characteristics to cover Coulomb material where c and ϕ can either remain constant [79] or vary throughout the stress field in some specified manner [3]. In the theory of limit equilibrium, the introduction of stress—strain relations was obviated by the restriction to the consideration of equations of equilibrium and a yield condition. This produces what appears to be and sometimes is static determinacy for the solutions of slip line field equations. However, in many soil—structure interaction problems, the boundary conditions involve rates of displacement and the slip line equations are generally statically indeterminate. The key to obtaining a valid solution for such cases requires the basic knowledge of the stress—strain relations. Otherwise, a so-called solution is merely a guess.

The general theory of limit analysis, developed in the early 1950s, considers the stress—strain relation of a soil in an idealized manner. This idealization, termed normality or the associated flow rule, establishes the limit theorems on which limit analysis is based. Although the applications of limit analysis to soil mechanics are relatively recent, there have been an enormous number of practical solutions available [13]. Many of the solutions obtained by the method are remarkably good when comparing with the existing results for which satisfactory solutions already exist. As a result of this development, the meaning of the limit equilibrium solutions in the light of the upper- and lower-bound theorems of limit analysis becomes clear.

The first major advance in the extension of metal plasticity to soil plasticity was made in the paper "Soil Mechanics and Plastic Analysis or

Limit Design" by Drucker and Prager [32]. In this paper, the authors extended the Coulomb criterion to three-dimensional soil mechanics problems. The Coulomb criterion was interpreted by Drucker [28] as a modified Tresca as well as an extended von Mises yield criterion. The yield criterion obtained by Drucker and Prager for the later case is now known as the Drucker—Prager model or the extended von Mises model.

One of the main stumbling blocks in the further development of the stress—strain relations of soil based on the Drucker—Prager type or Coulomb type of yield surfaces to define the limit of elasticity and beginning of a continuing irreversible plastic deformation was the excessive prediction of dilation, which was the result of the use of the associated flow rule. It became necessary, therefore, to extend classical plasticity ideas to a "non-associated" form in which the plastic potential and yield surfaces are defined separately [24]. However, this modification eliminated the validity of the use of limit theorems for bounding collapse loads and created doubts about the uniqueness of solutions. Attempts have been made to revise the bounding theorems and to resolve the uniqueness problem, but to date not much success has been achieved through this route [58].

In 1957, an important advance was made in the paper "Soil Mechanics and Work-Hardening Theories of Plasticity" by Drucker, Gibson and Henkel [30]. In this paper the authors introduced the concept of work-hardening plasticity into soil mechanics. There are two important innovations in the paper. The first is the introduction of the idea of a work-hardening cap to the perfectly plastic yield surface such as the Coulomb type or Drucker—Prager type of yield criterion. The second innovation is the use of current soil density (or voids ratio, or plastic compaction) as the state variable or the strain-hardening parameter to determine the successive loading cap surfaces.

These ideas have led in turn to the generation of many soil models, most notably the development of the critical state soil mechanics at Cambridge University, U.K. These new soil models have grown increasingly complex as additional experimental data have been gathered, interpreted, and matched. This extension marks the beginning of the modern development of a consistent theory of soil plasticity [13].

Instead of tracing the historical development of the theory of soil plasticity in this paper, I shall attempt to survey its present status. The present survey deals with three equally important aspects of progress: stress—strain modeling of soils, the application of limit analysis to stability problems in soil mechanics, and new numerical solutions to specific boundary value problems. In stressing the new solutions in soil mechanics, three broad groups of typical problems in soil mechanics should be tackled: (1) earth retaining structures (active and passive pressures); (2) foundation (bearing capacity); and (3) soil slopes. However, the discussion of most of these problems would require the presentation of a greater amount of details than is practical. Furthermore, almost all the interesting aspects of recent achievements can be brought out in the class of problems related to slopes. Accordingly, this paper is devoted primarily to a consideration of the concept of soil plasticity in relation to its applications to stability and progressive failures of slopes under static and earthquake loading conditions.

2. Stress—strain relations

Soil mechanics along with all other branches of mechanics of solids requires the consideration of geometry or compatibility and of equilibrium or dynamics. The essential set of equations that differentiate the soil from other solids is the relation between stress and strain. The behavior of soils is very complicated. The attempt to incorporate the various features of soil properties in a single mathematical model is not likely to be successful, but even if such a model could be constructed, it would be far too complex to serve as the basis for the solution of practical geotechnical engineering problems. Simplifications and idealizations are essential in order to produce simpler models that can represent those properties that are essential to the considered problem. Thus, any such simpler models should not be expected to be valid over a wide range of conditions [33, 35, 71].

The need for mathematical simplicity in the description of the mechanical properties of solids is understood quite well for metals where so much research effort has been expended by so many investigators [29]. Yet even for metals, the simple idealizations such as perfect plasticity, isotropic hardening, kinematic hardening, and mixed hardening are frequently used in solving practical problems [12, 21]. The same situation is to be expected for the stress—strain modeling of soil which is a far more complex material [34, 35, 70].

Drastic idealizations are valuable not only for the ease of treatment of practical engineering problems but also conceptually for a clear physical understanding of the essential features of the complex behavior of a material under certain conditions. Therefore, for soils, as for metals, perfect plasticity is still an excellent design assumption, while very complex stress—strain relations of soil which require an ever increasing elaboration in detail of a mathematical description may be approximated crudely by simple isotropic, kinematic, or mixed hardening models. Thus, the isotropic hardening cap models and Cambridge models, the kinematic hardening nested yield surfaces models, or the mixed hardening bounding surface models that have been proposed and developed in recent years are all within the realm of this simplification. In the sections that follow, these developments are described and, hopefully, unified within the same framework of physically and mathematically well-established theory of work-hardening plasticity.

It is important to note here that the path to reach the present state-of-the-art in soil plasticity is by no means easy and that much of the classical theory of plasticity for metals has been considerably modified in order to obtain reasonable results for soils. These achievements have been summarized in two earlier symposia [58, 61], as well as in two recent ASCE workshop and symposium proceedings [59, 81, 82]. An up to date summary will appear in the forthcoming book "Constitutive Equations for Engineering Materials — Vol. 2 Plasticity and Modeling" by Chen and Saleeb [21] and the Proceedings on "Constitutive Laws for Engineering Materials: Theory and Application" edited by Desai and Gallagher [25].

The use of strain- or work-hardening plasticity theories in soil mechanics has been developed for about twenty-five years, since publication of the classical paper by Drucker, Gibson and Henkel [30]. Most of the research has been conducted by engineers working in the area of soil statics. Recently, attention has been focused on the use of these models in soil dynamics [11]. The objective of this section is to set forth the state-of-the-art with respect to elastic-plastic stress—strain relations of soils. In doing so, it achieves not only the purpose of surveying the current research activity that has been going on very actively in this field in recent years, but also the survey gives the best indications of future problems that may result from the observations of the trend of recent developments.

One of the main problems in the theory of plasticity is to determine the nature of the subsequent yield surfaces. This post-yielding response is described by the hardening rule which specifies the rule for the evolution of the loading surfaces during the course of plastic deformations. Indeed, the assumption made concerning the hardening rule introduces a major distinction among various plasticity models developed for soils in recent years.

2.1 HARDENING (SOFTENING) RULES

There are several hardening rules that have been proposed to describe the growth of subsequent yield surfaces for strain-hardening (softening) materials. The choice of a specific rule depends primarily on the ease with which it can be applied and its ability to represent the hardening behavior of particular material. In general, three types of hardening rules have been commonly utilized [12]. These are: (1) Isotropic hardening; (2) Kinematic hardening; (3) Mixed hardening. In an isotropic hardening model, the initial yield surface is assumed to expand (or contract) uniformly without distortion as plastic flow continues. On the other hand, the kinematic hardening rule assumes that, during plastic deformations, the loading surface translates without rotation as a rigid body in the stress space, maintaining the size and shape of the initial yield surface. This rule provides a means of accounting for the Bauschinger effect, which refers to one particular type of directional anisotropy induced by plastic deformations; namely that an initial plastic deformation of one sign reduces the resistance of the material with respect to a subsequent plastic deformation of the opposite sign. Therefore, kinematic hardening models are particularly suitable for materials with pronounced Bauschinger effect such as soils under cyclic and reversed types of loading.

A combination of isotropic and kinematic hardening models leads to a more general hardening rule, and therefore provides for more flexibility in describing the hardening behavior of the material. For a mixed (combined) hardening model, the loading surface experiences translation as well as expansion (contraction) in all directions, and different degrees of Bauschinger effect may be simulated. Kinematic and mixed types of hardening rules are generally known as anisotropic hardening models.

In the last few years, several plasticity models with more complex hardening rules combining the concepts of kinematic and isotropic hardening have been developed and applied to describe the behavior of soils under cyclic loading. Some of these models will be discussed in the sections that follow.

2.2 PERFECT PLASTICITY MODELS

Perfect plasticity is an appropriate idealization for a structural metal because it captures the essential features of its behavior. This includes small tangent modulus when compared with elastic modulus, when loading in the plastic range, and the unloading response is elastic. However, perfect plasticity is not nearly appropriate for soils. Some of the troubles and their justifications for adoption of this idealization for practical use were discussed in the paper "Concepts of Path Independence and Material Stability for Soils" by Drucker [27].

For the most part, the concept of perfect plasticity has been used extensively in the past in conventional soil mechanics in assessing the collapse load in stability problems. Different widely known techniques have been employed to obtain numerical solutions in these cases; such as the slip-line method [79], and the limit equilibrium method [80]. For the later case, the simple ideas of perfect plasticity have found direct application in many practical geotechnical engineering problems.

In addition to these classical methods, the more rigorous approach of modern limit analysis of perfect plasticity has been applied to a wide variety of practical stability problems. Using the well-known Coulomb yield criterion and its associated flow rule, many solutions have been obtained [13]. Recently, the stability analysis has been extended to include the earthquake loading, employing the pseudo-static force method [11, 14, 15, 19]. It should be emphasized here that the useful application of these techniques has not been exhausted. New and striking applications are not only possible but to be encouraged strongly, because of their simplicity and power in helping us reach an understanding of, and feel for, a problem. Further, some predictions of this enormous idealization are very good. Much more of value will be uncovered as engineers who have need for particular results apply the methods of limit analysis and design to their own special problems. The power and simplicity of this method will be brought out in the subsequent discussions through the example of seismic stability analysis of slopes. This example will show not only the power and simplicity of the method, but also add insight to the practicing engineers' understanding of the complicated landslide problems [9, 42].

2.3 ISOTROPIC HARDENING MODELS (CONCEPT OF HARDENING CAP)

Drucker, Gibson and Henkel [30] were the first to suggest that soil might be modeled as an elastic-plastic isotropic hardening material. They proposed that successive yield surfaces might resemble extended von Mises (or Drucker—Prager) cones with convex end spherical caps (Fig. 1). As the soil strain-hardens, both the cone and the end cap expand. As mentioned previously, there are two important innovations in this work. The first is the introduction of the idea of a spherical cap fitted to the cone. The second is the use of current soil density (specific volume or void ratio) as the strain-hardening parameter to determine the successive loading surfaces, such as the surface marked I in Fig. 1 for a particular value of soil density. There will be a succession of such surfaces, all geometrically similar, but of different sizes, for different densities. This strain- or work-hardening model was a major step forward toward a more realistic representation of soil behavior. Two specific isotropic hardening models are discussed in the following.

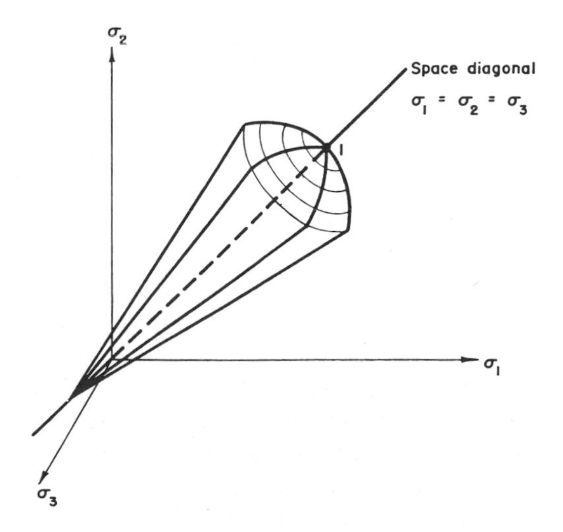

Fig. 1. A Drucker—Prager type of strain-hardening cap model.

(i) Cambridge models (concept of critical state)

The introduction of isotropic hardening plasticity into soil mechanics led in turn to the generation of the family of soil models of the strain-hardening cap type developed at Cambridge University by the late Professor Roscoe and his co-workers. Two of these models are now widely used. The first is known as the Cam-clay model [75] and was originally formulated by Roscoe, Schofield, and Thurairajah [69] for normally and lightly over-consolidated clays in the triaxial test. The second, known as the Modified Cam-clay model, was developed by Roscoe and Burland [67] as a modification and extension of the Cam-clay model to a general three-dimensional stress state. In both models, associated flow rule was used. The modified Cam-clay model is an isotropic, nonlinear elastic-plastic strain-hardening model. Only the volumetric strain is assumed to be partially recoverable; i.e. elastic shearing strain is assumed to be identically zero. The elastic volumetric strain is non-linearly dependent on the hydrostatic stress and is independent of the deviatoric (shear) stresses. For certain stress histories, strain-softening may occur. Extensive reviews of various types of Cambridge models have been given in two symposia [58, 61].

The important feature that has been the integral part of all Cambridge models is the concept of critical states proposed by Roscoe, Schofield, and Wroth [68], and independently conceived by Parry [60]. The critical state line is the locus of the failure points of all shear tests under both drained and undrained conditions. Its crucial property is that failure of initially isotropically compressed samples will occur once the states of stress in the samples reach the line, irrespective of the test path followed by the samples on their way to the critical state line [1]. At the critical state, large shear deformations occur with no change in stress or plastic volumetric strain.

Experimental results have indicated that, at failure, the behavior of soil is governed by the Coulomb failure criterion [2]. The failure (yield) surface corresponding to the Coulomb criterion in principal stress space is an irregular hexagon pyramid whose apex lies on the space diagonal as shown in Fig. 2. A strain-hardening cap intersects with the Coulomb surface in a line $ABCDEFA$. This line is called the critical state locus, and it is simply a line which separates the two surfaces corresponding to one fixed value of the specific volume (or void ratio). The complete failure surface is termed a state boundary surface [1, 67]. There will be similar surfaces of different sizes, but of the same shape, corresponding to different values of the specific volume. The geometric representation of the shape of the generalized state boundary surface requires four dimensions (i.e., three principal stress and one specific volume).

Since the development of Cambridge models, many attemps have been made to use various versions of these models in numerical solutions of

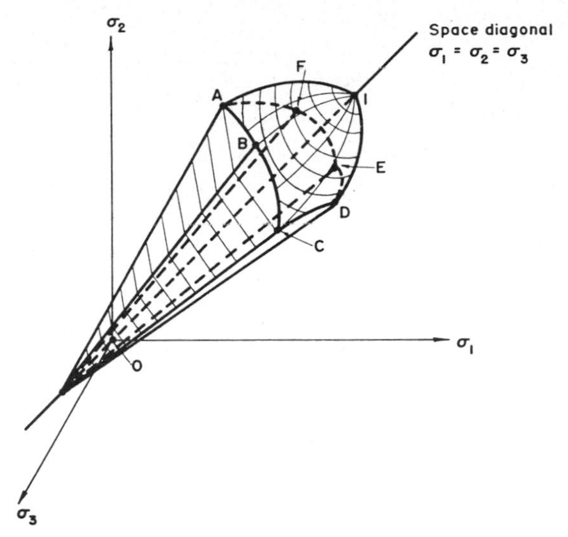

Fig. 2. The complete state boundary surface in principal stress space — Cambridge type of strain-hardening cap model [1].

boundary value problems [4] and predictions of soil behavior in the field (see for example, Palmer [58].)

(ii) Cap models

Dimaggio and Sandler [26] proposed a generalized cap model. The yield function they used consisted of a perfectly-plastic (failure) portion fitted to a strain-hardening elliptical cap. Associated flow rule was employed for the failure and cap functions. In this model, the functional forms for both the perfectly-plastic and the strain-hardening portions may be quite general and would allow for the fitting of a wide range of material properties. A simple version of the cap model will be used in the later part of the paper for a complete progressive failure stress analysis of earthquake-induced landslide problems.

The movement of the cap is controlled by the increase or decrease of the plastic volumetric strain. Strain-hardening in this model can therefore be reversed. It is this mechanism that leads to an effective control on dilatancy, which can be kept quite small (effectively zero) as required for many soils. Their model has also been adopted for rocks [73, 74] by allowing only expansion of the cap (i.e. hardening). In this variation of the model, the cap movement is assumed to depend only on the maximum previous value of the plastic volumetric strain, and consequently the cap is not reversible. In this way, the model allows representation of a relatively large amount of dilatancy, which is often observed during failure of rocks at low pressures.

These generalized cap models have also been expanded to include rate effects, and anisotropic behavior within the yield surface and viscoplastic behavior during yielding [54, 72]. Many of these variations of the generalized cap models are now widely used in ground shock computations [54—56, 72].

Using the idea of a cap fitted to the yield surface, Lade [44, 45] has developed a strain-hardening model to describe the behavior of different sands and normally consolidated clays. He used a spherical cap together with a modified version of the conical yield surface suggested earlier by Lade and Duncan [46]. In this model, both the conical yield surface and the cap are allowed to harden isotropically. However, for the cap, associated flow rule was employed, while non-associated rule was used in connection with the conical portion.

2.4 KINEMATIC HARDENING MODELS (CONCEPT OF NESTED YIELD SURFACE)

An alternative approach to the isotropic hardening type of models described above is provided by the kinematic type of strain-hardening rules. Recently, this approach has been employed by several researchers to provide for a more realistic representation of soil behavior under reversed, and particularly cyclic, loading conditions.

Iwan [38], following the related work of Masing [47], proposed one-dimensional plasticity models which consist of a collection of perfectly elastic and rigid-plastic or slip elements arranged in either a series—parallel or a parallel—series combination. The model can contain a very large number of elements, and the properties of these elements can be distributed such that they can match the particular form of hysteretic behavior of a certain type of soil. Such models are known as the overlay or mechanical sublayer models [83].

In order to extend the one-dimensional model to three-dimensional situations, an extended formulation of the classical incremental theory of plasticity has been proposed by Iwan [38]. Instead of using a single yield surface in stress space, he postulated a family (nest) of yield surface (Fig. 3) with each surface translating independently in a pure kinematic manner, or individually obeying a linear work-hardening model. Their combined action, in general, gives rise to a nonlinear work-hardening behavior for the material as a whole. The approach leads to a realistic Bauschinger effect of a type that could not be obtained by using a single yield surface and a nonlinear work-hardening rule even with kinematic hardening. The same concept of using a field of nested yield surfaces was also proposed independently by Mroz [51]. These models have been used recently for soils and are usually known as multi-surface plasticity models. In all the proposed models of this type, associated flow rule has been utilized.

Figure 3 demonstrates the qualitative behavior of a multisurface model

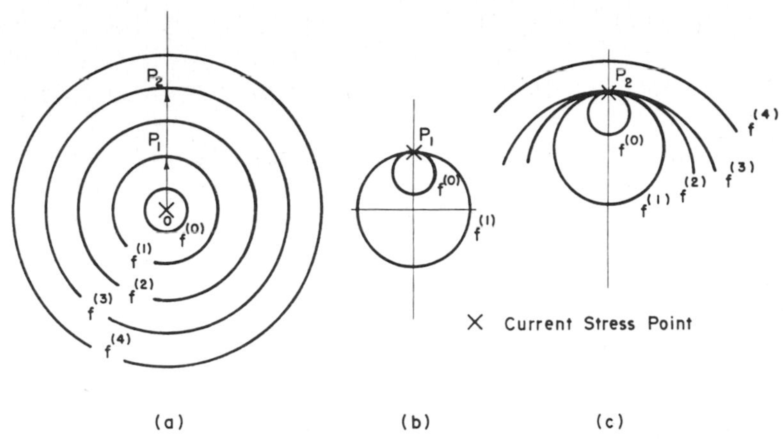

Fig. 3. Series of nested yield surfaces in Mroz/Iwan type of strain-hardening models.

with pure kinematic hardening. The initial positions of the yield surface $f^{(0)}$, $f^{(1)}$, $f^{(2)}$, and $f^{(3)}$ are shown in Fig. 3(a). When the stress point moves from O to P_1, elastic strains first occur until the surface $f^{(0)}$ is reached, where the plastic flow begins and the surface $f^{(0)}$ starts to move towards the surface $f^{(1)}$. Before their contact, the hardening modulus $H^{(0)}$ associated with $f^{(0)}$ governs the plastic flow according to the normality flow rule. However, when $f^{(0)}$ engages $f^{(1)}$ at P_1 (Fig. 3(b)), the first nesting surface $f^{(1)}$ becomes the active surface and, upon further loading, the hardening modulus $H^{(1)}$ applies in the flow rule. Both $f^{(0)}$ and $f^{(1)}$ are then translated by the stress point, and they remain tangent to each other on the stress path until they touch the yield surface $f^{(2)}$ which then becomes the active surface. For subsequent contacts of consecutive surfaces, the process is repeated with new corresponding values of hardening moduli applying. The situation when $f^{(3)}$ is reached is illustrated in Fig. 3(c).

Applications of the Iwan/Mroz model to study the seismic response of two-dimensional configurations of soils have been made by Joyner and Chen [40]. The model has been found to be particularly promising for use in calculating the response of earth dams subjected to earthquake ground shaking.

Recently, Prevost [63–65] has extended the Iwan/Mroz model for the undrained behavior of clays under monotonic and cyclic loading conditions. In this development, the basic strain-hardening rule is still of the kinematic type but a simultaneous isotropic hardening (or softening) is allowed. The rule used to govern the translation of the nested yield surfaces during plastic loading was that suggested by Mroz [51]. According to this rule, when the stress point P (Fig. 4) reaches the yield surface $f^{(m)}$, then, upon further loading (i.e. when the stress increment $\dot{\sigma}_{ij}$ is applied), the instantaneous translation of $f^{(m)}$ towards the next yield

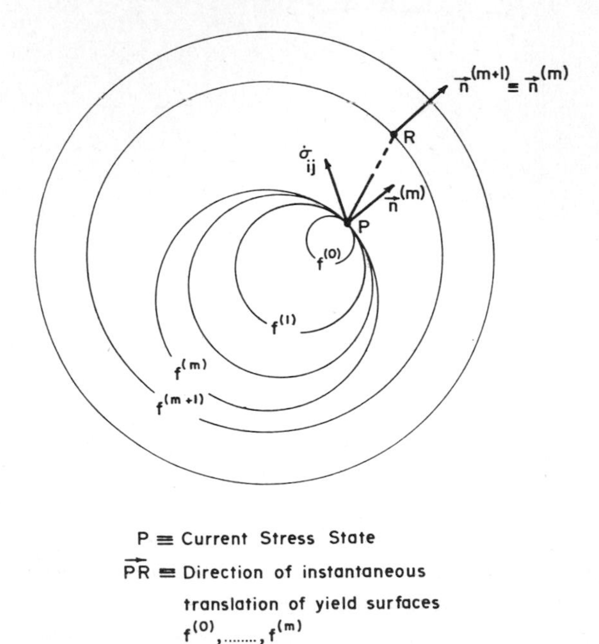

Fig. 4. Translation rule of nested yield surfaces in stress space — Prevost type of strain-hardening models.

surface $f^{(m+1)}$ will be along $\overline{PR}$, where R (known as the conjugate point) is the point on $f^{(m+1)}$ with outward normal in the same direction as the normal to $f^{(m)}$ at P (i.e., $\overline{n}^{(m+1)} = \overline{n}^{(m)}$ in Fig. 4).

This model has been adopted very recently by Prevost et al. [66] in the finite element analyses of soil-structure interaction of centrifugal models under both monotonic and cyclic loadings simulating the situation encountered in the analysis of offshore gravity structure foundations under wave forces. The results obtained from the analysis agree quite well with those measured experimentally in the centrifuge. This study has demonstrated the ability of the multi-surface model to provide realistic representation of soil behavior under complex loadings.

2.5 MIXED HARDENING MODELS (CONCEPT OF BOUNDING SURFACE)

Various types of strain-hardening plasticity models have been recently employed for soils based on the bounding (consolidation or limiting) surface concept introduced earlier for metals [22, 43]. A two-surface model of this type was proposed by Mroz, Norris and Zienkiewicz [52, 53] for clays. A bounding surface, $F = 0$, representing the consolidation history of the soil, and a yield surface, $f = 0$, defining the elastic domain within the bounding surface (Fig. 5) were employed in the model.

The bounding surface was assumed to expand or contract isotropically, but the yield surface was allowed to translate, expand or contract within

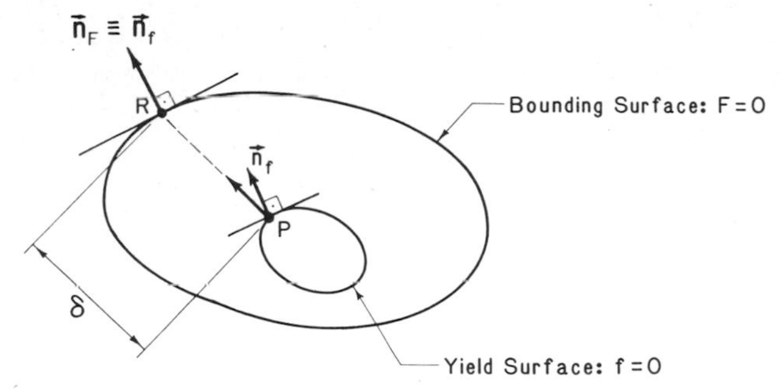

Fig. 5. Yield and bounding surfaces in stress space.

the domain enclosed by the bounding surface. The translation of the yield surface is governed by the same rule as the multi-surface models described earlier (i.e., f will translate towards the bounding surface along $\overline{PR}$ in Fig. 5).

The hardening modulus on the yield surface is assumed to vary depending on the relative configuration of the yield and bounding surfaces. It was this last assumption that distinguished the present formulation from the previous multi-surface models. Here, instead of using a field of nested yield surfaces with associated hardening moduli, an interpolation rule is utilized to define the variation of the hardening moduli between the yield and the bounding surfaces. In the interpolation rule used by Mroz et al. [52, 53], the hardening modulus, H, was taken as a function of the distance δ (Fig. 5) between the current stress point, P, on the yield surface and its conjugate point, R, on the bounding surface. Very detailed discussions of the present model and its application to represent the behavior of clays under monotonic and cyclic triaxial test conditions have been given by Mroz et al. [53].

Dafalias and Herrmann [23] have also used the bounding surface formulation to describe the behavior of clay under cyclic loadings. However, no explicit yield surface was postulated. Associated flow rule was utilized for the bounding surface. In this model, the variation of the plastic (hardening) modulus within the boundary surface is defined based on a very simple radial mapping rule. For each actual stress point within (or on) the bounding surface, a corresponding "image" point on the surface is specified as the intersection of the surface with the straight (radial) line connecting the origin with the current stress point (the origin was assumed to be always within the bounding surface). The actual hardening modulus is then assumed to be a function of the hardening modulus on the bounding surface, at the "image" point, and the distance between the actual stress point and its "image".

3. Limit analysis of perfect plasticity

Limit analysis is concerned with the development of efficient methods for computing the collapse load in a direct manner. It is therefore of intense practical interest to practicing engineers. There have been an enormous number of applications in metal structures. Applications of limit analysis to reinforced concrete structures are more recent and are given in a recent book by Chen [12] as well as a colloquium proceedings [37]. Applications to typical stability problems in soil mechanics have been the most highly developed aspect of limit analysis so that the basic techniques and many numerical results have been summarized in the book by Chen [13]. Extensive references to the work before 1975 are also given in the book cited. An up-to-date reference to recent work on the applications of limit analysis to earth pressure, bearing capacity and slope stability problems can be found in the ASCE Proceedings [81, 82], theses [5–7, 41] and papers [8, 15], among others.

It is true, as in most fields of knowledge, that many of the basic ideas of perfect plasticity and limit analysis have been used extensively and fruitfully in the past in conventional soil mechanics through experimental studies and engineering intuition. Here, the standard and widely-known techniques of the slip-line method and the limit equilibrium method, among others, come to mind immediately, but these methods also have been mentioned previously.

The slip-line method uses the Coulomb criterion as the yield condition for soil. From the basic slip-line differential equations the slip-line network can be constructed and the collapse load determined. Examples of this approach are the solutions presented in the book by Sokolovskii [79].

The limit equilibrium method can be best described as an approximate approach to the construction of the slip-line field. It generally entails the assumption of the failure surface of various simple configurations from which it is possible to solve problems by simple statics. Terzaghi [80] cites some examples of this approach.

Although these methods are widely used in geotechnical practice, they neglect altogether the important fact that the stress–strain relations constitute an essential consideration in a complete theory of any branch of the continuum mechanics of deformable solids. Modern limit analysis methods take into consideration in an idealized manner the stress–strain relations of the continuum, the soil in the present case. This idealization, termed normality or flow rule, establishes the limit theorems on which limit analysis is based. Within the framework of perfect plasticity and the associated flow rule assumption, the approach is rigorous and the techniques are competitive with those of limit equilibrium approach. In several instances, especially in slope stability analysis, such a level of reliability and completeness has been achieved and firmly established in recent years that the limit analysis method can be used as a working tool for design engineers to solve everyday problems. Examples of this approach are the solutions presented in the book by Chen [13].

However, most of the applications of limit analysis of perfect plasticity to geotechnical problems have been limited to soil statics. The on-going work discussed here attempts to extend this method to the area of soil dynamics, in particular to earthquake-induced landslide problems. Further details of the method of attack are given in the sections that follow.

3.1 THE PROBLEM OF EARTHQUAKE-INDUCED LANDSLIDES

The conventional method for evaluating the effect of an earthquake load on the stability of a slope is the so-called "pseudo-static method of analysis". In this analysis, the inertia force is treated as an equivalent concentrated horizontal force (the "pseudo-static force") at some critical point (usually the center of gravity) of the critical sliding mass. The inadequacies of this method for slope stability analysis are discussed in various papers by many authors, most notably by H. Bolton Seed of the University of California [76].

(1) The pseudo-static force is applied as a permanent force whereas in reality the reduction in the stability of the slope exists only during the short period of time for which the unfavorable direction of the inertia force is induced. As a result, the factor of safety may drop below unity for a very short duration and some permanent displacements can occur in this duration. But this may not cause the collapse of a slope.

(2) Since the soils are not rigid, the acceleration is not uniform throughout the slope. The distribution of seismic coefficients is therefore a function of the height of a slope. This has been the subject of study of several investigators, in particular, Seed and his colleagues who used the viscoelastic model together with the finite element method [76]. Suggested values for the seismic coefficients for design purposes are available but have not been incorporated into the present pseudo-static method of analysis.

(3) There is no strong basis for the value of the seismic coefficient for the pseudo-static force as it is specified in the building codes. The empirical value ranges from 0.05 to 0.15g for the design of earth dams in the United States. Somewhat higher values, 0.12—0.25g, are used in Japan. For building design, the specified value in Los Angeles County is 0.15g. In some parts of Japan, the value is as high as 0.30g. California is currently considering a inertia force of 1.0g.

Despite these limitations, the pseudo-static method continues to be used by consulting geotechnical engineers because it is required by the building codes, it is easier and less costly to apply, and satisfactory results have been obtained since 1933. This method will continue to be popular until an alternative method can be shown to be a more reasonable approach.

3.2 PROPOSED PROCEDURES FOR SEISMIC SAFETY ANALYSIS OF SLOPES

Limit analysis of perfect plasticity has been shown to be a powerful tool for geotechnical engineers in practice. Many solutions to static problems obtained by this method have been substantiated numerically by comparing these solutions with other existing results [13]. Herein, we suggest the following extension and modifications to the pseudo-static method of slope stability analysis, using the limit analysis techniques:

(1) Using the concept of superposition, the variation of the horizontal pseudo-static force throughout the depth of the slope can be incorporated. The factor of safety of a slope against collapse with nonuniform seismic coefficients can be determined. This work is currently in progress at Purdue University, where satisfactory results have been obtained [5—7]. Further extension to include such important items as pore water pressure, seepage forces, stratification, and three-dimensional effects will be made.

(2) In addition to the calculation of the factor of safety of a slope after a given earthquake shock, the effects of earthquakes on the displacements of a slope can be assessed. This is done following the concept of Newmark [57], again using the limit analysis techniques. This can be achieved in the following steps:

(a) Calculate the yield acceleration at which slippage will just begin to occur (or for which the factor of safety of the slope determined in (1) would reduce to 1.0).

(b) Instead of applying a permanent single, uniform value or several non-uniform values of the pseudo-static force to the slope, apply values from a discretized accelerogram of an actual or simulated earthquake.

(c) When the induced acceleration exceeds the yield acceleration, rigid body type of slope movement will occur. The magnitude of displacements can be evaluated by double integration of the part of the acceleration history above the yield acceleration. The acceleration and deceleration of the sliding mass are functions of the soil dynamics strength parameters.

(d) Determine the "stability" of the slope on the basis of this estimated total displacement by rigid body sliding.

(3) After several cycles of shaking, the potential failure surface may not be the one as originally assumed, because of the possible occurrence of liquefaction in the earth slopes. It is therefore necessary to find the most critical failure surface that reflects the change of shear strength of soil due to liquefaction. This can be achieved in the following steps.

(a) Determine the effective normal and shear stresses along the potential slip surfaces before and during a given earthquake shaking using the limit equilibrium method of slices.

(b) Evaluate the liquefaction potential for all points along the selected sliding surfaces by comparing the calculated shear stresses during earthquake shaking with the critical shear stresses required for liquefaction corresponding to a given initial stress condition and a given number of cycles.

(c) Calculate the total shear resistance against sliding for each of the selected potential failure surfaces.

(d) Determine the "reduced" yield acceleration corresponding to the critical slip surface that has been partially liquefied.

(e) Compute the total displacement by rigid body sliding of the critical mechanism.

Computer programs for the computation of safety factor, yield acceleration and displacements of slopes corresponding to a given earthquake loading have been developed at Purdue University [6]. A typical numerical example will be given in the following.

3.3 AN ILLUSTRATIVE EXAMPLE OF EARTHQUAKE-INDUCED LANDSLIDE PROBLEMS

A typical numerical example based on the present formulation is given below with the following input data (Fig. 6):

Design earthquake = El Centro, Dec. 30, 1934: c = cohesion = 200 psf, ϕ = friction angle = 24.5°, γ = unit weight of soil = 60 psf (submerged), β = slope angle = 44.5°, α = upper slope angle = 0, p = surcharge boundary load on AB = 120 psf, H = height of slope = 50 feet.

The yield acceleration of the slope at the beginning of the design earthquake is found to be $0.269g$. To obtain the updated or reduced yield acceleration at the end of each specified cycle of stress application, the empirical relationship proposed by Seed et al. [77] for liquefaction potential calculation is used. This results in the following information using a total of 20 slices in a typical limit equilibrium stress calculation.

(a) After 2 cycles of loading

Location of failure surface starting at $L = 4H$ on the top of the slope
Number of slices liquefied = 15
Yield acceleration after liquefaction = $0.192g$.

(b) After 4 cycles of loading

Location of failure surface starting at $L = 4H$ on the top of the slope
Number of slices liquefied = 17
Yield acceleration after liquefaction = $0.178g$

(c) After 7 cycles of loading

Location of failure surface starting at $L = 3H$ on the top of the slope
Number of slices liquefied = 18
Yield acceleration after liquefaction = $0.169g$

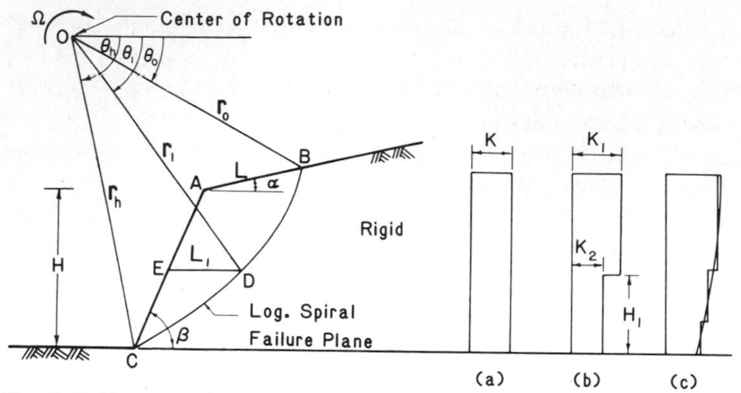

Fig. 6. Failure mechanism for the stability of an embankment and distributions of seismic coefficient.

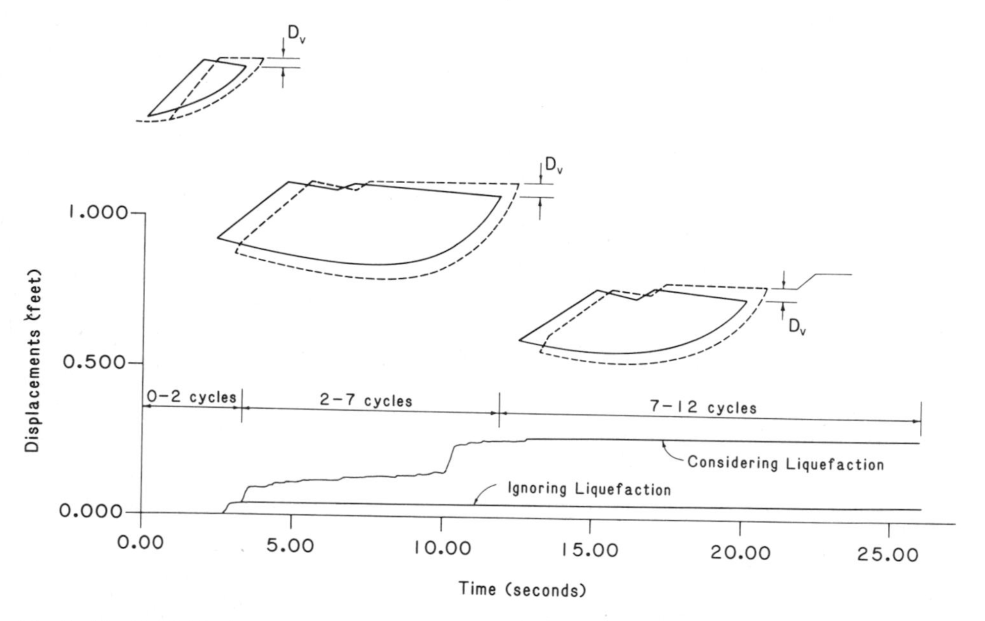

Fig. 7. Vertical displacement, D_v of point A of failure section (logspiral slip surface of local failure).

Once the critical sliding mechanism is identified and its reduced yield acceleration calculated, the rigid body type of slope movement can then be proceeded in a rather straightforward manner by double integration of that part of the acceleration history above the yield acceleration. A typical result is shown in Fig. 7 where the displacement—time relationships corresponding to the two cases: liquefaction considered and liquefaction ignored, are given. For the case of slope stability analysis ignoring the occurrence of liquefaction, the slope movement is generally small. However, including the effect of liquefaction, the slope movement can be significantly increased. The need for the determination of the "seismic stability" of the slope on the basis of this estimated total displacement is clearly demonstrated in this example calculation.

The influence of progressing liquefaction on the location of slip surfaces is shown in Fig. 8 where the displacements of sliding sections at the end of each cycle of loading are sketched. These calculated displacement patterns are seen to be similar to those observed in actual landslides due to earthquake shaking [57].

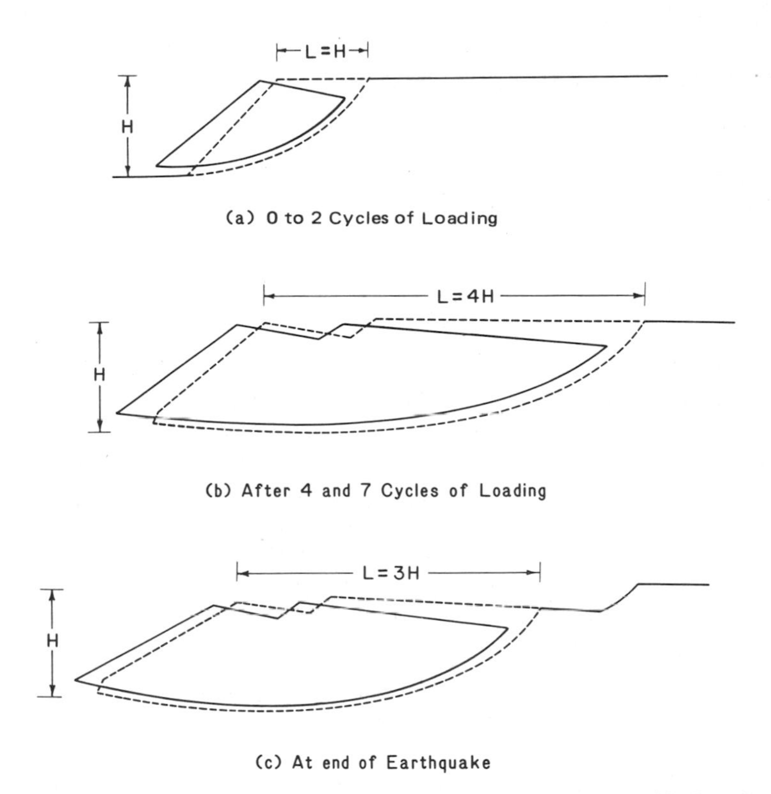

Fig. 8. Displacement patterns of failure sections after specified cycles of loading.

4. Finite element analysis of work-hardening plasticity

It is well-known that a complete analysis of stress and strain in a structure, as the load is increased to failure, is generally very complicated. This is particularly true in soil mechanics and soil–structure interaction problems where, unlike traditional structural engineering, the analysis almost always involves two- or three-dimensional continua. With the present state of development of finite element computer programs, we can confidently say that an almost unlimited range of solutions can now be obtained. These are not limited to linear elastic solid mechanics but can be extended to include problems of various kinds involving material and geometric non-linearities [10, 12, 16].

As has been indicated previously, a notable achievement in the stress strain modeling of soils based on work-hardening theories of plasticity has been made in recent years. What is really needed now in soil mechanics is to implement these theories in a finite element program for a proper study of the load—deformation behavior of soils at all stress levels leading up to failure in typical basic configurations of mixed boundary value problems under static and dynamic loading conditions. The objective of the work described in the forthcoming is therefore to show how the Drucker—Prager model with or without a hardening cap has been used for the solution of a complicated nonlinear boundary value problem. Attempts to obtain such computed solutions have shed much needed light on the suitability of such a drastic but extremely useful idealization of the soil as a perfectly plastic or isotropic hardening solid.

In the sections that follow, the development and verification of the "pseudo-static" method using finite element analysis with the elliptic cap model as well as the Drucker—Prager model are presented, and a comparative study is made with the limit analysis method. Details of the model formulations together with their computer implementations are given elsewhere [48—50]. Emphasis is placed here on the effect of large deformation on the evaluation of the overall slope stability problems. Further, of particular concern is the comparison of failure modes and limit loads with those assumed in the limit analysis method [18, 20].

4.1 ANALYSES OF SLOPES PRIOR TO SEISMIC LOADINGS

In order to simulate the stress condition inside the slope after the completion of the ground excavation, an elastic-plastic effective stress analysis of the slope is performed in this section. Sequential loading to simulate a cut-down or build-up process is not considered here. Instead, the final configuration of the slope, and the spreading of yielded zones are investigated qualitatively by increasing the internal force due to the weight of soil from zero to the natural weight of $\gamma = 120$ pcf (18.85 kN/m^3). A 30 ft (9.15 m) high vertical slope, as shown in Fig. 9 is considered. Here, the vertical boundaries and bottom boundary are placed respectively at 300 and 150 ft away from the toe of the expected slope surface. Movement on the vertical boundaries is constrained horizontally only and that along the bottom boundary is constrained in both directions. The same mesh as used by Snitbhan and Chen [78] consisting of 250 nodes and 216 rectangular elements is utilized in the present finite element analysis. Further, large deformation analysis is employed to

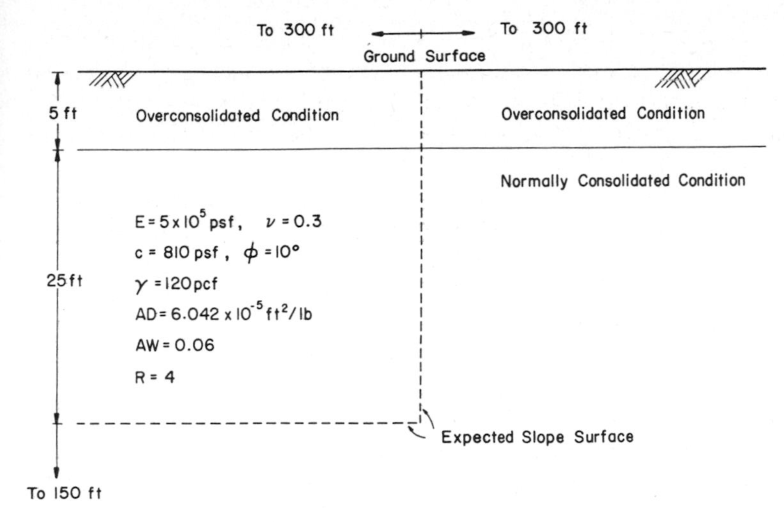

Fig. 9. Initial condition of ground.

reflect the effect of slope configuration on the limit load due to earthquake loading. In addition, the initial stress method [84] is utilized in the numerical analysis. Details of the analysis are given elsewhere [48].

In the Drucker—Prager model case, the ground surface behind the crest is found to settle approximately 3.2 ft (0.98 m) and the bulging extends 0.604 ft (0.184 m) from the original vertical slope line. For the elliptic cap model case, larger amounts of ground settlement (3.8 ft or 1.16 m) and bulging (0.804 ft or 0.245 m) in the horizontal direction are observed. However, the deformed geometries of the two slopes are very similar at this load level, regardless of the relatively large difference in ground settlement.

4.2 ANALYSES OF SLOPES DURING SEISMIC LOADINGS

Based on the analysis described in the previous section, the seismic large deformation analyses of the vertical slopes are now performed in this section by employing the pseudo-static method. The 1934 El Centro (S—N direction) accelerogram is used as input for the seismic loadings which are obtained as the product of soil mass, m, and acceleration, a. In the present analysis, the horizontal acceleration data between time $t = 1.5$ s and $t = 4.5$ s in the accelerogram are applied to the deformed shape of the slope after excavation. For simplicity, the distribution of acceleration is here assumed to be uniform throughout the slope. It acts to the right when its sign is positive. The response of the slopes during seismic loading, and the velocity fields at the collapse stage are discussed briefly as follows:

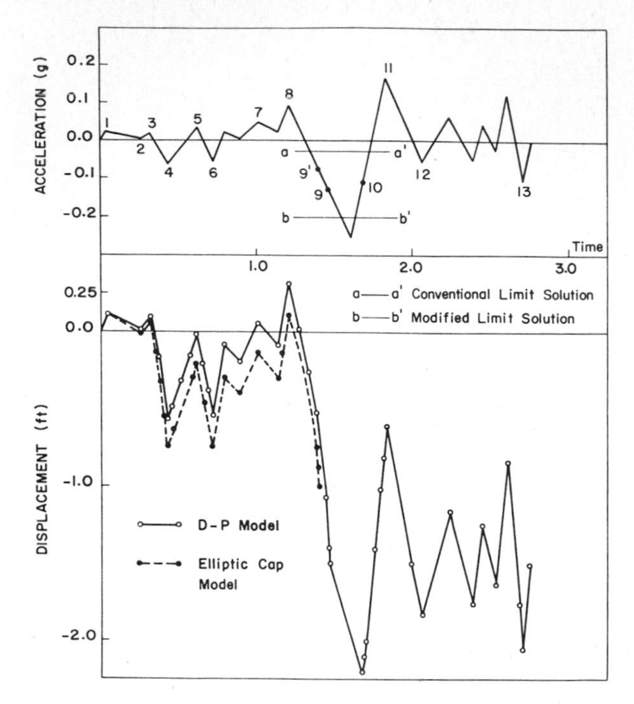

Fig. 10. Acceleration—time—displacement curves by the Drucker—Prager and elliptic cap models.

Response of Slopes

In Fig. 10, the acceleration—time—displacement curves are presented for both model cases. Note that the horizontal displacement at the nodal point above the toe is taken as the reference displacement in the figure, and that displacement is measured from the deformed shape of the slope after excavation, not from the original shape. The displacement response curves with the solid and broken lines are results from the analyses with the Drucker—Prager and elliptic cap models, respectively.

Based on the finite element calculation (Fig. 10), the sliding of the slope is estimated to occur at approximately $a = -0.135\,g$ and $-0.075\,g$, for the Drucker—Prager and elliptic cap model cases, respectively. On the other hand, the limit analysis method predicts the occurrence of sliding at $a = -0.0287\,g$ and $-0.196\,g$ by applying the full slope height of 30 ft (9.15 m) and the modified slope height of 25 ft (7.63 m), respectively. Thus, the finite element solutions lie between the two extreme solutions of the limit analysis method.

Relative velocity fields

The relative velocity is defined here as the difference between the displacement increment at a nodal point and that at the toe. The relative

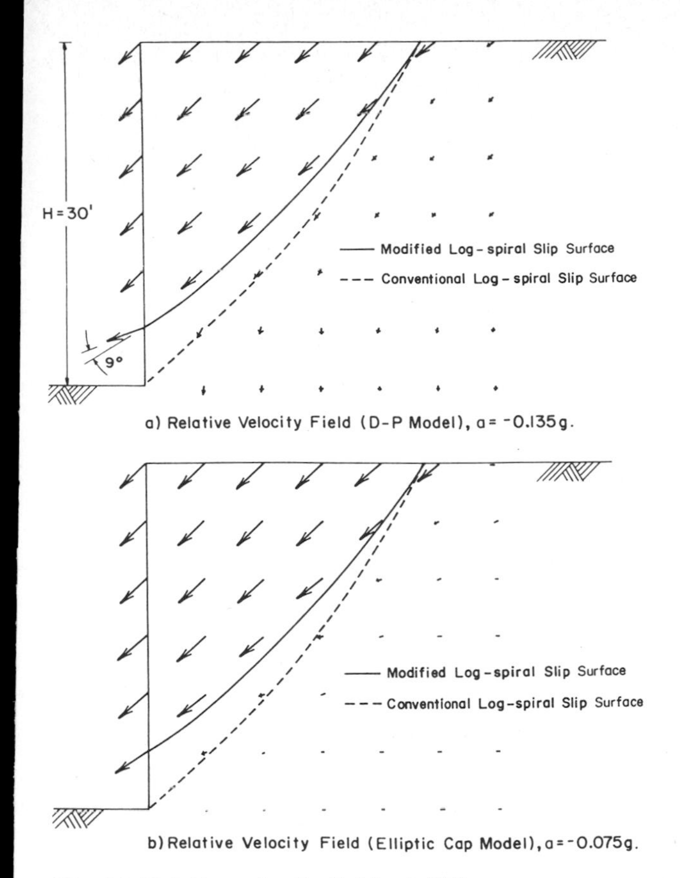

Fig. 11. Relative velocity fields at sliding.

velocity fields at the sliding stage are shown in Fig. 11, for both model cases. The magnitude and direction of each relative velocity vector is shown by an arrow, and the resultant of the relative-displacement increment at the nodal point above the toe is taken as a normalized unit length. Further, the modified and conventional log-spiral slip surface of the limit analysis method are illustrated with the solid and broken lines, respectively [17, 20]. The former and latter slip surfaces are obtained by using slope heights of 25 ft (7.63 m) and 30 ft (9.15 m) in the limit analysis solutions. Also, note that the modified and conventional slip surfaces correspond to failure mechanisms at $a = -0.196g$ and $-0.0287g$, respectively. Since the loss of ground appears to occur at a distance of approximately one-sixth of the total height above the toe as discussed previously, the relative velocity fields of both models agree quite well with the modified slip surfaces.

The angle between the modified slip surface and the relative velocity vector at the nodal point above the toe is measured from the figures. For the Drucker—Prager model case, it has an angle of 9°. It is expected that if the seismic load is increased slightly, this angle would approach 10° as would be predicted by the limit analysis method. On the other hand, the angle in the elliptic cap model case is nearly zero. This means that the dilatancy has not been developed along the slip surface. Referring to Fig. 11, most of the stress states along the slip surface are in the corner zone. Consequently, the dilatancy on the slip surface is prevented to some extent.

5. Conclusion

This paper gives a summary of recent progress and achievement of theoretical soil plasticity with emphasis on applications to earthquake-induced slope failure and landslide analysis. It is also intended to give a brief summary of the origins and development of soil plasticity from about 1950 onwards. To this end, I should like to conclude by emphasizing again the great impact of D.C. Drucker's ideas on the evolution and modern development of the theory of soil plasticity.

The mechanical behavior of soils is much more complex than that of ductile metals. Accordingly, a more elaborate plasticity theory is required for soils than is required for metals. The extension of metal plasticity to problems involving plastic deformations in soils was first made by Drucker and Prager [32] based on the concept of perfect plasticity. This formed the central and most extensively developed part of the early theory of soil plasticity. The general theory of limit analysis, due to Drucker, Greenberg and Prager [31], had shed much needed light on the foundation of the much older theory of earth pressure. An important defect in the theory of earth pressure lay in its development without reference to stress—strain relations, the theory being based upon the concept of limiting equilibrium satisfying Coulomb's failure criterion.

The modern development of soil plasticity did not copy this unsatisfactory feature of the theory of earth pressure nor continued the simplicity of perfect plasticity, but instead, pursued the difficult course of work-hardening theory of plasticity. As a result of this, the theory of soil plasticity is now in a position to lead that of metal plasticity theory, especially in connection with cyclic loading applications. It is therefore appropriate to conclude here that the pioneer work of Drucker et al. [30] on work-hardening theories of soil plasticity and also that of the subsequent developments of Roscoe and his co-workers (1958—1963) mark the beginning of the modern developments of a consistent theory of soil plasticity.

Acknowledgment

This material is based in part upon work supported by the National Science Foundation under Grant No. PFR-7809326 to Purdue University.

References

1 J.H. Atkinson and P.L. Bransby, The Mechanics of Soils — An Introduction to Critical State Soil Mechanics. McGraw-Hill, Maidenhead, England, 1978.
2 A.W. Bishop, The strength of soils as engineering materials, Geotechnique, 22 (1972) 509—513.
3 J.R. Booker and E.H. Davis, A note on a plasticity solution to the stability of slopes in homogeneous clay. Geotechnique, 22 (1972) 509—513.
4 J.B. Burland, A method of estimating the pore pressures and displacements beneath embankments on soft, natural clay deposits. In: R.H.G. Parry (Ed.), Proc. Roscoe Memorial Symp., Cambridge Univ., March 29—31, 1971, Foulis, Oxfordshire, U.K., 1972, pp. 505—536.
5 S.W. Chan, Perfect Plasticity Upper Bound Limit Analysis of the Stability of a Seismic-Infirmed Earthslope. M.S. Thesis, Sch. of Mech. Eng., Purdue Univ., Aug. 1980.
6 C.J. Chang, Seismic Safety Analysis of Slopes. Ph.D. Thesis, Sch. of Civ. Eng., Purdue Univ., Aug. 1981.
7 M.F. Chang, Static and Seismic Lateral Earth Pressures on Rigid Retaining Structures, Ph.D. Thesis, Sch. of Civ. Eng., Purdue Univ., Aug. 1981.
8 M.F. Chang and W.F. Chen, Lateral Earth Pressures on Rigid Retaining Walls Subjected to Earthquake Forces. Solid Mech. Archives, 7, Sijthoff & Noordhoff, The Netherlands, Sept., 1982, pp. 315—362.
9 W.F. Chen, Mechanics of slope failures and landslides. In: J. Penzien, S.T. Mau and Z. Yang (Eds.), Proc. Advisory Meeting on Earthquake Eng. and Landslides, U.S.-R.O.C. Coop. Sci. Prog., Taipei, August 29—September 2, 1977, Nat. Sci. Found., Washington, D.C., 1977, pp. 219—232.
10 W.F. Chen, Constitutive equations for concrete. In: Special Publ. of the Assoc. of Bridge and Str. Eng. Colloq. on Plasticity in Reinforced Concrete, Copenhagen, 1979, Vol. 2, pp. 11—34.
11 W.F. Chen, Plasticity in soil mechanics and landslides. J. Eng. Mech. Div., ASCE, 106 (1980) 443—464.
12 W.F. Chen, Plasticity in Reinforced Concrete. McGraw-Hill, New York, 1982.
13 W.F. Chen, Limit Analysis and Soil Plasticity, Elsevier, Amsterdam, 1975.
14 W.F. Chen, C.J. Chang and J.T.P. Yao, Limit analysis of earthquake-induced slope failure. In: R.L. Sierakowski, (Ed.), Proc. 15th Annual Meeting Soc. Eng. Sci., Dec. 4—6, Gainesville, Univ. of Florida, 1978, pp. 533—538.
15 W.F. Chen and M.F. Chang, Limit analysis in soil mechanics and its applications to lateral earth pressure problems. Solid Mech. Archives, 6 (3) (1981) 331—399.
16 A.C.T. Chen and W.F. Chen, Constitutive relations for concrete, J. Eng. Mech. Div., ASCE, 101 (1975) 465—481.
17 W.F. Chen, and M.W. Giger, Limit analysis of stability of embankments. J. Soil Mech. and Found. Div., ASCE, 97 (1971) 19—26.
18 W.F. Chen, M.W. Giger and H.Y. Fang, On the limit analysis of stability of slopes, soils and foundations. Jap. Soc. Soil Mech. and Found. and Eng., 9 (1969) 23—32.
19 W.F. Chen and S.L. Koh, Earthquake-induced landslide problems. Proc. Central American Conf. on Earthquake Eng., 1, Universidad Centroamericana Jose Simeon Canas and the Ministry of Public Works of El Salvador and Lehigh Univ., San Salvador, Jan. 9—12, 1978, Envo, PA., pp. 665—685.
20 W.F. Chen, N. Snitbhan and H.Y. Fang, Stability of slopes in anisotropic nonhomogenous soils. Canadian Geot. J., 12 (1975) 146—152.
21 W.F. Chen and A.F. Saleeb, Constitutive Equations for Engineering Materials, Vol. 1 — Elasticity and Modeling, Feb., 1982; Vol. 2 — Plasticity and Modeling, 1984. John Wiley Interscience, New York.
22 Y.F. Dafalias and E.P. Popov, A model of nonlinearly hardening materials for complex loadings. Acta Mech., 21 (1975) 173—192.

23 Y.F. Dafalias and R. Herrmann, A boundary surface soil plasticity model. Int. Symp. Soils Under Cyclic and Transient Loading, Swansea, U.K., 1980, Vol. 1, pp. 335–345.
24 E.H. Davis, Theories of plasticity and the failure of soil masses. In: I.K. Lee (Ed.), Soil Mechanics: Selected Topics, Butterworths, London, 1968, pp. 341–380.
25 C.S. Desai and R.H. Gallagher (Eds.), Constitutive Laws for Engineering Materials: Theory and Application. John Wiley, U.K., 1983.
26 F.L. Dimaggio and I.S. Sandler, Material model for granular soils. J. Eng. Mech. Div., ASCE, 97 June (1971) 935–950.
27 D.C. Drucker, Concepts of path independence and material stability for soils. In: J. Kravtchenko and P.M. Sirieys (Ed.), Rheol. Mecan. Soils Proc. IUTAM Symp. Grenoble, Springer, Berlin, 1966, pp. 23–43.
28 D.C. Drucker, Limit analysis of two- and three-dimensional soil mechanics problems. J. Mech. and Phys. Solids, 1 (1953) 217–226.
29 D.C. Drucker, Plasticity. In: J.N. Goodier and N.J. Hoff (Eds.), Structural Mechanics, Pergamon Press, London, 1960, pp. 407–455.
30 D.C. Drucker, R.E. Gibson and D.J. Henkel, Soil mechanics and work hardening theories of plasticity. Trans. ASCE, 122 (1957) 338–346.
31 D.C. Drucker, J.H. Greenberg and W. Prager, Extended limit design theorems for continuous media. Quart. Appl. Math., 9 (1952) 381–389.
32 D.C. Drucker and W. Prager, Soil mechanics and plastic analysis or limit design. Quart. Appl. Math. 10 (1952) 157–165.
33 J.M. Duncan and C.Y. Chang, Nonlinear analysis of stress and strain in soils. J. Soil Mech. Found. Div. ASCE, 96 (Sept. 1970) 1629–1653.
34 B.O. Hardin, The nature of stress–strain behavior of soils, Earthquake Eng. and Soil Dynamics, ASCE, 1 (1978) 3–90.
35 B.O. Hardin and V.P. Drnevich, Shear modulus and damping in soils: design equation and curves. J. Soil Mech. Found. Div., ASCE, 98 (1972) 667–692.
36 R. Hill, The Mathematical Theory of Plasticity, Oxford, Clarendon Press, 1950.
37 IABSE Proc. Colloq. on Plasticity in Reinforced Concrete, Copenhagen, May 21–23, IABSE Pub., Zurich, 1979.
38 W.D. Iwan, On a class of models for the yielding behavior of continuous and composite systems. J. Appl. Mech., 34 (1967) 612–617.
39 N. Janbu, Soil compressibility as determined by odometer and triaxial tests. Proc. Europ. Conf. on Soil Mech. and Found. Eng., Wiesbaden, Germany, 1963, Vol. 1, pp. 19–25.
40 W.B. Joyner and A.T.F. Chen, Calculation of nonlinear ground response in earthquakes. Bulletin Seismological Soc. of America, 65 (5) (1976).
41 K. Karal, Energy Method for Soil Stability Analyses. River and Harbour Lab., Norwegian Inst. Tech., Trondheim, Norway, 1979.
42 S.L. Koh and W.F. Chen, The Prevention and Control of Landslides. Presented at Sept. 1977, Joint U.S.–Southeast Asia Symp. on Eng. for Natural Hazards Protection, Manila.
43 R.D. Krieg, A practical two-surface plasticity theory. J. Appl. Mech., 42 (1975) 641–646.
44 P.V. Lade, Elasto-plastic stress–strain theory for cohesionless soil with curved yield surfaces. Int. J. Solids and Str., 13 (1977) 1014–1035.
45 P.V. Lade, Stress–strain theory for normally consolidated clay. Proc. 3rd Int. Conf. on Numer. Methods in Geomech., Aachen, Germany, 1979, Vol. 4, pp. 1325–1337.
46 P.V. Lade and J.M. Duncan, Elastoplastic stress–strain theory for cohesionless soil. J. Geot. Eng. Div., ASCE, 101 (1975) 1037–1053.
47 G. Masing, Eigenspannungen and Verfestigung beim Messing. Proc. 2nd Int. Cong. of Applied Mech., Zurich, 1926.

48 E. Mizuno, Plasticity Modeling of Soils and Finite Element Applications. Ph.D. Thesis, Sch. of Civ. Eng., Purdue Univ., Dec. 1981.

49 E. Mizuno and W.F. Chen, Analysis of soil response with different plasticity models. In: R.N. Yang and E.T. Selig (Eds.), Proc. of the Symp. Applications of Plasticity and Generalized Stress—Strain in Geotechnical Engineering, Hollywood, Florida, Oct. 27—31, 1980, ASCE, New York, 1982, pp. 115—138.

50 E. Mizuno and W.F. Chen, Plasticity models and finite element implementation. Proc. Symp. on Implementation of Computer Procedures and Stress—Strain Laws in Geot. Eng., Chicago, Aug. 3—6, 1981, Acorn Press, Durham, N.C., 1981, pp. 519—534.

51 Z. Mroz, On the description of anisotropic hardening. J. Mech. Phys. of Solids, 15 (1967) 163—175.

52 Z. Mroz, V.A. Norris and O.C. Zienkiewicz, An anisotropic hardening model for soils and its application to cyclic loading. Int. J. for Numer. and Anal. Meth. in Geomech., 2 (1978) 203—221.

53 Z. Mroz, V.A. Norris and O.C. Zienkiewicz, Application of an anisotropic hardening model in the analysis of elastoplastic deformation of soils. Geotech. 29 (1979) 1—34.

54 I. Nelson, Constitutive models for use in numerical computations. Proc. Int. Symp. on Dynamic Methods in Soil and Rock Mech., Balkema, Rotterdam, 1978, Vol. 2, pp. 45—97.

55 I. Nelson and G.Y. Baladi, Outrunning ground shock computed with different models, J. Eng. Mech. Div. ASCE, 103 (1977) 377—393.

56 I. Nelson, M.L. Baron and I. Sandler, Mathematical Models for Geological Materials for Wave Propagation Studies, Shock Waves and the Mechanical Properties of Solids. Syracuse Univ. Press, Syracuse, N.Y., 1971.

57 N.M. Newmark, Effects of earthquakes on dams and embankments. Geotech. 15 (1965) 139—160.

58 A.C. Palmer (Ed.), Proc. Symp. on the Role of Plasticity in Soil Mechanics, Cambridge Univ. Press, 1973.

59 G.N. Pand and O.C. Zienkiewicz (Eds.), Proc. Int. Symp. Soils Under Cyclic and Transient Loading, Balkema, Rotterdam, 1980.

60 R.H.G. Parry, correspondence on "On the Yielding of Soils". Geotech. 8 (1958) 185—186.

61 R.H.G. Parry (Ed.), Roscoe Memorial Symp.: Stress—Strain Behavior of Soils, Henly-on-Thames, Cambridge Univ., 1972.

62 W. Prager and P.G. Hodge, Theory of Perfectly Plastic Solids, Wiley, New York, 1950.

63 J. Prevost, Mathematical modelling of monotonic and cyclic undrained clay behavior. Int. J. for Numer. and Anal. Meth. in Geomech., 1 (1977) 195—216.

64 J.H. Prevost, Anisotropic undrained stress—strain behavior of clays. J. Geotech. Eng. Div. ASCE, 104 (1978) 1075—1090.

65 J.H. Prevost, Plasticity theory for soil stress behavior, J. Eng. Mech. Div. ASCE, 104 (1978) 1177—1194.

66 J.H. Prevost, B. Cuny, T.J.R. Hughes and R.F. Scott, Offshore gravity structures: analysis. J. Geotech. Div. ASCE, 107 (1981) 143—165.

67 K.H. Roscoe and J.B. Burland, On the generalized stress—strain behavior of "wet" clay. In: J. Heyman and F.A. Leckie (Eds.), Engineering Plasticity, Cambridge Univ. Press, 1968, pp. 535—609.

68 K.H. Roscoe, A.N. Schofield and C.P. Wroth, On the yielding of soils. Geotech. 8 (1958) 22—52.

69 K.H. Roscoe, A.N. Schofield and A. Thurairajah, An Evaluation of Test Data for Selecting a Yield Criterion for Soils. ASTM Spec. Tech. Pub. No. 361, ASTM, 1963, pp. 111—128.

70 A.F. Saleeb, Constitutive Models for Soils in Landslides, Ph.D. Thesis, Sch. Civ. Eng. Purdue Univ., May, 1981.
71 A.F. Saleeb and W.F. Chen, Nonlinear hyperelastic (Green) constitutive models for soils, Part I — Theory and Calibration; Part II — Predictions and Comparisons. In: R.K. Yong and H.Y. Ko (Eds.), Proc. North American Workshop on Limit Equilibrium, Plasticity, and Generalized Stress—Strain In Geotech. Eng., McGill Univ., Montreal, May 28—30, 1980; ASCE, New York, 1981.
72 I.S. Sandler and M.L. Baron, Recent development in the constitutive modeling of geological materials. In: Proc. 3rd Int. Conf. on Num. Methods in Geomechanics, Aachen, Germany, 1979, pp. 363—376.
73 I.S. Sandler and M.L. Baron, Material models of geological materials in ground shock. In: C.S. Desai (Ed.), Proc. of the Second Int. Conf. on Num. Methods in Geomech., Blacksburg, Virginia, June, 1976, ASCE, 1976, pp. 219—231.
74 I.S. Sandler, F.L. DiMaggio and G.Y. Baladi, Generalized cap model for geological materials. J. Geotech. Eng. Div. ASCE, 102 (1976) 683—699.
75 A. Schofield and P. Wroth, Critical State Soil Mechanics, McGraw-Hill, New York, 1968.
76 H.B. Seed, A method for earthquake resistant design of earth dams. J. Soil Mech. and Found. Div. ASCE, 92 (1966) 13—41.
77 H.B. Seed, K.L. Lee and I.M. Idriss, Analysis of Sheffield dam failure. J. Soil Mech. Found. Div., ASCE, 95 (1969) 1453—1490.
78 N. Snitbhan and W.F. Chen, Elastic-plastic large deformation analysis of slopes. Computers and Structures, 8 (1978) 567—577.
79 V.V. Sokolovskii, Statics of Granular Media. Pergamon, New York, 1965.
80 K. Terzaghi, Theoretical Soil Mechanics. John Wiley, New York, 1943.
81 R.K. Yong and H.Y. Ko, (Eds.), Limit Equilibrium, Plasticity and Generalized Stress—Strain in Geotechnical Engineering. ASCE, New York, 1981.
82 R.N. Yong and E.T. Selig (Eds.), Application of Plasticity and Generalized Stress—Strain in Geotechnical Engineering. ASCE, New York, 1982.
83 O.C. Zienkiewicz, V. Norris and D.J. Naylor, Plasticity and viscoplasticity in soil mechanics with special reference to cyclic loading problems. In: P.G. Bergan et al. (Eds.), Finite Elements in Non-Linear Mechanics, Tapir, Trondheim, 1977, Vol. 2, pp. 455—485.
84 O.C. Zienkiewicz, S. Valliapan and I.P. King, Elasto-plastic solutions of engineering problems: initial stress, finite element approach. Int. J. Numer. Meth. Eng. 1 (1969) 75—100.

Restrictions on Quasi-Statically Moving Surfaces of Strong Discontinuity in Elastic-Plastic Solids

W.J. DRUGAN

Department of Engineering Mechanics, University of Wisconsin, Madison, Wisconsin 53706 (U.S.A.)

J.R. RICE

Division of Applied Sciences, Harvard University, Cambridge, Massachusetts 02138 (U.S.A.)

Abstract

This paper educes the restrictions on a hypothesized quasi-statically propagating planar surface of strong discontinuity in an elastic-plastic solid. The derivation treats a solid under general three-dimensional conditions, employing a "small strain" formulation; the principal constitutive assumptions are that the elastic part of total strain be linearly related (although with arbitrary anisotropy) to the stress state, that the material possess a positive definite elastic strain energy function, and that inelastic deformation occur in accordance with the principle of maximum plastic resistance. It is proved that under these conditions all components of stress must be continuous across such a propagating surface. Further, the only components of strain which may suffer discontinuities across such a surface are shown to be the plastic components which do not deform elements in the plane of the surface, and these may be discontinuous only if the stress state at the moving surface meets certain specific conditions.

1. Introduction

Many problems in the mechanics of solids involve quasi-static changes in loading or geometry, and in the context of elastic-plastic solids these changes often correspond to quasi-static motion of zones of active plasticity through the material. The solution of such problems necessarily involves a knowledge of the conditions which must be satisfied across the moving surfaces which separate such zones of active plastic deformation from regions of purely elastic response or from other zones of different types of plastic response.

An example motivating our own interest is the asymptotic analysis of crack tip fields during quasi-static crack growth. A recent analysis (Rice [1]) revealed that four distinct types of response are possible in different angular sectors at a plane strain crack tip in a general class of elastic-ideally plastic solids. The proper assembly of these sectors, accomplished recently by Drugan et al. [2] for the isotropic Huber—Mises material, and addressed by several previous workers for the special case of Poisson ratio $\nu = 1/2$, requires specification of the necessary continuity conditions, or of permissible discontinuities, at the moving interfaces between sectors.

The following discussion addresses this issue by determining the requirements placed by the governing equations on planar surfaces of strong discontinuity which propagate quasi-statically through an elastic-plastic solid. Strong discontinuities are discontinuities in components of stress or strain, as opposed to discontinuities in their derivatives (generally referred to as weak discontinuities).

The analysis is formulated for a solid under general three-dimensional conditions, employing the assumption of "small strain"; that is, strain is taken to be the symmetric part of the displacement gradient, and the distinction between the undeformed and deformed configurations is neglected in writing equilibrium equations and in formulating stress rates. The class of elastic-plastic solids treated is rather broad, with most of the results relying only on the constitutive assumptions that the elastic part of the total strain be linearly related (although with arbitrary anisotropy) to the stress state, that stress be derivable from a positive definite elastic strain energy function, and that inelastic deformation be governed by the principle of maximum plastic resistance.

Accordingly, consider a planar surface of discontinuity, Σ, in a general three-dimensional elastic-plastic solid, which moves through the solid with normal velocity $V > 0$. The Cartesian coordinate system x_1, x_2, x_3 moves with the discontinuity surface and is chosen so that x_1 is normal to Σ and points in the direction of propagation, as illustrated in Fig. 1.

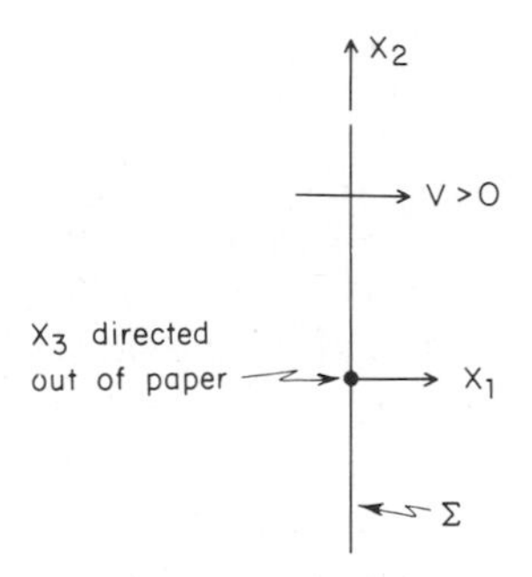

Fig. 1. Quasi-statically moving surface of discontinuity.

Values of a field quantity, say $g(x_1, x_2, x_3, t)$, where t is time, directly ahead of and directly behind the moving surface Σ will be denoted as

$$g^+ \equiv \lim_{\mu \to 0} \{g(\mu, x_2, x_3, t)\} = \lim_{\mu \to 0} \{g(x_1, x_2, x_3, t_a - \mu)\},$$

$$g^- \equiv \lim_{\mu \to 0} \{g(-\mu, x_2, x_3, t)\} = \lim_{\mu \to 0} \{g(x_1, x_2, x_3, t_a + \mu)\},$$

respectively, where $\mu > 0$ and t_a is the time at which Σ arrives at a particular material point. The jump in a field quantity across Σ will be denoted as

$$[\![g]\!] \equiv g^+ - g^-.$$

2. Governing equations and assumptions

2.1 EQUILIBRIUM

Since attention is limited to quasi-static processes, the inertia terms in the equations of motion are negligible, and hence the equations of equilibrium govern:

$$\int_S T_i \mathrm{d}S + \int_V f_i \mathrm{d}V = 0. \tag{2.1}$$

Here, V is an arbitrary subregion of the body with boundary S, $T_i = \sigma_{ij} n_j$ is the traction vector, n_i is the unit normal to S, $\sigma_{ij} (= \sigma_{ji})$ is the stress tensor and f_i is the body force vector. Here and throughout the paper, components of tensors with respect to the Cartesian coordinate system of Fig. 1 are indicated either by the Latin indices i, j, k, l which have range 1, 2, 3 or by the Greek indices $\alpha, \beta, \gamma, \delta$ which have range 2, 3 only and thus refer to tensor components in planes parallel to Σ. Both types of index follow the summation convention.

Choosing V to be a subregion containing the discontinuity surface Σ, then taking the limit as V shrinks to zero in such a manner that S collapses onto Σ, the well-known requirement of traction continuity results

$$[\![T_i]\!] = [\![\sigma_{ij} n_j]\!] = 0. \tag{2.2}$$

For the geometry illustrated in Fig. 1,

$$n_i = \delta_{1i}, \tag{2.3}$$

where δ_{ij} is the Kronecker delta; (2.2) thus requires

$$[\![\sigma_{1i}]\!] = 0. \tag{2.4}$$

2.2 DISPLACEMENT CONTINUITY

The solid is assumed to remain coherent throughout the deformation process; along Σ this is the requirement that

$$[\![u_i]\!] = 0, \tag{2.5}$$

where u_i is the displacement vector. Further justification for enforcing (2.5) is provided by the fact that it is necessary for finite total plastic work rate in any subregion of the body traversed by Σ since Σ is non-stationary.

Assuming that $\partial u_i/\partial x_j$ exist in the neighborhood of Σ and tend to finite limits as Σ is approached, differentiation $\partial/\partial x_\alpha$ of (2.5) immediately implies (e.g., Hill [3])

$$[\![\partial u_i/\partial x_\alpha]\!] = 0. \tag{2.6}$$

This permits only the component of displacement gradient in the direction normal to Σ to undergo a jump across Σ:

$$[\![\partial u_i/\partial x_j]\!] = \lambda_i n_j, \tag{2.7}$$

where λ_i is an arbitrary vector function of position. In terms of the components of strain,

$$\epsilon_{ij} = \tfrac{1}{2}(\partial u_i/\partial x_j + \partial u_j/\partial x_i), \tag{2.8}$$

(2.6) can be restated as

$$[\![\epsilon_{1\alpha}]\!] = \tfrac{1}{2}[\![\partial u_\alpha/\partial x_1]\!], \tag{2.9a}$$

$$[\![\epsilon_{\alpha\beta}]\!] = 0, \tag{2.9b}$$

(2.9b) being the requirement that all strain components in the discontinuity plane be continuous.

Jumps in the components of material velocity are determined by taking the time derivative of u_i following Σ, and employing (2.5). Thus (e.g., Hill [3])

$$[\![v_i]\!] + V[\![\partial u_i/\partial x_j]\!]n_j = 0; \tag{2.10}$$

using (2.3) this becomes

$$[\![v_i]\!] = -V[\![\partial u_i/\partial x_1]\!], \tag{2.11}$$

so that in terms of strains, via (2.8) and (2.9a),

$$[\![v_1]\!] = -V[\![\epsilon_{11}]\!], \tag{2.12a}$$

$$[\![v_\alpha]\!] = -2V[\![\epsilon_{1\alpha}]\!]. \tag{2.12b}$$

The jump restrictions derived thus far, independent of material constitution, permit evaluation of the total work accumulated discontinuously at a material point due to the passage of Σ, which is given by

$$W \equiv \int_{\epsilon_{ij}^+}^{\epsilon_{ij}^-} \sigma_{ij} d\epsilon_{ij}. \tag{2.13}$$

Such an integral is dependent (for the type of elastic-plastic materials considered) on the path in strain space, from $\boldsymbol{\varepsilon}^+$ to $\boldsymbol{\varepsilon}^-$, traversed at the discontinuity. Since σ_{1i} must be continuous across Σ via (2.4), (2.13) is

$$W = -\sigma_{11}^{\Sigma} [\![\epsilon_{11}]\!] - 2\sigma_{1\alpha}^{\Sigma} [\![\epsilon_{1\alpha}]\!] + \int_{\epsilon_{\alpha\beta}^+}^{\epsilon_{\alpha\beta}^-} \sigma_{\alpha\beta} d\epsilon_{\alpha\beta}, \tag{2.14}$$

where the notation h^{Σ} indicates that the continuous field quantity $h(x_1, x_2, x_3, t)$ has been evaluated on Σ. Restriction (2.9b) requires that $d\epsilon_{\alpha\beta} = 0$ across Σ, so the integral in (2.14) vanishes, leaving the simple result

$$W \equiv \int_{\epsilon_{ij}^+}^{\epsilon_{ij}^-} \sigma_{ij} d\epsilon_{ij} = -\sigma_{11}^{\Sigma} [\![\epsilon_{11}]\!] - 2\sigma_{1\alpha}^{\Sigma} [\![\epsilon_{1\alpha}]\!]$$
$$= -\tfrac{1}{2} (\sigma_{ij}^+ + \sigma_{ij}^-) [\![\epsilon_{ij}]\!].$$

In deriving this result we have obviously restricted the path from $\boldsymbol{\varepsilon}^+$ to $\boldsymbol{\varepsilon}^-$ by requiring that (2.4) and (2.9b) be satisfied by all states traversed along the path.

2.3 CONSTITUTIVE ASSUMPTIONS

Most of the conclusions to follow will be arrived at on the basis of a rather general constitutive model. A principal assumption is that the elastic part of the total strain be linearly related to the stress state:

$$\epsilon_{ij}^{e} = M_{ijkl} \sigma_{kl}. \tag{2.15}$$

Here, M_{ijkl} is the (constant) elastic compliance tensor having the following symmetries

$$M_{ijkl} = M_{jikl} = M_{ijlk} = M_{klij} \tag{2.16}$$

due to stress symmetry, strain symmetry and the assumed existence of an elastic strain energy function, respectively.

The total strain is taken to be the sum of the elastic and plastic parts

$$\epsilon_{ij} = \epsilon^{e}_{ij} + \epsilon^{p}_{ij} = M_{ijkl}\sigma_{kl} + \epsilon^{p}_{ij}, \tag{2.17}$$

and arbitrary anisotropy is evidently included.

The plastic behavior of the solids considered is assumed to proceed in accordance with the maximum plastic work inequality

$$(\sigma_{ij} - \sigma^{0}_{ij})\mathrm{d}\epsilon^{p}_{ij} \geqslant 0, \tag{2.18}$$

where σ_{ij} is the stress state (at yield) corresponding to the plastic strain increment $\mathrm{d}\epsilon^{p}_{ij}$, and σ^{0}_{ij} is any other stress state which is at or below yield.

The maximum plastic work inequality (2.18) results from a number of different viewpoints as discussed by Rice [4, 5]. In its small strain form, as employed here, (2.18) can be seen to result from Drucker's material stability postulate (Drucker [6, 7]) which states that any additional stresses imposed on a stressed body must do non-negative work on the strains they produce. Alternatively, (2.18) follows from a postulate of Il'yushin [8] that the work produced by a strain cycle originating from an arbitrary deformed state must be non-negative. Yet another line of reasoning was begun by Bishop and Hill [9] who considered a rigid-plastic polycrystal and showed that if conditions for plastic straining of a particular slip system depend only on the resolved shear stress in the slip direction, (2.18) can be derived for macroscopic stress and strain. Later investigators extended this derivation of Bishop and Hill to the elastic-plastic polycrystal; see e.g. Mandel [10], Hill [11].

Since we seek restrictions on possible jumps in stress and strain across Σ, an integrated form of (2.18) would seem most informative. In particular, integration of (2.18) at a material point just during the passage of the discontinuity surface Σ results in

$$\int_{\epsilon^{p+}_{ij}}^{\epsilon^{p-}_{ij}} (\sigma_{ij} - \sigma^{0}_{ij})\mathrm{d}\epsilon^{p}_{ij} \geqslant 0, \tag{2.19}$$

where here σ^{0}_{ij} is understood to be a stress state at or below yield for all states along the strain path from $\boldsymbol{\varepsilon}^{+}$ to $\boldsymbol{\varepsilon}^{-}$.

3. General impossibility of a stress jump across a quasi-statically propagating surface

We shall now prove that the restrictions of Section 2 rule out the possibility of a stress jump across a quasi-statically propagating (non-stationary) planar surface in an elastic-plastic solid under general three-dimensional conditions. To this end, note that the first term of (2.19) is the *plastic* work accumulated discontinuously at a material point due to the passage of Σ:

$$W^{\rm p} \equiv \int_{\epsilon_{ij}^{\rm p+}}^{\epsilon_{ij}^{\rm p-}} \sigma_{ij} {\rm d}\epsilon_{ij}^{\rm p}. \tag{3.1}$$

Since σ_{1i} must be continuous across Σ due to (2.4), (3.1) is

$$W^{\rm p} = -\sigma_{11}^{\Sigma} [\![\epsilon_{11}^{\rm p}]\!] - 2\sigma_{1\alpha}^{\Sigma} [\![\epsilon_{1\alpha}^{\rm p}]\!] + \int_{\epsilon_{\alpha\beta}^{\rm p+}}^{\epsilon_{\alpha\beta}^{\rm p-}} \sigma_{\alpha\beta} {\rm d}\epsilon_{\alpha\beta}^{\rm p}. \tag{3.2}$$

Restriction (2.9b) coupled with (2.17) yields the requirement that ${\rm d}\epsilon_{\alpha\beta}^{\rm p} = -{\rm d}\epsilon_{\alpha\beta}^{\rm e}$ across Σ; thus the integral in (3.2) becomes

$$\int_{\epsilon_{\alpha\beta}^{\rm p+}}^{\epsilon_{\alpha\beta}^{\rm p-}} \sigma_{\alpha\beta} {\rm d}\epsilon_{\alpha\beta}^{\rm p} = -\int_{\epsilon_{\alpha\beta}^{\rm e+}}^{\epsilon_{\alpha\beta}^{\rm e-}} \sigma_{\alpha\beta} {\rm d}\epsilon_{\alpha\beta}^{\rm e}. \tag{3.3}$$

This integral in terms of elastic strains is easily evaluated using (2.15) and the symmetry properties (2.16) of $M_{\alpha\beta\gamma\delta}$, along with the fact that ${\rm d}\sigma_{1i} = 0$ along all admissible paths from $\boldsymbol{\varepsilon}^+$ to $\boldsymbol{\varepsilon}^-$ due to (2.4):

$$\begin{aligned}
\int_{\epsilon_{\alpha\beta}^{\rm e+}}^{\epsilon_{\alpha\beta}^{\rm e-}} \sigma_{\alpha\beta} {\rm d}\epsilon_{\alpha\beta}^{\rm e} &= \int_{\sigma_{\alpha\beta}^{+}}^{\sigma_{\alpha\beta}^{-}} \sigma_{\alpha\beta} M_{\alpha\beta\gamma\delta} {\rm d}\sigma_{\gamma\delta} = \tfrac{1}{2} \int_{\sigma_{\alpha\beta}^{+}}^{\sigma_{\alpha\beta}^{-}} {\rm d}(\sigma_{\alpha\beta} {\rm M}_{\alpha\beta\gamma\delta}\, \sigma_{\gamma\delta}) \\
&= \tfrac{1}{2} (\sigma_{\alpha\beta}^{+} + \sigma_{\alpha\beta}^{-}) M_{\alpha\beta\gamma\delta} (\sigma_{\gamma\delta}^{-} - \sigma_{\gamma\delta}^{+}) \\
&= \tfrac{1}{2} (\sigma_{\alpha\beta}^{+} + \sigma_{\alpha\beta}^{-}) M_{\alpha\beta kl} (\sigma_{kl}^{-} - \sigma_{kl}^{+}) \\
&= -\tfrac{1}{2} (\sigma_{\alpha\beta}^{+} + \sigma_{\alpha\beta}^{-}) [\![\epsilon_{\alpha\beta}^{\rm e}]\!] = \tfrac{1}{2} (\sigma_{\alpha\beta}^{+} + \sigma_{\alpha\beta}^{-}) [\![\epsilon_{\alpha\beta}^{\rm p}]\!],
\end{aligned} \tag{3.4}$$

the last step making use of (2.9b) in the form

$$[\![\epsilon_{\alpha\beta}^{\rm e}]\!] = -[\![\epsilon_{\alpha\beta}^{\rm p}]\!]. \tag{3.5}$$

Thus, from (3.3) and (3.4), (3.2) becomes

$$W^{\rm p} = -\sigma_{11}^{\Sigma} [\![\epsilon_{11}^{\rm p}]\!] - 2\sigma_{1\alpha}^{\Sigma} [\![\epsilon_{1\alpha}^{\rm p}]\!] - \tfrac{1}{2} (\sigma_{\alpha\beta}^{+} + \sigma_{\alpha\beta}^{-}) [\![\epsilon_{\alpha\beta}^{\rm p}]\!], \tag{3.6}$$

and noting that (2.4) requires

$$\sigma_{1i}^{+} = \sigma_{1i}^{-} = \sigma_{1i}^{\Sigma}, \tag{3.7}$$

we obtain finally

$$W^{\rm p} \equiv \int_{\epsilon_{ij}^{\rm p+}}^{\epsilon_{ij}^{\rm p-}} \sigma_{ij} {\rm d}\epsilon_{ij}^{\rm p} = -\tfrac{1}{2} (\sigma_{ij}^{+} + \sigma_{ij}^{-}) [\![\epsilon_{ij}^{\rm p}]\!]. \tag{3.8}$$

This has a deceptively similar appearance to the result given earlier for W; however, the derivation of (3.8) relies on the constitutive assumptions relating to $\boldsymbol{\varepsilon}^{\mathrm{e}}$, while the result for W is independent of material constitution.

As regards the second term of (2.19), recall from Section 2.3 that σ_{ij}^0 is any stress state at or below yield. We make the choice

$$\sigma_{ij}^0 \equiv \sigma_{ij}^+, \tag{3.9}$$

observing that $\boldsymbol{\sigma}^+$ is necessarily at or below yield throughout the transition from $\boldsymbol{\varepsilon}^+$ to $\boldsymbol{\varepsilon}^-$ for ideally plastic materials and also for many types of hardening materials as discussed later in this Section. Thus

$$\int_{\epsilon_{ij}^{\mathrm{p}+}}^{\epsilon_{ij}^{\mathrm{p}-}} \sigma_{ij}^0 \mathrm{d}\epsilon_{ij}^{\mathrm{p}} = \int_{\epsilon_{ij}^{\mathrm{p}+}}^{\epsilon_{ij}^{\mathrm{p}-}} \sigma_{ij}^+ \mathrm{d}\epsilon_{ij}^{\mathrm{p}} = -\sigma_{ij}^+ [\![\epsilon_{ij}^{\mathrm{p}}]\!]. \tag{3.10}$$

Substituting (3.8—3.10) into (2.19), we find

$$\int_{\epsilon_{ij}^{\mathrm{p}+}}^{\epsilon_{ij}^{\mathrm{p}-}} (\sigma_{ij} - \sigma_{ij}^+) \mathrm{d}\epsilon_{ij}^{\mathrm{p}} = -\tfrac{1}{2}(\sigma_{ij}^+ + \sigma_{ij}^-) [\![\epsilon_{ij}^{\mathrm{p}}]\!] + \sigma_{ij}^+ [\![\epsilon_{ij}^{\mathrm{p}}]\!] \geqslant 0,$$

which simplifies to

$$\tfrac{1}{2} [\![\sigma_{ij}]\!] [\![\epsilon_{ij}^{\mathrm{p}}]\!] \geqslant 0. \tag{3.11}$$

Use of (2.4) further reduces this to

$$\tfrac{1}{2} [\![\sigma_{\alpha\beta}]\!] [\![\epsilon_{\alpha\beta}^{\mathrm{p}}]\!] \geqslant 0,$$

which is expressible via (3.5) as

$$\tfrac{1}{2} [\![\sigma_{\alpha\beta}]\!] [\![\epsilon_{\alpha\beta}^{\mathrm{e}}]\!] \leqslant 0. \tag{3.12}$$

Now

$$[\![\epsilon_{\alpha\beta}^{\mathrm{e}}]\!] = M_{\alpha\beta kl} [\![\sigma_{kl}]\!] = M_{\alpha\beta\gamma\delta} [\![\sigma_{\gamma\delta}]\!]$$

from (2.15) and (2.4), respectively, and this permits (3.12) to be written as a restriction on stress jumps only:

$$\tfrac{1}{2} [\![\sigma_{\alpha\beta}]\!] M_{\alpha\beta\gamma\delta} [\![\sigma_{\gamma\delta}]\!] \leqslant 0. \tag{3.13}$$

Noting that since the full compliance tensor M_{ijkl} is positive definite $M_{\alpha\beta\gamma\delta}$ is also positive definite (it is the "plane stress" compliance tensor, for plane stress in the x_2, x_3 plane), (3.13) can be satisfied only if

$$[\![\sigma_{\alpha\beta}]\!] = 0.$$

Coupling this with the result of equilibrium, (2.4), gives finally

$$[\![\sigma_{ij}]\!] = 0. \tag{3.14}$$

That is, *all* components of stress *must* be continuous across Σ, provided that Σ is indeed moving and that the material response is quasi-static.

The preceding proof of the necessity of full stress continuity across Σ applies for materials which satisfy the constitutive assumptions outlined in Section 2.3. These assumptions permit arbitrary anisotropy of elastic response, as well as arbitrary anisotropy of the initial yield surface in stress space provided that it is convex. (Convexity is required by (2.18).) The proof is clearly valid for non-hardening materials; its validity also extends to a fairly broad class of strain-hardening materials, delineated by one additional restriction. This restriction, implicit in the use of the integrated form (2.19) of the maximum plastic work inequality, is that the material harden in such a manner that σ_{ij}^{+} (the value chosen for σ_{ij}^{0} in the proof) remain on or inside the current yield surface during the passage of Σ. One type of strain hardening which evidently satisfies this additional restriction, and which is thus included in our stress continuity proof, is isotropic hardening. Many types of anisotropic hardening would be included as well; namely, any type of hardening in which the current yield locus fully incorporates all preceding yield loci. Examples include some cases of hardening by the formation of a vertex on the yield surface, and hardening comprised of appropriate combinations of kinematic and isotropic hardening.

4. Restrictions on propagating surfaces of strain discontinuity

4.1 RESTRICTIONS IN GENERAL

The fact that all components of stress must be continuous across Σ has a number of immediately evident consequences as regards strain jumps across Σ. From (2.15) we observe that the elastic part of every strain component must be continuous across Σ, i.e.,

$$[\![\epsilon_{ij}^{\mathrm{e}}]\!] = 0. \tag{4.1}$$

Condition (3.5) then shows that all components of plastic strain in the plane of Σ must be continuous

$$[\![\epsilon_{\alpha\beta}^{\mathrm{p}}]\!] = 0. \tag{4.2}$$

Thus in general the only strain components which may suffer discontinuities across Σ are $\epsilon^{\mathrm{p}}_{1i}$.

A restriction on possible jumps in these components is provided by (2.19) (using (3.8)) by choosing $\sigma^0_{ij} \equiv 0$ (which can always be considered to be below yield for the class of materials just discussed) and realizing that due to full stress continuity $\sigma^+_{ij} = \sigma^-_{ij} = \sigma^{\Sigma}_{ij}$; the resulting restriction is

$$-\sigma^{\Sigma}_{11} [\![\epsilon^{\mathrm{p}}_{11}]\!] - 2\sigma^{\Sigma}_{1\alpha}[\![\epsilon^{\mathrm{p}}_{1\alpha}]\!] \geqslant 0. \tag{4.3}$$

This is the requirement that Σ produce non-negative plastic work.

Conditions (4.1) through (4.3) are general restrictions on possible strain jumps across Σ which must be satisfied by the full class of materials for which stress continuity was proved, as discussed at the end of Section 3.

4.2 FURTHER RESTRICTIONS FOR MORE SPECIFIC CONSTITUTIVE MODELS

All of the conclusions drawn thus far have been based on the rather general constitutive assumptions described in Section 2.3. If this constitutive model is made more specific, further restrictions on possible jumps in plastic strain are seen to result.

4.2.1 Plastically incompressible solids

For solids which are plastically incompressible, the following additional condition applies

$$\epsilon^{\mathrm{p}}_{ii} = 0. \tag{4.4}$$

Along Σ, this relates the jumps in the normal plastic strain components as

$$[\![\epsilon^{\mathrm{p}}_{11}]\!] = -[\![\epsilon^{\mathrm{p}}_{22}]\!] - [\![\epsilon^{\mathrm{p}}_{33}]\!],$$

so that via (4.2),

$$[\![\epsilon^{\mathrm{p}}_{11}]\!] = 0. \tag{4.5}$$

Combining this with (4.1) and (2.17) yields the conclusion

$$[\![\epsilon_{11}]\!] = 0. \tag{4.6}$$

Going back to (2.12a), (4.6) shows that

$$[\![v_1]\!] = 0. \tag{4.7}$$

Thus in a plastically incompressible solid, only components of velocity parallel to Σ (i.e., v_α), and plastic strain components $\epsilon^{\mathrm{p}}_{1\alpha}$, may suffer discontinuities, and then only provided that the restriction

$$-\sigma^{\Sigma}_{1\alpha}[\![\epsilon^{\mathrm{p}}_{1\alpha}]\!] \geqslant 0 \tag{4.8}$$

is satisfied.

4.2.2 Elastic-ideally plastic solids with smooth yield surfaces

Solids of this type are assumed to have a yield condition which is a function of the stress state only, i.e.

$$f(\sigma_{ij}) = 0 \tag{4.9}$$

(with $f < 0$ for σ_{ij} below yield). It is easily proved from the maximum plastic work inequality, (2.18), that $f(\sigma_{ij})$ must be convex, and that $\mathrm{d}\epsilon^{\mathrm{p}}_{ij}$ must be normal to $f(\sigma_{ij})$. The latter result can be stated mathematically as an associated flow rule

$$\mathrm{d}\epsilon^{\mathrm{p}}_{ij} = \mathrm{d}\Lambda P_{ij}, \tag{4.10}$$

where it is assumed that f depends symmetrically on σ_{ij} and σ_{ji} so that

$$P_{ij} \equiv \partial f/\partial\sigma_{ij}$$

is a symmetric tensor having the interpretation of outer normal to the yield surface in stress space, and $\mathrm{d}\Lambda \geqslant 0$ is an undetermined scalar function of position; note that $\sigma_{ij}P_{ij} > 0$. Restricting consideration now to yield functions $f(\sigma_{ij})$ which are smooth (no vertices), and noting that this limitation coupled with (3.14) means that

$$[\![P_{ij}]\!] = 0, \tag{4.11}$$

integration of (4.10) at a material point just during the passage of the discontinuity surface Σ gives

$$-[\![\epsilon^{\mathrm{p}}_{ij}]\!] = \eta P^{\Sigma}_{ij}, \tag{4.12}$$

where $\eta = \eta(x_2, x_3, t) \geqslant 0$ is an undetermined scalar ($\eta \equiv \int_{\Lambda^+}^{\Lambda^-}\mathrm{d}\Lambda = -[\![\Lambda]\!]$).

As concluded in Section 4.1, only the components $\epsilon^{\mathrm{p}}_{1i}$ may suffer discontinuities across Σ. Relation (4.12) shows that in the case of elastic-ideally plastic solids, additional conditions must be met along Σ for such a discontinuity to exist. In particular, since (4.2) must be met, it is apparent from (4.12) that *either*

(i) $\eta \equiv 0$

or

(ii) $P^{\Sigma}_{\alpha\beta} \equiv 0$.

That is, if $P^{\Sigma}_{\alpha\beta} \not\equiv 0$, we *must* have $\eta \equiv 0$, and (4.12) shows then that all $[\![\epsilon^{\mathrm{p}}_{ij}]\!] = 0$; this result together with (3.14) and (4.1) means that Σ cannot be a surface of strong discontinuity when $P^{\Sigma}_{\alpha\beta} \not\equiv 0$.

If however $P^{\Sigma}_{\alpha\beta} \equiv 0$, η *may* be nonzero, in which case a jump in $\epsilon^{\mathrm{p}}_{1i}$ is not precluded so long as $[\![\epsilon^{\mathrm{p}}_{1i}]\!] = -\eta P^{\Sigma}_{1i}$. If we assume further that the solid is plastically incompressible, (4.5) and (4.12) impose the additional restriction $P^{\Sigma}_{11} = 0$. Thus an elastic-ideally plastic solid which is plastically incompressible may sustain a jump in one or both of $\epsilon^{\mathrm{p}}_{1\alpha}$ across Σ only if the following conditions are met:

$$[\![\epsilon^{\mathrm{p}}_{1\alpha}]\!] = -\eta P^{\Sigma}_{1\alpha}, \eta \geqslant 0, \tag{4.13a}$$

$$P^{\Sigma}_{11} = P^{\Sigma}_{\alpha\beta} = 0. \tag{4.13b}$$

If (4.13) are satisfied, a jump in one or both of v_α across Σ may exist; from (2.12b), (4.1) and (4.13a), these will be given by

$$[\![v_\alpha]\!] = -2V[\![\epsilon^{\mathrm{p}}_{1\alpha}]\!] = 2V\eta P^{\Sigma}_{1\alpha}. \tag{4.13c}$$

4.2.3 Huber—Mises solids

We examine finally a special case of the class analyzed in Section 4.2.2: elastic-ideally plastic solids which obey the Huber—Mises yield condition. This condition is

$$f(\sigma_{ij}) = \tfrac{1}{2} s_{ij}s_{ij} - k^2 = 0, \tag{4.14}$$

where $s_{ij} = \sigma_{ij} - \delta_{ij}\sigma_{kk}/3$ is the deviatoric stress tensor and k is the shear strength. From (4.14), evidently,

$$P_{ij} \equiv \partial f/\partial \sigma_{ij} = s_{ij}. \tag{4.15}$$

Restrictions (4.13) thus specify to

$$[\![\epsilon^{\mathrm{p}}_{1\alpha}]\!] = -\eta\sigma^{\Sigma}_{1\alpha}, \quad \eta \geqslant 0, \tag{4.16a}$$

$$s^{\Sigma}_{11} = s^{\Sigma}_{22} = s^{\Sigma}_{33} = \sigma^{\Sigma}_{23} = 0, \tag{4.16b}$$

$$[\![v_\alpha]\!] = -2V[\![\epsilon^{\mathrm{p}}_{1\alpha}]\!] = 2V\eta\sigma^{\Sigma}_{1\alpha}; \tag{4.16c}$$

condition (4.16b) is the requirement that the stress state on Σ consist of pure shear plus equal normal stresses in all directions, and that Σ be a plane of maximum shearing stress.

5. Discussion

Gao and Hwang [12], utilizing a very different approach from the one employed above (they examine a higher-order differential equation in an Airy stress function), have investigated the conditions of continuity across a quasi-statically propagating surface for the special case of plane strain deformation of an elastic-ideally plastic Huber–Mises solid. They conclude that *all* components of stress and strain must be continuous for Poisson ratio $\nu < 0.5$ (elastic compressibility), in contrast to our conclusion that a jump in plastic shear strain cannot be ruled out under certain conditions — namely those of (4.16) for the material model which they considered. However, the proof of full strain continuity supplied by Gao and Hwang is faulted because these authors make the *a priori* assumption that a certain combination of terms, involving second derivatives of the Airy function, does not vanish. In this case the governing equations are shown to be elliptic and discontinuity-free in plastic as well as elastic domains. But elementary analysis suffices to show that the non-vanishing of this combination of terms is equivalent to non-vanishing of the deviatoric stress component s_{33} (when x_1, x_2 define the plane of straining). Hence the *a priori* (and never justified) assumption of Gao and Hwang rules out one of the conditions that we have just shown in (4.16b) to be necessary in order that a strain discontinuity exist.

The motivation of our study in the elastic-plastic growing crack problem was mentioned at the outset. It is interesting to note that the asymptotic crack tip field derived by Drugan et al. [2], for quasi-static crack growth in isotropic ideally plastic Huber–Mises materials, involves (at least asymptotically, as the crack tip is approached) a tangential velocity discontinuity and strain discontinuity, of just the type allowed by satisfaction of all the conditions (4.16). This occurs at the moving interface between what are referred to as "constant stress" and "centered fan" asymptotic plastic angular sectors about the crack tip; this is an interface on which the stress state satisfies (4.16b). Drugan et al. show further that at least for the sequence of arrangement of asymptotic sectors for which they construct a solution, no acceptable alternate solution can be found which eliminates the velocity and strain discontinuity at the interface referred to above.

It is well known that asymptotic stress fields at *stationary* tensile crack tips, say, in bodies on which monotonic deformations are imposed, are not unique for ideally plastic solids. Typically, different fields exist for well-contained yielding and for fully plastic general yielding. This is because fields with $[\![\sigma_{\alpha\beta}]\!] \neq 0$ across a stationary surface cannot be ruled out and also because rather general types of sub-yield stress distributions may exist at a crack tip. By contrast, it is evident from what we have proven here that σ_{ij} must be fully continuous at a quasi-statically *growing* crack tip. Further, the admissible asymptotic fields at, say, a growing plane strain crack tip consist of only four different types of angular sectors. Given that these sectors must be assembled with full continuity of σ_{ij} at their boundaries, and given also that attempts thus far reported have resulted in only one acceptable asymptotic stress distribution for the growing crack in an isotropic Huber—Mises material, we pose the following open question: Could it be that there is a *unique* asymptotic stress distribution at a growing crack tip in an elastic-ideally plastic solid (on the class of distributions having symmetry appropriate to tensile loading and transmitting tension across the prolongation of the crack plane)? The answer has relevance to the description of ductile crack growth, because if the asymptotic stress field is unique, then in all cases for which it prevails over a size scale that is large compared to that of fracture micro-mechanisms, the crack growth criterion presumably reduces to the requirement that an adequate intensity of near tip deformation be maintained to sustain growth. While there may be some ambiguity as to how the deformation field should be characterized, the result resembles the type of "one parameter" elastic-plastic crack growth criterion which is, in fact, usually argued against on the basis of the well known non-uniqueness of ideally plastic near tip stress fields for stationary cracks.

Acknowledgements

This study was supported by the U.S. Department of Energy under Contract DE-AC02-80ER/10556 with Brown University, with additional support (JRR) from the NSF Materials Research Laboratory at Harvard University. We acknowledge gratefully discussions with Professors L.B. Freund, R.D. James and A.C. Pipkin of Brown University.

References

1 J.R. Rice, in: H.G. Hopkins and M.J. Sewell (Eds.), Mechanics of Solids: The R. Hill 60th Anniversary Volume, Pergamon Press, Oxford, 1982, p. 539.
2 W.J. Drugan, J.R. Rice and T.-L. Sham, J. Mech. Phys. Solids, 30 (1982) 447—473 and 31 (1983) 191.
3 R. Hill, in: I.N. Sneddon and R. Hill (Eds.), Progress in Solid Mechanics, Vol. II, North-Holland, Amsterdam, 1961, Ch. 6.
4 J.R. Rice, J. Appl. Mech., 37 (1970) 728—737.
5 J.R. Rice, J. Mech. Phys. Solids, 19 (1971) 433—455.

6 D.C. Drucker, in: E. Sternberg (Ed.), Proc. First U.S. Nat. Congr. Appl. Mech., A.S.M.E., New York, 1951, pp. 487—491.
7 D.C. Drucker, in: J.N. Goodier and N.J. Hoff (Eds.), Structural Mechanics, Pergamon Press, Oxford, 1960, pp. 407—455.
8 A.A. Il'yushin, Prikl. Mat. Mekh., 25 (1961) 503—507.
9 J.F.W. Bishop and R. Hill, Phil. Mag., 42 (1951) 414—427.
10 J. Mandel, in: H. Gortler (Ed.), Proc. Eleventh Int. Congr. Appl. Mech., Springer, Berlin, 1964, p. 502.
11 R. Hill, J. Mech. Phys. Solids, 15 (1967) 79—95.
12 Y.-C. Gao and K.-C. Hwang, On the formulation of plane strain problems for elastic-perfectly plastic medium, Int. J. Eng. Sci., in press.

Temperature Measurements during the Formation of Shear Bands in a Structural Steel

J. DUFFY

Brown University, Providence, RI 02912 (U.S.A.)

Abstract

Under dynamic conditions, localized plastic flow in metals is accompanied by a local rise in the temperature. In a series of experiments with steel specimens deformed in shear, the surface temperature was measured in the shear band and to either side of it as the deformation proceeded. This measurement was effected by employing a high speed infra-red radiation detection system. The results are compared to predictions based on analyses of other investigators.

1. Introduction

Many investigations currently are under way into the causes of shear bands, or more generally of localized plastic flow. The phenomenon has been observed for quite some time, occurring in many applications as for instance in machining or in penetration and perforation. The importance of shear bands lies in the fact that they weaken metal components rapidly and often precede fracture. Reviews of the subject have been presented by Rogers [9] and by Clifton [2].

In the case of steel, shear bands frequently are divided into two classes: deformed bands and transformed bands. The latter generally appear white when etched with a nital solution and are generally narrower than the deformed bands. Examples of both types of shear bands are given by Backman and Finnegan [1]. Considerable evidence, see for instance Rogers and Shastri [10], indicates that the transformed band is preceded by a deformation band. For either type of shear band, when high rates of deformation are involved then the formation of the shear band is accompanied by a local temperature rise which serves to soften the metal locally. When large enough it may produce the phase transformation seen in the transformed bands. Measurements of the increase in temperature are of considerable significance but are difficult to effect due to the speed with which the whole process occurs. As a result only a few investigators have attempted temperature measurements with high speed equipment; in all these cases a radiant energy detection system was employed. Moss and Pond [8] performed tensile tests on specimens of copper at strain rates from 0.56 to 172 s^{-1}. They measured the radiant energy emitted from neighboring points on the metal surface by employing a photoconductive detection system with a germanium crystal. They calibrated their system so as to account for the changes in surface finish occurring as a result of the plastic strain, since these produce an increase in the metal's emissivity. Hayashi et al. [5] made use of an indium-antimonide detector to measure the surface temperature of metals during plastic deformation. However, in their experiment the observed surface underwent a uniform strain distribution. They did find, however, that the changes in surface finish due to the increasing plastic strain increased somewhat the emissivity of the metal surface. They also found a time-lag of about 10 microseconds between deformation and the consequent temperature rise.

2. Description of experiment

More recently, Costin et al. [3] observed shear bands in thin-walled tubular specimens of a cold-rolled steel deformed in torsion in a Kolsky bar. They found that a low-carbon cold-rolled steel (CRS) produced shear bands at high strain rates whereas a hot-rolled steel (HRS) of similar composition did not, nor did aluminum or copper specimens. Their investigation was directed toward determining the reasons for the contrasting behavior of these metals and in particular of the two steels.

The torsional Kolsky bar in which these experiments were conducted, has been described by Senseny et al. [11]. By the sudden release of a previously stored torque, thin-walled specimens can be loaded in shear at strain rates ranging from 200 s^{-1} to 1000 s^{-1}. The apparatus can also be employed for quasi-static loading of the specimen. Figure 1 shows the results of quasi-static and dynamic tests with the two steels in question. For the HRS the work-hardening rate is always positive for either strain rate and at all temperatures except the lowest, where an instability occurs when the shear strain γ is equal approximately to 10% or 15%. For CRS at the quasi-static strain rate the flow stress remains nearly constant up to large strains, but the work-hardening rate though small is always positive up to strains as high as 70%. On the other hand, CRS deformed dynamically shows a positive strain-hardening rate only to a strain of about 8% or 10% (at room temperature) while for larger strains the material softens. Aluminum and copper specimens deformed in similar tests show only a positive strain hardening rate.

The records of strain and strain rate provided by Kolsky's method represent average values over the whole length of the specimen; simply

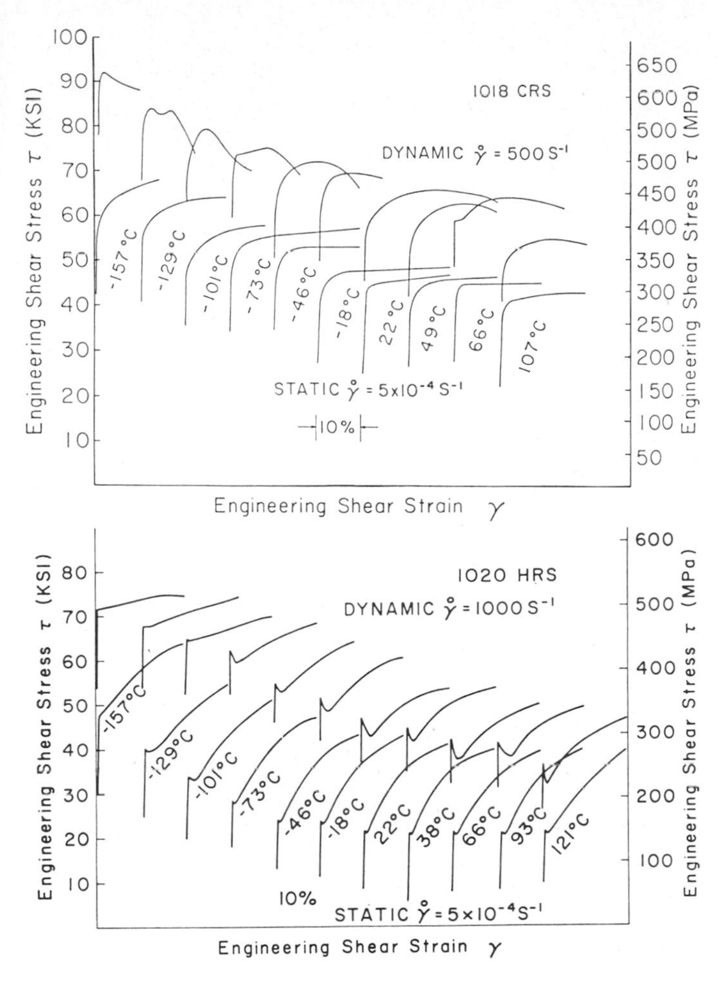

Fig. 1. Engineering shear stress vs. engineering shear strain at various temperatures for 1018 CRS and 1020 HRS (Costin et al. [3]).

on the basis on strain rate records, there is no way of determining whether the strain distribution within the specimen is homogeneous or not. However, the permanent strain distribution in the specimen can be measured from the inclination of fine lines scribed on the inside surface of the specimens. These lines are scribed before testing and originally lie parallel to the tube's axis. After testing, the inclination of the lines provides a direct measure of permanent strain. For the HRS, the constant inclination of the line within the gage length region indicates a homogeneous strain distribution, Fig. 2(b). Outside this region the line remains axial. For dynamic tests with CRS the strain distribution is inhomogeneous, Fig. 2(a), and this becomes evident because the inclination of the scribed line is not constant and the presence of the shear bands results in a very

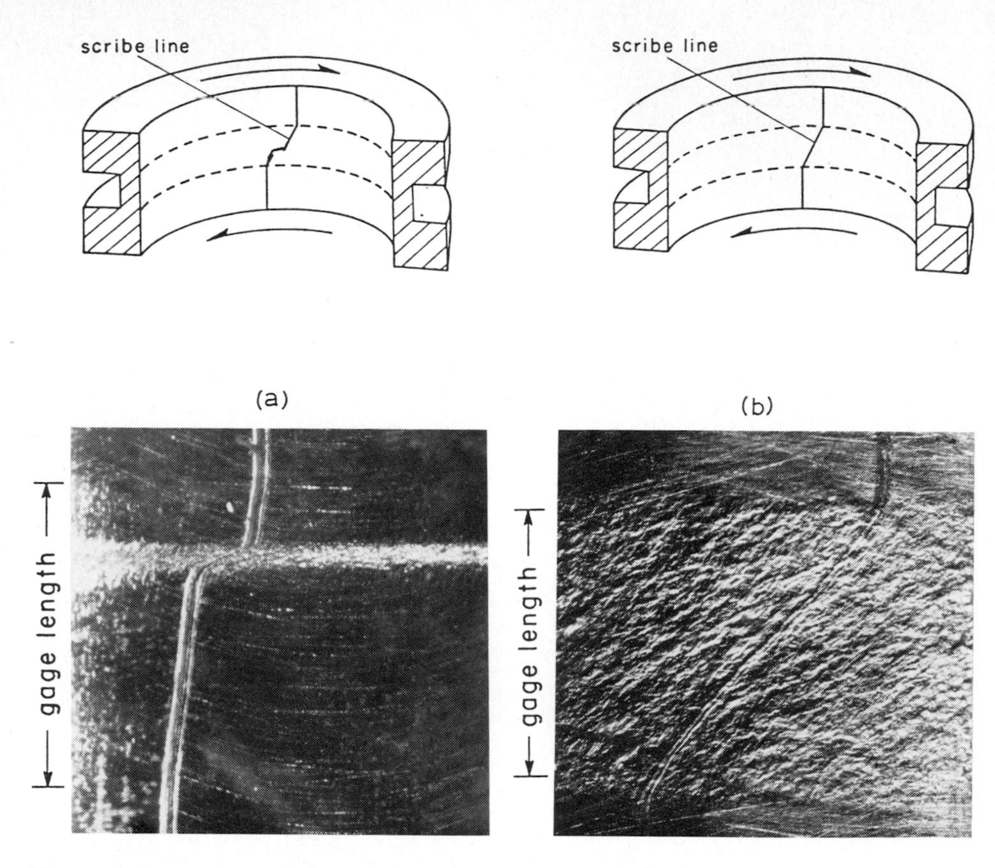

Fig. 2. Schematic drawings and photographs showing scribed lines deformed as a result of dynamic loading in shear: (a) cold-rolled steel and (b) hot rolled steel (Costin et al. [3]).

large local inclination. However, if the test with CRS is stopped short at 8% average strain, then the strain distribution always remains homogeneous. This is consistent with the slope of the line in Fig. 2(a) which shows some permanent strain along the entire gage length of the specimen. This result also indicates that the formation of the shear band does not start when initial yield is first attained.

During several tests conducted at room temperature, a high-speed infrared radiation detection system, whose essential element was an indium-antimonide photovoltaic detector, was used to measure the change in the surface temperature of the specimen as it was undergoing dynamic loading. The temperature of a small area (about 1 mm square) was measured by focussing the heat radiated by this surface area onto the element of the infrared detector. The detector was calibrated by measuring its output when focussed onto an undeformed 1018 CRS specimen heated to various temperatures by a small heater inserted within the specimen. The output of the detector was then compared to the specimen's temperature as measured by a thermocouple. In tests where a shear band formed within the observed region a significant temperature increase was observed. Its precise value, however, depends on the estimated width of the shear band, since the shear band is narrower than the observation area. If the width of the shear band is taken as 0.01 inches, the observed temperature increase would be 80 to 130°C*; for an estimated width of 0.02 inches, the observed values are half as great. A calculation of the temperature based on an estimated strain of about 200% indicates that the temperature rise should be about 200°C, on the assumption that all plastic work is converted to heat. However, the area covered by the detector includes both the shear band and some of the more slowly deforming material on either side of it. When allowance is made for this, the measured temperature increase agrees reasonably well with the predicted adiabatic temperature increase. It thus appears that the localization of plastic strain into a shear band involves a nearly adiabatic process.

Costin et al. sought further evidence of thermal effects during the formation of shear bands in CRS. For this purpose, specimens which had first been tested dynamically, as described above, and in which a shear band had appeared, were retested as follows. These specimens

*The temperatures presented here are corrected values. They differ from those originally reported by Costin et al. [3].

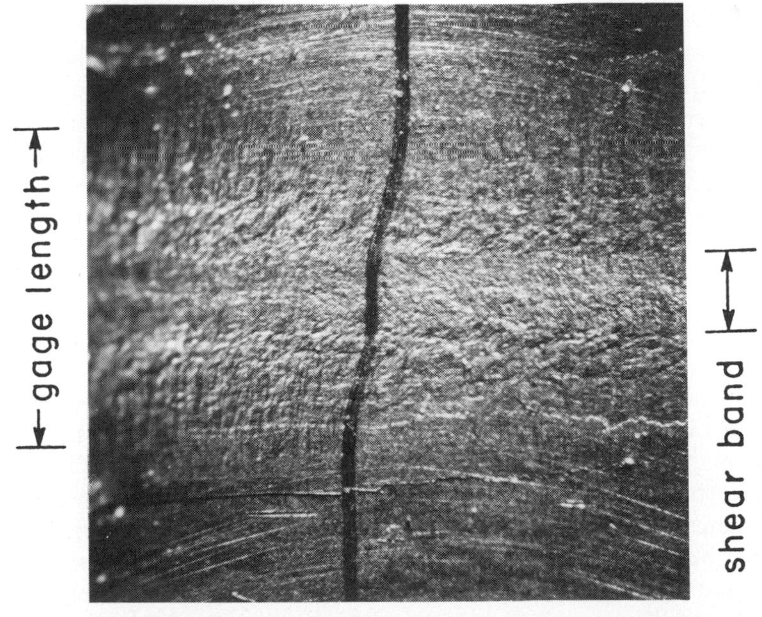

Fig. 3. Photograph of scribed line through shear band in CRS specimen, showing no further straining in the shear band as a result of quasi-static plastic deformation (Costin et al. [3]).

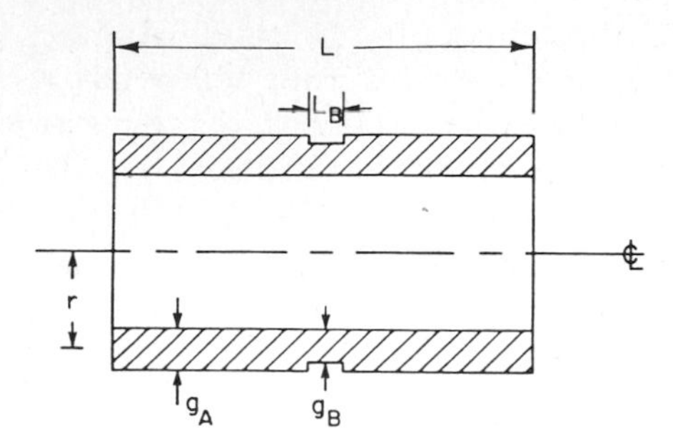

Fig. 4. Cross-section of torsional tube in Litonski's model (Costin et al. [3]).

were removed from the apparatus and new axial lines were inscribed along their inner surface. After this, they were remounted and the deformation continued, but quasi-statically. The result, Fig. 3, shows that the shear band material did not deform plastically on reloading: all strain accumulated in the remainder of the specimen. Thus after cooling the shear band is harder than the material to either side of it. It seems likely, therefore, that during the dynamic tests plastic flow occurs within the shear band because of a local softening due to a higher temperature within the band.

3. Comparison to analyses

A calculation was made to estimate the expected plastic strain and temperature within the shear band as compared to conditions during homogeneous deformation. For this purpose, it is convenient to use an analysis of torsion in thin-walled tubes presented by Litoński [7] in which the tube has a defect represented by a slightly thinner region, designated by B in Fig. 4. The constitutive relation adopted by Litonski gives the shear stress τ as a function of the average plastic shear strain γ and also of the strain rate $\dot{\gamma}$, as follows

$$\tau = c(1 - aT)(1 + b\dot{\gamma})^m \gamma^n \quad (1)$$

where a, b, c, m and n are parameters evaluated on the basis of experimental results. In this equation, the linear variation of stress with temperature T approximates the temperature dependence observed in Fig. 1. The constants in the equation were evaluated from these curves for both HRS and CRS. The expected plastic strain and the increase in temperature are found by a numerical integration of eqn. (1). The results, shown in Fig. 5, are qualitatively similar to those observed experimentally: in CRS at $\gamma = 10\%$ or 15% there is a rapid increase in strain and temperature within the narrow region of the defect designated by B. In HRS there is no appreciable difference in strain or temperature in this region. Evidently,

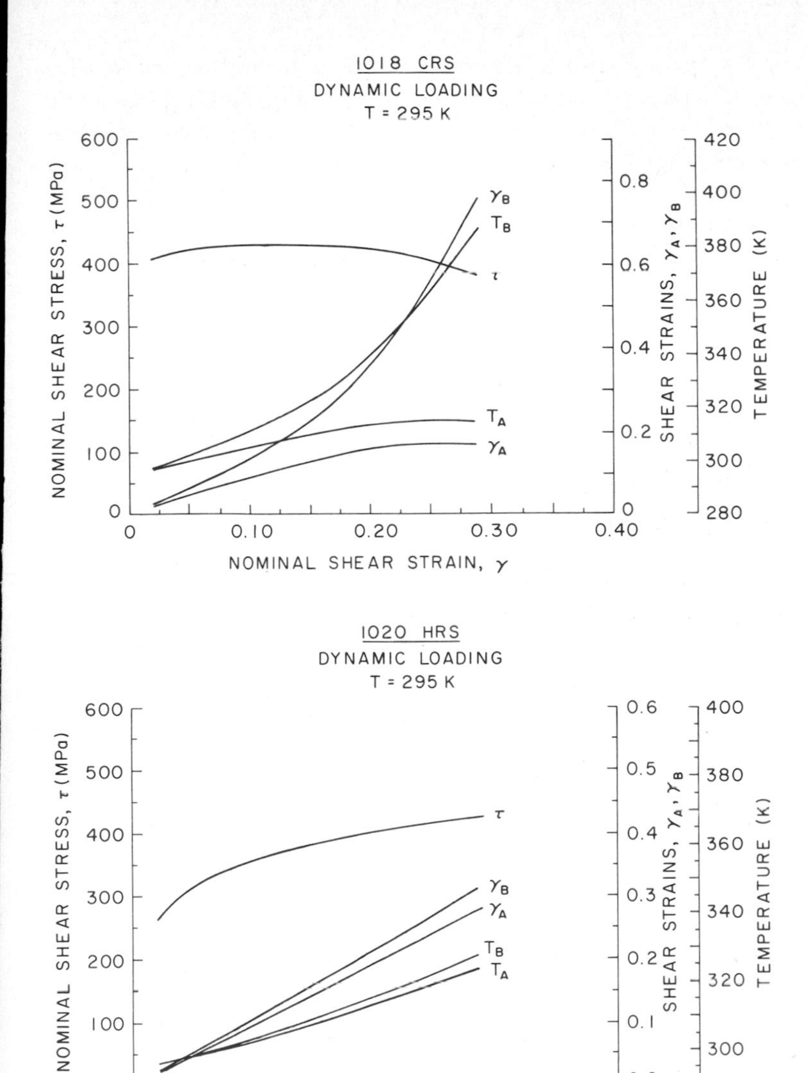

Fig. 5. Results of calculations based on Litonski's model showing strain and temperature within the thinned section, *B*, compared to that in the remainder of the specimen, *A* (Costin et al. [3]).

in the cold-rolled steel the strain and temperature within the thinned region begin to increase rapidly at a nominal strain of about 10% or 15%. In the remainder of the specimen the increase continues nearly linearly. For the hot-rolled steel, whose strain hardening rate is much steeper, this analysis predicts an almost equal rise in strain and temperature inside and outside the defect region.

The results of Costin et al. are in general agreement with those of Culver [4] and with the analysis presented by Clifton [2]. In his analysis, Clifton imposes a simple shear stress as a boundary condition on a body made of a rate-sensitive material and containing a geometric inhomogeneity. The shear stress is applied at the boundary by a traction $\tau^0(t)$, which is a specified function of time. It results in a shear strain rate $\dot{\gamma}(t)$ which may be divided into elastic and plastic components, $\dot{\gamma}^{\mathrm{e}}$ and $\dot{\gamma}^{\mathrm{p}}$. In addition there is a thermoelastic volume change. For the plastic shear strain rate, Clifton takes an Arrhenius type relation

$$\dot{\gamma}^{\mathrm{p}} = \dot{\gamma}_0 \exp\left(-H(\tau, \tau_{\mathrm{r}}, p)/KT\right) \tag{2}$$

where $\dot{\gamma}_0$ is a constant, H is an activation enthalpy, K is Boltzmann's constant and T is absolute temperature. The activation enthalpy is assumed to depend on the applied shear stress τ, a shear stress τ_{r} that represents the resistance of the material to plastic deformation, and the hydrostatic pressure p which is related to the change in temperature. Clifton then employs a perturbation technique, simultaneously satisfying equations of compatibility of displacements, heat conduction as well as heat production due to the plastic deformation. By this method he shows that the deformation process is stable whenever the following quantity is positive

$$m\dot{\gamma}^{0\mathrm{p}} \left[\frac{1}{\tau_{\mathrm{r}}} \frac{\partial \tau_{\mathrm{r}}}{\partial \gamma^{\mathrm{p}}} + \frac{\beta}{\rho c} \frac{\partial \tau}{\partial T} \right] + \frac{k\epsilon^2}{\rho c} \tag{3}$$

In this expression, m is a strain rate sensitivity parameter defined by

$$m = \frac{\partial \ln \dot{\gamma}^{\mathrm{p}}}{\partial \ln \tau} \tag{4}$$

while c is the specific heat of the material, k its thermal conductivity, ρ its mass density, $\dot{\gamma}^{0\mathrm{p}}$ the homogeneous plastic strain rate, $1/\epsilon$ is the wavelength of the inhomogeneity, and β is the fraction of plastic work converted into heat; β generally lies in the range 0.9—1.0. It should be noted also that $\partial \tau / \partial T < 0$ for most materials so that the two terms within the brackets are of opposite sign. The expression presented above is a simplification of Clifton's original criterion applicable to a material for which the pressure sensitivity of the plastic resistance is assumed to be negligible. The first two terms in this criterion are the ones most likely to vary depending on the material and its condition. Their relative magnitude is thus most important, and the quantity most likely to vary by a large amount is the strain hardening rate, $\partial \tau_{\mathrm{r}} / \partial \gamma^{\mathrm{p}}$. A large strain hardening rate, according to this criterion, will tend to a stable deformation process, while a vanishingly small strain hardening rate may make the second term dominant. Hence, shear bands are more likely in cold-rolled steel than in hot-rolled steel or than in aluminium or copper. The value of the product $m\dot{\gamma}^{0\mathrm{p}}$ is also important, since it multiplies two terms in expression (3). Thus a large value of $m\dot{\gamma}^{0\mathrm{p}}$ combined with

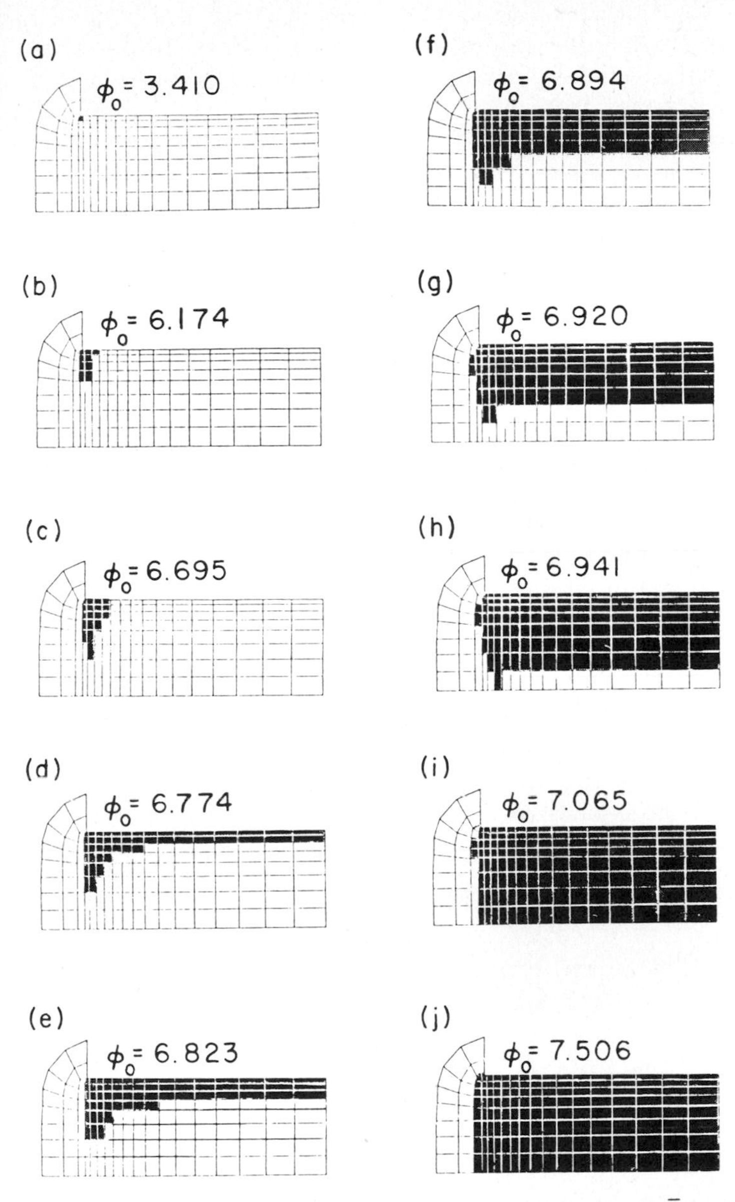

Fig. 6. Spread of the plastic zone at various loading stages $\bar{\sigma}_y = 6.25 \times 10^3$ psi, $H' = 1.13 \times 10^5$ psi (Leung [6]).

a relatively small strain hardening rate would imply instability. In the experiments described above, shear bands did not form at the reentrant corners of the specimen where the thin-walled tube joins the flange. It is evident that these corners are sites of stress concentration and it might have been expected that the plastic deformation starting there would quickly extend across the specimen's wall and thus form a region of uncontained plastic flow leading, with further deformation, to a shear band. Indeed, when the imposed strain rate drops below about $200\,s^{-1}$, then no shear band is formed unless straining is continued far enough, when shear bands form at the corners. Similarly, in aluminum and copper for large enough strains a shear band always forms at the corners, at least for strain rates up to $1000\,s^{-1}$. In the experiments with CRS it was not possible to predict the location of the shear band before testing, unless the specimen was first thinned in a particular region. However, shear bands never were observed at the reentrant corners. A finite element analysis of quasi-static deformation in these specimens was performed by Leung [6]. For a material with a bi-linear stress—strain relation, he showed that plastic deformation indeed starts at the stress concentration. However, for the work-hardening rates considered, it extends along the whole outer surface of the thin-walled tube before reaching the inside surface. Furthermore, when the inside surface becomes plastic as in Fig. 6(j), then the average plastic strain in the specimen is not yet quite equal to the elastic strain at first yield.

It is clear from the above that the infrared detector used by Costin et al. [3] focussed on too large a surface area of the specimen. Furthermore, it is important to determine the temperature--time history to either side of the shear band as well as within the band area. Accordingly, the design of the temperature detector is now being revised. We have constructed a five-element detector to measure the temperature—time history simultaneously at five equally spaced points along the specimen's 0.1 inch gage length. For this purpose, we are using a set of very fine thin-walled tubes as waveguides to bring the radiation from the specimen surface to the detector. At present each tube has an internal diameter

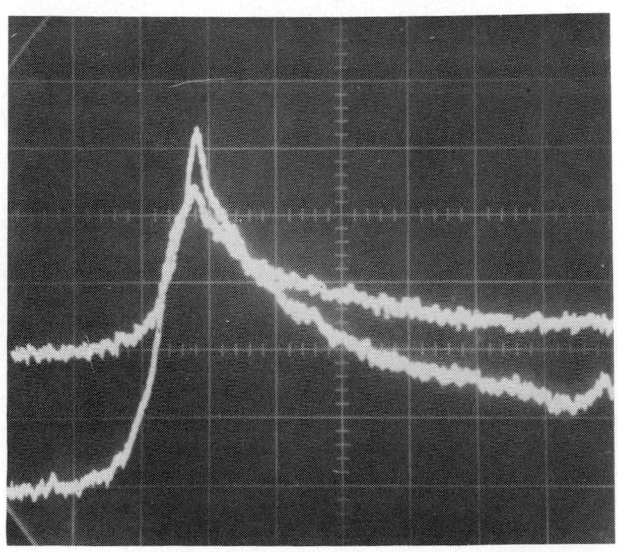

Fig. 7. Temperature--time history during formation of a shear band. Vertical scale represents temperature (1 division = 2 mV) and horizontal scale is time (1 division = 200 μsec).

of 0.016 inches. To improve the reflectivity of the tube walls, their interior surfaces are given a fine gold coating. Typical records obtained with two of the detectors are shown in Fig. 7, which indicates a maximum temperature rise of 210°C recorded by the detector element that monitored the temperature within the zone of the shear band. However, this value is tentative, since the surface finish of the specimen plays an important role in determining the amplitude of the detector's output, and we have not yet satisfied ourselves that our calibration method takes into account properly the changes in surface finish that occur during deformation. Furthermore, the diameter of the tubes is still somewhat too large so that we are now experimenting with a set of ten 0.010 inch tubes. The result should provide a finer spatial resolution of the temperature history.

4. Conclusions

More work needs to be done to determine the conditions that are necessary and sufficient for the formation of shear bands. On the analytic side, a stability analysis for the deformation of thin-walled tubes, both with and without a flaw, would be extremely interesting. Experimentally, the influence of strain rate on the width of the shear band needs to be investigated, since evidence suggests that higher strain rates produce finer shear bands. This, in turn, leads to a requirement of an ever finer temperature resolution. Numerous other aspects of the problem require examination. Shear bands are known frequently to lead to fracture, as we have found with the thin-walled tubes, Fig. 8. Microscopic investigations

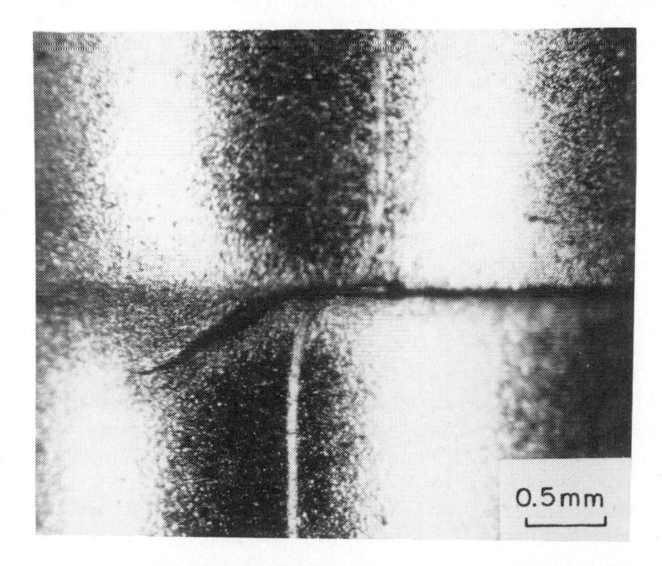

Fig. 8. A shear band leading to fracture in AISI 1018 cold-rolled steel.

also are needed and may lead to relations between the substructure or perhaps the occurrence of twinning and the initiation of shear bands. Eventually, stability criteria should be found to allow for the prediction of shear bands under various temperature and loading conditions and for quasi-static as well as dynamic deformation.

Acknowledgements

The results presented in this paper were obtained in the course of research supported by the National Science Foundation under Grant CME 79-23742. Thanks are due to K.A. Hartley and E.E. Crisman for their contributions to the development of the multiple element temperature detection system. The technical assistance of G.J. LaBonte is much appreciated.

References

1 Backman, M.E. and Finnegan, S.A., 1973. The propagation of adiabatic shear. In: R.W. Rohde, B.M. Butcher, J.R. Holland, and C.H. Karnes (Eds.), Proc. Conf. on Metallurgical Effects at High Strain Rates, Plenum Press, New York, pp. 531—544.
2 Clifton, R.J., 1980. Adiabatic Shear Banding. Report No. NMAB-356, National Materials Advisory Board Committee, *8*, pp. 129—142.
3 Costin, L.S., Crisman, E.E., Hawley, R.H. and Duffy, J., 1979. On the localisation of plastic flow in mild steel tubes under dynamic torsional loading. In: J. Harding (Ed.), Proc. 2nd Conference on Mechanical Properties at High Rates of Strain, The Institute of Physics, Bristol and London, Conf. Series No. 17, pp. 90—100.
4 Culver, R.S., 1973. Thermal instability strain in dynamic plastic deformation. In: R.W. Rohde, B.M. Butcher, J.R. Holland, and C.H. Karnes, (Eds.), Proc. Conf. on Metallurgical Effects at High Strain Rates, Plenum Press, New York, pp. 519—530.
5 Hayashi, T., Yamamura, H. and Okano, S., 1977. Temperature measurement of metals under high velocity deformation. Proc., Twentieth Japan Cong. on Mat. Res., Soc. of Mat. Sci., Kyoto, Japan, pp. 94—98.
6 Leung, E.K.C., 1980. An elestic-plastic stress analysis of the specimen used in the torsional Kolsky bar. J. Appl. Mech., 47: 278—282.
7 Litonski, J., 1977. Plastic flow of a tube under adiabatic torsion. Bulletin de l'Academie Polonaise des Sciences, 25 (1): 7—14.
8 Moss, G.L. and Pond, R.B., 1975. Inhomogeneous thermal changes in copper during plastic elongation. Met. Trans., 6A: 1223—1235.
9 Rogers, H.C., 1974. Adiabatic Shearing; A review. Drexel University Report for U.S. Army Research Office.
10 Rogers, H.C. and Shastry, C.V., 1981. Material Factors in Adiabatic Shearing in Steels. In: M.A. Meyers and L.E. Murr (Eds.), Proc. Conf. on Shock Waves and High-Strain-Rate Phenomena in Metals, Plenum Press, New York, Chapter 18, pp. 285—298.
11 Senseny, P.E., Duffy, J. and Hawley, R.H., 1978. Experiments on strain rate history and temperature effects during the plastic deformation of close packed metals. J. Appl. Mech., 45: 60—66.

Thermoplasticity of Unidirectional Metal Matrix Composites

G.J. DVORAK and C.J. WUNG

College of Engineering, University of Utah, Salt Lake City, UT 84112 (U.S.A.)

Abstract

Elastic-plastic deformation of fibrous metal matrix composites is analyzed for combined loading by axisymmetric mechanical loads, uniform thermal changes and variations in matrix yield stress. The composite cylinder model is used. New equilibrium and constraint equations for phase stress and strain averages are derived which make the model analysis adaptable to any phase constituent properties. Specific results are obtained for composites with transversely isotropic fibers and a kinematically hardening matrix. Instantaneous phase stress concentration factors and overall compliances are derived for the three types of loading. Loading by thermal and yield stress changes is converted to equivalent mechanical loading. Explicit expressions are obtained for the overall yield surface, hardening rule, and for the instantaneous overall compliance, which permit simple numerical evaluation of deformation response to any combined loading path.

1. Introduction

Design requirements for certain space structures indicate the need for light, stiff, and strong materials which are dimensionally stable under cyclic thermal changes. Aluminum or magnesium matrix composites reinforced by high-modulus graphite fibers are suitable for such applications. The current state of metal matrix composites technology, as well as the large span of many space structures, suggest the use of planar and spatial trusses constructed of unidirectionally reinforced comosite tubes with special end attachments for pin joints. Such an arrangement permits only axial mechanical loading of the tubes, in combination with thermal changes which can be usually regarded as nearly uniform.

The combination of stiff and dimensionally stable graphite fibers with relatively compliant and thermally expansive metal matrices leads to large thermal stresses in the constituents and to plastic flow in the matrix after small thermal changes. In most systems a thermal change of only 50—100°F from a stress-free state will cause onset of yielding, whereas the thermal cycle in outer space spans approximately 550°F. Also, axial mechanical loads as small as 20% of the composite strength may cause plastic straining of the matrix.

Matrix yielding does not necessarily impair the mechanical properties of the composite. Axial stiffness may decrease somewhat, but the axial thermal expansion coefficient is reduced almost to the level of the fiber coefficient, i.e., by a factor of ten or so. Thus plastic yielding improves dimensional stability of composite structures. However, a certain part of the total composite strain generated in a plastic loading cycle will be permanent. If such permanent strains accumulate, they may cause distortion of composite structures subjected to cyclic thermal and/or mechanical load histories.

Under such circumstances, plastic response of unidirectional composites must be analyzed and accounted for in design. It is clear from the outset that the strain range of interest is small. Indeed, the ultimate fiber strain seldom exceeds 0.01. However, the plastic deformation range will be large relative to the total strength range because the matrix may yield at strains as small as ± 0.001. Therefore, thermomechanical response of the constituents, and the effects caused by their mutual constriant must be accounted for rather accurately. Various simplifying assumptions which are often adopted in plasticity of metals, or in large strain theories of composite media are not admissible.

Deformation of unidirectional composites subjected to uniform thermal changes and axisymmetric mechanical loads can be evaluated from composite cylinder models of the fibrous medium. Several plasticity studies of such models have been made [1–6], but simplifying assumptions regarding constituent behavior make them unsuitable for the stated purpose. The present paper develops a novel approach to the composite cylinder problem. In particular, new equilibrium and constraint equations are derived in Section 2 for volume averages of stresses and strains in the phases. These permit evaluation of instantaneous stress concentration factors and overall instantaneous compliances of the composite medium from thermomechanical properties and volume fractions of the constituents. The modelling procedure imposes no limitations on material behavior. Specific results are obtained for composites with elastic, transversely isotropic fibers and elastic-plastic matrices which exhibit kinematic hardening. This type of hardening appears in aluminum at small strains [7, 8]. The load types considered include axisymmetric overall stresses, uniform thermal changes, and changes in matrix yield stress which cause contraction of the overall yield surface and additional plastic strains when mechanical loading remains constant. These are discussed in Section 3, together with other possible load types, such as local differential dilatation caused by an allotropic transformation in the matrix. The procedure can easily incorporate other practically significant effects, such as variability in phase thermal expansion coefficients.

Phase stress concentration factors for mechanical loading are derived in Section 4, and Section 5 presents the concentration factors for thermal loading and yield stress changes. The latter are derived from a new decomposition scheme that converts the corresponding loading into equivalent mechanical loading which is superimposed upon uniform local fields. Section 6 extends the scheme to the derivation of overall instantaneous compliances corresponding to mechanical loads. Finally, Section 7 is concerned with evaluation of the overall yield surface and hardening rule. A remarkable feature of the results is that both the instantaneous local concentration factors and the overall compliances depend only on local material properties. Therefore, the macroscopic response of the composite medium is given by explicit differential forms which can be readily integrated along any prescribed loading path.

The notation is somewhat similar to that adopted in References [6], [9] and [10].

2. Stress and strain averages

Consider a cylindrical composite element of infinite length which has the crossection shown in Fig. 1. In the coordinate system $r\phi z$ the matrix (m) occupies the region $b \geqslant r > a$, the fiber (f) is in $a > r \geqslant 0$. The current stresses applied to the composite cylinder consist of a uniform radial stress $\bar{\sigma}_r$, and a uniform axial normal stress $\bar{\sigma}_z$; when index m or f is not used the top bars indicate overall averages. Instantaneous stress increments $d\bar{\sigma}_r$, $d\bar{\sigma}_z$ are superimposed upon the current state while the element is kept at a constant temperature.

Our first objective is to derive relationships between overall and local average stress and strain increments. These will follow from certain equilibrium and compatibility conditions which must be satisfied in the composite cylinder.

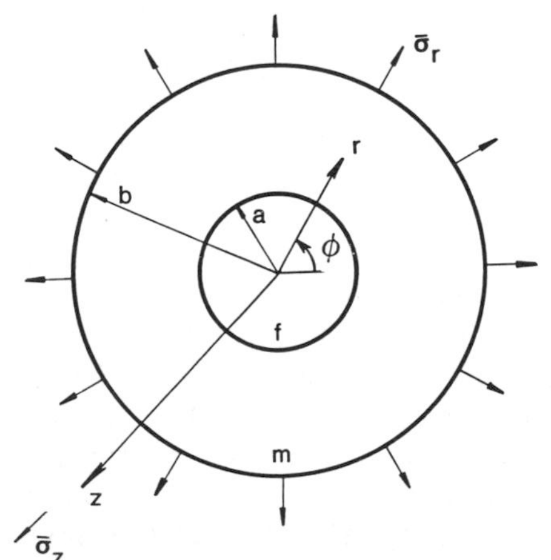

Fig. 1. Crossection of composite cylinder.

Under the axisymmetric applied stress, both overall and local shear stresses vanish. Accordingly, local and overall averages of the stress and strain increment vectors will be denoted as

$$\mathrm{d}\boldsymbol{\sigma}_{\mathrm{m}} = [\mathrm{d}\sigma_r^{\mathrm{m}}\ \mathrm{d}\sigma_\phi^{\mathrm{m}}\ \mathrm{d}\sigma_z^{\mathrm{m}}]^{\mathrm{T}}, \quad \mathrm{d}\boldsymbol{\varepsilon}_{\mathrm{m}} = [\mathrm{d}\epsilon_r^{\mathrm{m}}\ \mathrm{d}\epsilon_\phi^{\mathrm{m}}\ \mathrm{d}\epsilon_z^{\mathrm{m}}]^{\mathrm{T}} \tag{1}$$

$$\mathrm{d}\boldsymbol{\sigma}_{\mathrm{f}} = [\mathrm{d}\sigma_r^{\mathrm{f}}\ \mathrm{d}\sigma_z^{\mathrm{f}}]^{\mathrm{T}}, \quad \mathrm{d}\boldsymbol{\varepsilon}_{\mathrm{f}} = [2\mathrm{d}\epsilon_r^{\mathrm{f}}\ \mathrm{d}\epsilon_z^{\mathrm{f}}]^{\mathrm{T}} \tag{2}$$

$$\mathrm{d}\bar{\boldsymbol{\sigma}} = [\mathrm{d}\bar{\sigma}_r\ \mathrm{d}\bar{\sigma}_z]^{\mathrm{T}}, \quad \mathrm{d}\bar{\boldsymbol{\varepsilon}} = [2\mathrm{d}\bar{\epsilon}_r\ \mathrm{d}\bar{\epsilon}_z]^{\mathrm{T}}. \tag{3}$$

The equalities which follow from cylindrical symmetry are

$$\mathrm{d}\sigma_r^{\mathrm{f}} = \mathrm{d}\sigma_\phi^{\mathrm{f}}, \qquad \mathrm{d}\epsilon_r^{\mathrm{f}} = \mathrm{d}\epsilon_\phi^{\mathrm{f}}, \qquad \mathrm{d}\bar{\epsilon}_r = \mathrm{d}\bar{\epsilon}_\phi.$$

An alternative definition of stress and strain averages which will be useful in the sequel is

$$\mathrm{d}\bar{\boldsymbol{\sigma}}_{\mathrm{m}} = [\tfrac{1}{2}(\mathrm{d}\sigma_r^{\mathrm{m}} + \mathrm{d}\sigma_\phi^{\mathrm{m}})\ \ \mathrm{d}\sigma_z^{\mathrm{m}}]^{\mathrm{T}}, \qquad \mathrm{d}\bar{\boldsymbol{\varepsilon}}_{\mathrm{m}} = [(\mathrm{d}\epsilon_r^{\mathrm{m}} + \mathrm{d}\epsilon_\phi^{\mathrm{m}})\ \ \mathrm{d}\epsilon_z^{\mathrm{m}}]^{\mathrm{T}}$$

$$\mathrm{d}\bar{\boldsymbol{\sigma}}_{\mathrm{f}} = \mathrm{d}\boldsymbol{\sigma}_{\mathrm{f}}, \qquad \mathrm{d}\bar{\boldsymbol{\varepsilon}}_{\mathrm{f}} = \mathrm{d}\boldsymbol{\varepsilon}_{\mathrm{f}}. \tag{4}$$

Top bars in these equations do *not* indicate overall quantities.

The volume fractions of the phases c_{f}, c_{m} can be evaluated from the dimensions shown in Fig. 1. It is convenient to introduce the ratio

$$\eta = (b - a)/a \tag{5}$$

and to write

$$c_{\mathrm{f}} = (1+\eta)^{-2}, \qquad c_{\mathrm{m}} = \eta(2+\eta)(1+\eta)^{-2},$$

and $c_{\mathrm{f}} + c_{\mathrm{m}} = 1$.

To obtain a relationship between overall and local stress averages, we adapt the results (A-2) and (A-3) in the Appendix to volume averages of stress increments. From (A-3a), (A-3b) we obtain

$$q_1\mathrm{d}\bar{\sigma}_r + q_2\mathrm{d}\sigma_r^{\mathrm{f}} - \mathrm{d}\sigma_\phi^{\mathrm{m}} = 0, \tag{6}$$

and from (A-2)

$$2\mathrm{d}\bar{\sigma}_r = 2c_{\mathrm{f}}\mathrm{d}\sigma_r^{\mathrm{f}} + c_{\mathrm{m}}(\mathrm{d}\sigma_r^{\mathrm{m}} + \mathrm{d}\sigma_\phi^{\mathrm{m}}). \tag{7}$$

Similarly, for volumes averages of local and overall strain increments, from (A-5) and (A-6a), or (A-7b)

$$2\mathrm{d}\bar{\epsilon}_r = 2c_\mathrm{f}\mathrm{d}\epsilon_r^\mathrm{f} + c_\mathrm{m}(\mathrm{d}\epsilon_r^\mathrm{m} + \mathrm{d}\epsilon_\phi^\mathrm{m}) \tag{8}$$

$$p_1\mathrm{d}\bar{\epsilon}_r + p_2\mathrm{d}\epsilon_r^\mathrm{f} - \mathrm{d}\epsilon_\phi^\mathrm{m} = 0. \tag{9}$$

Equations (6) to (9) can be recast into a more convenient form

$$\mathrm{d}\sigma_r^\mathrm{m} = c_1\mathrm{d}\bar{\sigma}_r + c_2\mathrm{d}\sigma_r^\mathrm{f} \tag{10}$$

$$\mathrm{d}\sigma_\phi^\mathrm{m} = c_3\mathrm{d}\bar{\sigma}_r + c_4\mathrm{d}\sigma_r^\mathrm{f} \tag{11}$$

$$\mathrm{d}\sigma_z^\mathrm{m} = \frac{1}{c_\mathrm{m}}\mathrm{d}\bar{\sigma}_z - \frac{c_\mathrm{f}}{c_\mathrm{m}}\mathrm{d}\sigma_z^\mathrm{f} \tag{12}$$

$$\mathrm{d}\epsilon_r^\mathrm{m} = c_5\mathrm{d}\bar{\epsilon}_r + c_6\mathrm{d}\epsilon_r^\mathrm{f} \tag{13}$$

$$\mathrm{d}\epsilon_\phi^\mathrm{m} = c_7\mathrm{d}\bar{\epsilon}_r + c_8\mathrm{d}\epsilon_r^\mathrm{f} \tag{14}$$

$$\mathrm{d}\epsilon_z^\mathrm{m} = \mathrm{d}\bar{\epsilon}_z = \mathrm{d}\epsilon_z^\mathrm{f}. \tag{15}$$

Equations (12) and (15) are self-evident and were added for completeness. The constants c_1 to c_8 are

$$c_1 = 2/c_\mathrm{m} - q_1, \quad c_2 = -(2c_\mathrm{f}/c_\mathrm{m} + q_2), \quad c_3 = q_1, \quad c_4 = q_2,$$

$$c_5 = 2/c_\mathrm{m} - p_1, \quad c_6 = -(2c_\mathrm{f}/c_\mathrm{m} + p_2), \quad c_7 = p_1, \quad c_8 = p_2.$$

Approximate values of q_1 and q_2 are given by (A-3a) or (A-3b), those of p_1 and p_2 by (A-6a) or (A-7b). The choice between the various approximations can apparently be made on the basis of trial numerical solutions of the thermoplasticity problem of the composite cylinder.

Equations (10) to (15) provide equilibrium and constraint relations for overall and phase volume averages in the composite cylinder model. Their utility will be seen in derivation of stress concentration factors in Section 3. These equations could also be used to relate area averages of the local stress and strain increments, if the coefficients c_1 to c_8 were evaluated from equations (A-3), (A-7), and from similar area averages of (A-1) and (A-4). Except in the elastic case, where $(\epsilon_r + \epsilon_\phi) = \text{const.}$, and $(\sigma_r + \sigma_\phi) = \text{const.}$, area averages must again be evaluated from approximate local distributions of certain stress and displacement components.

Approximations of local fields should be taken with certain care. Specifically, (10) to (15) must be valid when fiber and matrix properties are identical. This suggests that the sum of coefficients in each equation is equal to unity.

3. Governing equations

In addition to the applied stress increments $\mathrm{d}\bar{\boldsymbol{\sigma}}$, the composite element may be loaded by a uniform thermal change $\mathrm{d}\theta$, subjected to an allotropic transformation in the matrix which causes local dilation $\Delta V_{\mathrm{m}}/V_{\mathrm{m}}$, and in the course of plastic loading the matrix may experience a sudden, or a gradual, change in yield stress $\mathrm{d}k$, which may be caused by annealing, or by a thermal change, respectively. The change $\mathrm{d}k < 0$ causes an isotropic contraction of the matrix and overall yield surfaces. Consequently, when $\bar{\boldsymbol{\sigma}} = \text{const.}$ is the current plastic loading, it is necessary to translate the center of the overall yield surface to keep the surface at the prescribed $\bar{\boldsymbol{\sigma}}$. These effects will influence the macroscopic or overall response of the composite material. The corresponding constitutive relations are

$$\mathrm{d}\bar{\boldsymbol{\sigma}} = L\mathrm{d}\bar{\boldsymbol{\varepsilon}} - \mathbf{l}\mathrm{d}\theta - \mathbf{l}'\mathrm{d}k, \qquad \mathrm{d}\bar{\boldsymbol{\varepsilon}} = M\mathrm{d}\bar{\boldsymbol{\sigma}} + \mathbf{m}\mathrm{d}\theta + \mathbf{m}'\mathrm{d}k, \tag{16}$$

where L, M are instantaneous stiffness and compliance matrices, $\mathbf{l}$, $\mathbf{m}$ are thermal stress and strain vectors, and $\mathbf{l}'$, $\mathbf{m}'$ are stress and strain vectors describing response to an instantaneous change $\mathrm{d}k$ in the matrix yield stress in shear. Local phase dilatation $\Delta V_{\mathrm{m}}/V_{\mathrm{m}}$, or $\Delta V_{\mathrm{f}}/V_{\mathrm{f}}$, need not be included explicitly in (16). It can be readily expressed in terms of a virtual thermal change $\mathrm{d}\theta$, with appropriately chosen local thermal expansion coefficients in (A-8), (A-9), or (A-11), (A-25), and (A-35) in the Appendix.

In order to evaluate the overall stiffnesses and compliances in (16), we assume that the local constitutive equations (A-8) and (A-11) are valid for stress and strain volume averages in the phases which appear in (10)–(15). When both phases are elastic, local properties are constant within each phase, and therefore, (A-8) and (A-11) with $\mathbf{l}'_{\mathrm{m}} = \mathbf{m}'_{\mathrm{m}} = \mathbf{0}$, are exact constitutive relations between the averages. However, when the matrix starts to yield, the homogeneity of local properties is lost and (A-11) is an approximation of the matrix response.

The admissibility of this approximation should be considered in the context of actual dimensions of the composite microstructure. The magnitudes of a and b in Fig. 1 are of the same order, and the fiber diameter a is of the order of $1\,\mu\mathrm{m}$ for graphite fibers, and $100\,\mu\mathrm{m}$ for boron or silicon carbide filaments. The matrix is usually made of aluminum or magnesium with grain sizes which are often comparable to the fiber diameter. Under these circumstances, the matrix itself may be regarded as a heterogeneous medium with average properties (A-11). On these grounds, the approximation is acceptable, and it may in fact convey a more appropriate image of physical reality than the conventional continuum model.

To find the overall stiffnesses and compliances which appear in (16), we assume that overall and local phase averages of stress and strain increments are related in the following way

$$d\boldsymbol{\sigma}_f = B_f d\bar{\boldsymbol{\sigma}} + \mathbf{b}_f d\theta + \mathbf{b}'_f dk, \tag{17}$$

$$d\boldsymbol{\sigma}_m = B_m d\bar{\boldsymbol{\sigma}} + \mathbf{b}_m d\theta + \mathbf{b}'_m dk, \tag{18}$$

and

$$d\boldsymbol{\varepsilon}_f = A_f d\bar{\boldsymbol{\varepsilon}} - \mathbf{a}_f d\theta - \mathbf{a}'_f dk, \tag{19}$$

$$d\boldsymbol{\varepsilon}_m = A_m d\bar{\boldsymbol{\varepsilon}} - \mathbf{a}_m d\theta - \mathbf{a}'_m dk, \tag{20}$$

where A_r, B_r, $\mathbf{a}_r$, $\mathbf{b}_r$, and $\mathbf{a}'_r$, $\mathbf{b}'_r$, $(r = \mathrm{f}, \mathrm{m})$ are concentration factors similar to those introduced by Hill [9] and Laws [10].

When both phases are elastic, it is possible to find exact values of the first four concentration factors for the composite cylinder element, and, with the procedure that follows, obtain exact overall elastic stiffnesses and compliances.

As a consequence of the definitions (1) to (3), B_m and A_m are 3×2 matrices, $\mathbf{b}_m$, $\mathbf{b}'_m$, $\mathbf{a}_m$, $\mathbf{a}'_m$ are 3×1 column vectors; B_f and A_f, and also M, L, M_f, L_f are 2×2 matrices; $\mathbf{b}_f$, $\mathbf{b}'_f$, $\mathbf{a}_f$, $\mathbf{a}'_f$, $\mathbf{l}$, $\mathbf{l}'$, $\mathbf{m}$, $\mathbf{m}'$ are 2×1 vectors and M_m, L_m in (A-24), (A-34) are 3×3 matrices. To facilitate some derivations which will be made in the sequel, we introduce the compliances $\bar{M}_m$, $\bar{\mathbf{m}}_m$, $\bar{\mathbf{m}}'_m$, and stiffnesses $\bar{L}_m$, $\bar{\mathbf{l}}_m$, $\bar{\mathbf{l}}'_m$, which conform with (4)

$$d\bar{\boldsymbol{\sigma}}_m = \bar{L}_m d\boldsymbol{\varepsilon}_m - \bar{\mathbf{l}}_m d\theta - \bar{\mathbf{l}}'_m dk; \qquad d\bar{\boldsymbol{\varepsilon}}_m = \bar{M}_m d\boldsymbol{\sigma}_m + \bar{\mathbf{m}}_m d\theta + \bar{\mathbf{m}}' dk \tag{21}$$

such that

$$\bar{L}_m = \frac{1}{2}\begin{bmatrix} (L_{11}^m + L_{12}^m) & (L_{12}^m + L_{22}^m) & (L_{13}^m + L_{23}^m) \\ 2L_{13}^m & 2L_{23}^m & 2L_{33}^m \end{bmatrix};$$

$$\bar{M}_m = \begin{bmatrix} (M_{11}^m + M_{12}^m) & (M_{12}^m + M_{22}^m) & (M_{13}^m + M_{23}^m) \\ M_{13}^m & M_{23}^m & M_{33}^m \end{bmatrix} \tag{22}$$

and

$$\bar{\mathbf{l}}_m = [\tfrac{1}{2}(l_1^m + l_2^m) \;\; l_3^m]^T, \qquad \bar{\mathbf{m}}_m = [(m_1^m + m_2^m) \;\; m_3^m]^T \tag{23}$$

with analogous forms for $\bar{\mathbf{l}}'_m$, $\bar{\mathbf{m}}'_m$. The L_{ij}^m, M_{ij}^m, l_i^m, m_i^m are taken from (A-34), (A-24), (A-35 or 36), and (A-25 or 26), respectively.

Also, (18) and (20) can be written in terms of (4)

$$d\bar{\boldsymbol{\sigma}}_m = \bar{B}_m d\bar{\boldsymbol{\sigma}} + \bar{\mathbf{b}}_m d\theta + \bar{\mathbf{b}}'_m dk, \tag{24}$$

$$\mathrm{d}\bar{\boldsymbol{\varepsilon}}_{\mathrm{m}} = \bar{A}_{\mathrm{m}}\,\mathrm{d}\boldsymbol{\varepsilon} - \bar{\mathbf{a}}_{\mathrm{m}}\,\mathrm{d}\theta - \bar{\mathbf{a}}'_{\mathrm{m}}\,\mathrm{d}k, \tag{25}$$

where

$$\bar{A}_{\mathrm{m}} = \begin{bmatrix} (A^{\mathrm{m}}_{11} + A^{\mathrm{m}}_{21}) & (A^{\mathrm{m}}_{12} + A^{\mathrm{m}}_{22}) \\ A^{\mathrm{m}}_{31} & A^{\mathrm{m}}_{32} \end{bmatrix};$$

$$\bar{B}_{\mathrm{m}} = \begin{bmatrix} \tfrac{1}{2}(B^{\mathrm{m}}_{11} + B^{\mathrm{m}}_{21}) & \tfrac{1}{2}(B^{\mathrm{m}}_{12} + B^{\mathrm{m}}_{22}) \\ B^{\mathrm{m}}_{31} & B^{\mathrm{m}}_{32} \end{bmatrix} \tag{26}$$

and

$$\bar{\mathbf{a}}_{\mathrm{m}} = [(a^{\mathrm{m}}_1 + a^{\mathrm{m}}_2)\;\; a^{\mathrm{m}}_3]^{\mathrm{T}}, \qquad \bar{\mathbf{b}}_{\mathrm{m}} = [\tfrac{1}{2}(b^{\mathrm{m}}_1 + b^{\mathrm{m}}_2)\;\; b^{\mathrm{m}}_3]^{\mathrm{T}}, \tag{27}$$

with analogous expressions for $\bar{\mathbf{a}}'_{\mathrm{m}}, \bar{\mathbf{b}}'_{\mathrm{m}}$.

Now, the overall stress and strain increments can be expressed in terms of phase averages (4)

$$\mathrm{d}\bar{\boldsymbol{\sigma}} = c_{\mathrm{f}}\mathrm{d}\bar{\boldsymbol{\sigma}}_{\mathrm{f}} + c_{\mathrm{m}}\,\mathrm{d}\bar{\boldsymbol{\sigma}}_{\mathrm{m}}, \qquad \mathrm{d}\bar{\boldsymbol{\varepsilon}} = c_{\mathrm{f}}\mathrm{d}\bar{\boldsymbol{\varepsilon}}_{\mathrm{f}} + c_{\mathrm{m}}\,\mathrm{d}\bar{\boldsymbol{\varepsilon}}_{\mathrm{m}}. \tag{28}$$

With $\mathrm{d}\bar{\boldsymbol{\sigma}}_{\mathrm{f}}$, and $\mathrm{d}\bar{\boldsymbol{\sigma}}_{\mathrm{m}}$ from (A-8), (4), and (21), and the definitions (19) and (20), eqn. (28) assumes a form comparable to (16). Thus, the instantaneous overall stiffnesses are

$$L = c_{\mathrm{f}}L_{\mathrm{f}}A_{\mathrm{f}} + c_{\mathrm{m}}\bar{L}_{\mathrm{m}}A_{\mathrm{m}}, \tag{29}$$

$$\mathbf{l} = c_{\mathrm{f}}(\mathbf{l}_{\mathrm{f}} + L_{\mathrm{f}}\mathbf{a}_{\mathrm{f}}) + c_{\mathrm{m}}(\bar{\mathbf{l}}_{\mathrm{m}} + \bar{L}_{\mathrm{m}}\mathbf{a}_{\mathrm{m}}), \tag{30}$$

$$\mathbf{l}' = c_{\mathrm{f}}L_{\mathrm{f}}\mathbf{a}'_{\mathrm{f}} + c_{\mathrm{m}}(\bar{\mathbf{l}}'_{\mathrm{m}} + \bar{L}_{\mathrm{m}}\mathbf{a}'_{\mathrm{m}}). \tag{31}$$

An analogous sequence of substitutions into (28) gives instantaneous overall compliances

$$M = c_{\mathrm{f}}M_{\mathrm{f}}B_{\mathrm{f}} + c_{\mathrm{m}}\bar{M}_{\mathrm{m}}B_{\mathrm{m}}, \tag{32}$$

$$\mathbf{m} = c_{\mathrm{f}}(\mathbf{m}_{\mathrm{f}} + M_{\mathrm{f}}\mathbf{b}_{\mathrm{f}}) + c_{\mathrm{m}}(\bar{\mathbf{m}}_{\mathrm{m}} + \bar{M}_{\mathrm{m}}\mathbf{b}_{\mathrm{m}}), \tag{33}$$

$$\mathbf{m}' = c_{\mathrm{f}}M_{\mathrm{f}}\mathbf{b}'_{\mathrm{f}} + c_{\mathrm{m}}(\bar{\mathbf{m}}'_{\mathrm{m}} + \bar{M}_{\mathrm{m}}\mathbf{b}'_{\mathrm{m}}). \tag{34}$$

Also, the following equalities relate the concentration factors (c.f., (A-7), (A-2))

$$c_f A_f + c_m \bar{A}_m = I, \qquad c_f B_f + c_m \bar{B}_m = I, \tag{35}$$

$$c_f \mathbf{a}_f + c_m \bar{\mathbf{a}}_m = \mathbf{0}, \qquad c_f \mathbf{b}_f + c_m \bar{\mathbf{b}}_m = \mathbf{0}, \tag{36}$$

$$c_f \mathbf{a}_f' + c_m \bar{\mathbf{a}}_m' = \mathbf{0}, \qquad c_f \mathbf{b}_f' + c_m \bar{\mathbf{b}}_m' = \mathbf{0}. \tag{37}$$

Evaluation of the overall response of the composite consists in integration of (16), with the instantaneous stiffnesses given by (29) to (31), and instantaneous compliances by (32) to (34). Thus, the problem has been reduced to finding the concentration factors which appear in eqns. (29)–(34).

4. Stress concentration factors B_f, B_m

When the composite is loaded only by applied stresses $\mathrm{d}\bar{\boldsymbol{\sigma}}$, with $\mathrm{d}\theta = \mathrm{d}k = 0$, equations (17) and (18) become

$$\mathrm{d}\boldsymbol{\sigma}_f = B_f \mathrm{d}\bar{\boldsymbol{\sigma}}, \qquad \mathrm{d}\boldsymbol{\sigma}_m = B_m \mathrm{d}\bar{\boldsymbol{\sigma}}. \tag{38}$$

The stress concentration factors B_f, B_m for the composite cylinder model can be found from the equilibrium and constraint equations (10)–(15), when the local constitutive relations (A-8), (A-11) are taken to be valid for phase averages of stress and strain increments.

For convenience we rewrite (A-8) and (A-9) in the form

$$2\mathrm{d}\epsilon_r^f = (n_f \mathrm{d}\sigma_r^f - l_f \mathrm{d}\sigma_z^f)/(k_f E_A^f),$$

$$\mathrm{d}\epsilon_z^f = (-l_f \mathrm{d}\sigma_r^f + k_f \mathrm{d}\sigma_z^f)/(k_f E_A^f), \tag{39}$$

and (A-11) and (A-24) as

$$\mathrm{d}\epsilon_r^m = M_{11}^m \mathrm{d}\sigma_r^m + M_{12}^m \mathrm{d}\sigma_\phi^m + M_{13}^m \mathrm{d}\sigma_z^m,$$

$$\mathrm{d}\epsilon_\phi^m = M_{12}^m \mathrm{d}\sigma_r^m + M_{22}^m \mathrm{d}\sigma_\phi^m + M_{23}^m \mathrm{d}\sigma_z^m, \tag{40}$$

$$\mathrm{d}\epsilon_z^m = M_{13}^m \mathrm{d}\sigma_r^m + M_{23}^m \mathrm{d}\sigma_\phi^m + M_{33}^m \mathrm{d}\sigma_z^m.$$

The particular values of $M_{ij}^m = M_{ji}^m$ will depend on the loading history which is controlled, in part, by the local yield condition (A-14), the hardening rule (A-12), and the loading and unloading criteria (A-18). Therefore, the coefficients of (40) must be evaluated at each step of plastic loading.

To find the concentration factors, it is necessary to evaluate the local stresses $\mathrm{d}\boldsymbol{\sigma}_f$, $\mathrm{d}\boldsymbol{\sigma}_m$ in terms of applied stress $\mathrm{d}\bar{\sigma}_r$, $\mathrm{d}\bar{\sigma}_z$. This requires the solution of eqns. (10)–(15), (39), and (40), a system of twelve equations for the five local stresses, five local strains, and the overall strains $\mathrm{d}\bar{\epsilon}_r$, $\mathrm{d}\bar{\epsilon}_z$. The two unknowns we seek can be evaluated in closed form. We omit the heavy algebra involved and write the final result

$$\mathrm{d}\sigma_r^{\mathrm{f}} = B_{11}^{\mathrm{f}} \mathrm{d}\bar{\sigma}_r + B_{12}^{\mathrm{f}} \mathrm{d}\bar{\sigma}_z, \tag{41}$$

where

$$B_{11}^{\mathrm{f}} = d_1/g, \quad B_{12}^{\mathrm{f}} = d_2/g, \tag{42}$$

and

$$d_1 = c_1 f_1 + c_3 f_2,$$

$$d_2 = \frac{h(c_5 M_{23}^{\mathrm{m}} - c_7 M_{13}^{\mathrm{m}})/M_{33}^{\mathrm{m}} + \nu_{\mathrm{A}}^{\mathrm{f}}}{c_{\mathrm{m}}/M_{33}^{\mathrm{m}} + c_{\mathrm{f}} E_{\mathrm{A}}^{\mathrm{f}}},$$

$$g = 1/(2k_{\mathrm{f}}) - c_2 f_1 - c_4 f_2 + 2c_{\mathrm{f}} d_2 \nu_{\mathrm{A}}^{\mathrm{f}},$$

$$f_1 = h[(c_5 M_{12}^{\mathrm{m}} - c_7 M_{11}^{\mathrm{m}}) - M_{13}^{\mathrm{m}}(c_5 M_{23}^{\mathrm{m}} - c_7 M_{13}^{\mathrm{m}} - c_{\mathrm{m}} d_2/h)/M_{33}^{\mathrm{m}}],$$

$$f_2 = h[(c_5 M_{22}^{\mathrm{m}} - c_7 M_{12}^{\mathrm{m}}) - M_{23}^{\mathrm{m}}(c_5 M_{23}^{\mathrm{m}} - c_7 M_{13}^{\mathrm{m}} - c_{\mathrm{m}} d_2/h)/M_{33}^{\mathrm{m}}],$$

$$h = \frac{1}{c_5 c_8 - c_6 c_7} = \frac{1}{c_5 - c_7}.$$

Also

$$\mathrm{d}\sigma_z^{\mathrm{f}} = B_{21}^{\mathrm{f}} \mathrm{d}\bar{\sigma}_r + B_{22}^{\mathrm{f}} \mathrm{d}\bar{\sigma}_z, \tag{43}$$

where

$$B_{21}^{\mathrm{f}} = e_2 + e_3 B_{11}^{\mathrm{f}}, \quad B_{22}^{\mathrm{f}} = e_1 + e_3 B_{12}^{\mathrm{f}} \tag{44}$$

and

$$e_1 = E_{\mathrm{A}}^{\mathrm{f}}/(c_{\mathrm{m}}/M_{33}^{\mathrm{m}} + c_{\mathrm{f}} E_{\mathrm{A}}^{\mathrm{f}}), \quad e_2 = e_1 c_{\mathrm{m}}(c_1 M_{13}^{\mathrm{m}} + c_3 M_{23}^{\mathrm{m}})/M_{33}^{\mathrm{m}},$$

$$e_3 = e_1[c_{\mathrm{m}}(c_2 M_{13}^{\mathrm{m}} + c_4 M_{23}^{\mathrm{m}})/M_{33}^{\mathrm{m}} - 2c_{\mathrm{f}}\nu_{\mathrm{A}}^{\mathrm{f}}] + 2\nu_{\mathrm{A}}^{\mathrm{f}}.$$

This completes the evaluation of B_{f}.

Now, (41) can be substituted into (10) and (11), and (43) into (12). The result is the concentration factor B_{m}

$$B_{11}^{\mathrm{m}} = c_1 + c_2 B_{11}^{\mathrm{f}}, \qquad B_{12}^{\mathrm{m}} = c_2 B_{12}^{\mathrm{f}},$$

$$B_{21}^{\mathrm{m}} = c_3 + c_4 B_{11}^{\mathrm{f}}, \qquad B_{22}^{\mathrm{m}} = c_4 B_{12}^{\mathrm{f}}, \tag{45}$$

$$B_{31}^{\mathrm{m}} = -c_{\mathrm{f}} B_{21}^{\mathrm{f}}/c_{\mathrm{m}}, \qquad B_{32}^{\mathrm{m}} = (1 - c_{\mathrm{f}} B_{22}^{\mathrm{f}})/c_{\mathrm{m}}.$$

It can be verified that in the elastic case, when $M_{\mathrm{m}} = M_{\mathrm{me}}$ in (A-24), the concentration factors $B_{\mathrm{f}} = B_{\mathrm{fe}}$, $B_{\mathrm{m}} = B_{\mathrm{me}}$ in (41), (43), and (45) are exactly equal to those obtained from the elasticity solution of the composite cylinder. Specifically, if the coefficients c_1 to c_8 in (10) to (15) are derived from approximations (A-3a) and (A-6a) and exact relations (A-2) and (A-5) for volume averages, then $B_{\mathrm{me}} = B_{\mathrm{me}}^{\mathrm{V}}$ gives exactly the volume averages of elastic fields. This is to be expected, since (A-3a) and (A-6a) were found from elastic fields. Furthermore, if coefficients c_1 to c_8 are taken from area averages of local fields in (A-3b) and (A-7), and if (A-2) and (A-5) are regarded as relations between area averages, which is correct in the elastic case, then $B_{\mathrm{me}} = B_{\mathrm{me}}^{\mathrm{A}}$ gives exact area averages of elastic fields. Of course, the coefficients of B_{me} are different in each case, whereas B_{fe} remain unchanged because the stress and strain fields in the fiber are uniform. Finally, one can find that

$$\bar{M}_{\mathrm{me}} B_{\mathrm{me}}^{\mathrm{A}} = \bar{M}_{\mathrm{me}} B_{\mathrm{me}}^{\mathrm{V}},$$

where $\bar{M}_{\mathrm{me}}$ is given by (22). Hence, it follows from (32) that either $B_{\mathrm{me}}^{\mathrm{V}}$ or $B_{\mathrm{me}}^{\mathrm{A}}$ found from the procedure discussed above will give the exact value of M. In the inelastic case, the $B_{\mathrm{f}}, B_{\mathrm{m}}$ are influenced by the approximations used to derive c_1 to c_8 in (10) to (15), and also by assumptions leading to (17) and (18).

Note that the concentration factors $B_{\mathrm{f}}, B_{\mathrm{m}}$ depend only on phase moduli, and not on overall instantaneous stiffnesses or compliances. Therefore, (32) is an explicit equation for M.

5. Concentration factors $\mathbf{b}_{\mathrm{f}}$, $\mathbf{b}_{\mathrm{m}}$, $\mathbf{b}_{\mathrm{f}}'$, $\mathbf{b}_{\mathrm{m}}'$

These factors can be derived in a similar way as $B_{\mathrm{m}}, B_{\mathrm{f}}$. However, it is advantageous to use a variant of the recent results by Dvorak [11], which leads to expressions relating the $\mathbf{b}_r$ and $\mathbf{b}_r'$ factors to the $B_r (r = \mathrm{f, m})$. The response of the composite to $\mathrm{d}\theta$ and $\mathrm{d}k$ is then evaluated as in the case of mechanical loading by $\mathrm{d}\bar{\boldsymbol{\sigma}}$, with equivalent surface tractions which depend on $\mathrm{d}\theta$ and $\mathrm{d}k$. The latter term causes plastic loading only if the yield surface contracts at constant applied stress $\bar{\boldsymbol{\sigma}}$, and constant temperature θ, i.e., only if $\mathrm{d}k < 0$.

Consider again the matrix and fiber stress—strain relations (A-11) and (A-8), in the explicit form (A-24) to (A-26)

$$\begin{Bmatrix} d\epsilon_r^m \\ d\epsilon_\phi^m \\ d\epsilon_z^m \end{Bmatrix} = \begin{bmatrix} M_{11}^m & M_{12}^m & M_{13}^m \\ & M_{22}^m & M_{23}^m \\ \text{SYM.} & & M_{33}^m \end{bmatrix} \begin{Bmatrix} d\sigma_r^m \\ d\sigma_\phi^m \\ d\sigma_z^m \end{Bmatrix} +$$

$$\begin{Bmatrix} \alpha_m \\ \alpha_m \\ \alpha_m \end{Bmatrix} d\theta - \frac{3}{2Hk} \begin{Bmatrix} s_r^* \\ s_\phi^* \\ s_z^* \end{Bmatrix}_m dk, \tag{46}$$

and rewrite (A-8) and (A-9) as

$$\begin{Bmatrix} d\epsilon_r^f \\ d\epsilon_\phi^f \\ d\epsilon_z^f \end{Bmatrix} = \begin{bmatrix} 1/E_T^f & -\nu_T^f/E_T^f & -\nu_A^f/E_A^f \\ & 1/E_T^f & -\nu_A^f/E_A^f \\ \text{SYM.} & & 1/E_A^f \end{bmatrix} \begin{Bmatrix} d\sigma_r^f \\ d\sigma_\phi^f \\ d\sigma_z^f \end{Bmatrix} + \begin{Bmatrix} \alpha_T^f \\ \alpha_T^f \\ \alpha_A^f \end{Bmatrix} d\theta. \tag{47}$$

Next, separate the phases and apply the changes $d\theta$ and dk to each phase, together with an unknown hydrostatic stress increment in the matrix

$$d\sigma_r^m = d\sigma_z^m = dS_T, \tag{48}$$

and a stress increment

$$d\sigma_r^f = dS_T, \quad d\sigma_z^f = (dS_A - c_m dS_T)/c_f \tag{49}$$

in the fiber. Also, tractions corresponding to current local stresses $\boldsymbol{\sigma}_m^0, \boldsymbol{\sigma}_f^0$ are added to the separated phases to preserve the current strain states $\boldsymbol{\varepsilon}_m^0, \boldsymbol{\varepsilon}_f^0$.

The resulting strain increments in the separated phases are: In the matrix

$$\begin{Bmatrix} d\epsilon_r^m \\ d\epsilon_\phi^m \\ d\epsilon_z^m \end{Bmatrix} = M_{me} \begin{Bmatrix} dS_T \\ dS_T \\ dS_T \end{Bmatrix} + \begin{Bmatrix} \alpha_m \\ \alpha_m \\ \alpha_m \end{Bmatrix} d\theta - \frac{3}{2Hk} \begin{Bmatrix} s_r^* \\ s_\phi^* \\ s_z^* \end{Bmatrix}_m dk. \tag{50}$$

Here we regard the matrix as elastic under isotropic stress increments. The s_m^* components are calculated from the current local stresses $\boldsymbol{\sigma}_m^0$, $dk < 0$. For $dk \geqslant 0$ the last term in (50) vanishes. In the fiber, from (47)

$$\begin{Bmatrix} d\epsilon_r^f \\ d\epsilon_\phi^f \\ d\epsilon_z^f \end{Bmatrix} = M_{fe} \begin{Bmatrix} dS_T \\ dS_T \\ (dS_A - c_m dS_T)/c_f \end{Bmatrix} + \begin{Bmatrix} \alpha_T^f \\ \alpha_T^f \\ \alpha_A^f \end{Bmatrix} d\theta. \quad (51)$$

Now, adjust dS_A, dS_T to satisfy constraint equations (13) to (15) which can be written as

$$d\epsilon_r^f = h(c_5 d\epsilon_\phi^m - c_7 d\epsilon_r^m), \qquad d\epsilon_z^f = d\epsilon_z^m, \quad (52)$$

where as in (42)

$$h = 1/(c_5 - c_7).$$

This adjustment requires that

$$a_1 dS_T + a_2 dS_A + a_3 d\theta + a_4 dk = 0,$$

$$b_1 dS_T + b_2 dS_A + b_3 d\theta + b_4 dk = 0, \quad (53)$$

where

$$a_1 = (n_f + l_f c_m/c_f)/(k_f E_A^f) - (n_m - l_m)/(k_m E_m),$$

$$a_2 = -l_f/(c_f k_f E_A^f), \qquad a_3 = 2(\alpha_T^f - \alpha_m),$$

$$a_4 = 3h(c_5 s_{\phi m}^* - c_7 s_{rm}^*)/(Hk),$$

$$b_1 = -(l_f + k_f c_m/c_f)/(k_f E_A^f) + (l_m - k_m)/(k_m E_m),$$

$$b_2 = (c_f E_A^f)^{-1}, \qquad b_3 = (\alpha_A^f - \alpha_m), \qquad b_4 = 3s_{zm}^*/(2Hk).$$

The solution of (53) is

$$dS_T = s_T^\theta d\theta + s_T^k dk, \qquad dS_A = s_A^\theta d\theta + s_A^k dk, \quad (54)$$

where

$$s_{\mathrm{T}}^{\theta} = (a_2 b_3 - a_3 b_2)/(a_1 b_2 - a_2 b_1), \qquad s_{\mathrm{T}}^{\mathrm{k}} = (a_2 b_4 - a_4 b_2)/(a_1 b_2 - a_2 b_1),$$
$$s_{\mathrm{A}}^{\theta} = -(a_1 b_3 - a_3 b_1)/(a_1 b_2 - a_2 b_1),$$
$$s_{\mathrm{A}}^{\mathrm{k}} = -(a_1 b_4 - a_4 b_1)/(a_1 b_2 - a_2 b_1). \tag{55}$$

Now, the phases are compatible and can be reassembled. The tractions (48) and (49) are in equilibrium at the fiber matrix interface, and in the z-direction. However, there remain tractions $\mathrm{d}S_{\mathrm{T}}$, $\mathrm{d}S_{\mathrm{A}}$ on the surface of the composite element. These need to be removed.

The final stresses caused in the phases by $\mathrm{d}\theta$ and $\mathrm{d}k < 0$ are

$$\mathbf{d\sigma}_{\mathrm{f}} = [\mathrm{d}\sigma_r^{\mathrm{f}}\ \mathrm{d}\sigma_z^{\mathrm{f}}]^{\mathrm{T}} = \mathbf{b}_{\mathrm{f}}\mathrm{d}\theta + \mathbf{b}_{\mathrm{f}}'\mathrm{d}k, \tag{17}$$

where

$$\mathbf{b}_{\mathrm{f}} = [s_{\mathrm{T}}^{\theta}\ \ (s_{\mathrm{A}}^{\theta} - c_{\mathrm{m}} s_{\mathrm{T}}^{\theta})/c_{\mathrm{f}}]^{\mathrm{T}} - B_{\mathrm{f}}[s_{\mathrm{T}}^{\theta}\ s_{\mathrm{A}}^{\theta}]^{\mathrm{T}} \tag{56}$$
$$\mathbf{b}_{\mathrm{f}}' = [s_{\mathrm{T}}^{\mathrm{k}}\ \ (s_{\mathrm{A}}^{\mathrm{k}} - c_{\mathrm{m}} s_{\mathrm{T}}^{\mathrm{k}})/c_{\mathrm{f}}]^{\mathrm{T}} - B_{\mathrm{f}}[s_{\mathrm{T}}^{\mathrm{k}}\ s_{\mathrm{A}}^{\mathrm{k}}]^{\mathrm{T}} \tag{57}$$

and

$$\mathrm{d}\boldsymbol{\sigma}_{\mathrm{m}} = [\mathrm{d}\sigma_r^{\mathrm{m}}\ \mathrm{d}\sigma_\theta^{\mathrm{m}}\ \mathrm{d}\sigma_z^{\mathrm{m}}]^{\mathrm{T}} = \mathbf{b}_{\mathrm{m}}\mathrm{d}\theta + \mathbf{b}_{\mathrm{m}}'\mathrm{d}k, \tag{18}$$

where

$$\mathbf{b}_{\mathrm{m}} = [1\ 1\ 1]^{\mathrm{T}} s_{\mathrm{T}}^{\theta} - B_{\mathrm{m}}[s_{\mathrm{T}}^{\theta}\ s_{\mathrm{A}}^{\theta}]^{\mathrm{T}} \tag{58}$$
$$\mathbf{b}_{\mathrm{m}}' = [1\ 1\ 1]^{\mathrm{T}} s_{\mathrm{T}}^{\mathrm{k}} - B_{\mathrm{m}}[s_{\mathrm{T}}^{\mathrm{k}}\ s_{\mathrm{A}}^{\mathrm{k}}]^{\mathrm{T}}. \tag{59}$$

Local stresses caused in the composite cylinder by $\mathrm{d}\theta$ and $\mathrm{d}k$ can now be evaluated directly from (56)–(59). The $\mathrm{d}k$ effect is present only in the plastic range. Indeed $\mathbf{m}_{\mathrm{m}}' = \mathbf{m}_{\mathrm{me}}' = \mathbf{0}$ in (A-26) when the matrix is elastic, or when $\mathrm{d}k \geqslant 0$; then both $s_{\mathrm{T}}^{\mathrm{k}}$ and $s_{\mathrm{A}}^{\mathrm{k}}$ in (55) vanish; i.e., $\mathbf{b}_{\mathrm{me}}' = \mathbf{b}_{\mathrm{fe}}' \equiv \mathbf{0}$. On the other hand, s_{T}^{θ}, s_{A}^{θ} are not dependent on the local stress state, therefore, thermal loading is present whenever $\mathrm{d}\theta \neq 0$. The distinction between elastic and plastic $\mathbf{b}_{\mathrm{f}}$, $\mathbf{b}_{\mathrm{m}}$ in (56), (58) is in taking $B_{\mathrm{m}} = B_{\mathrm{me}}$ and $B_{\mathrm{f}} = B_{\mathrm{fe}}$ in the elastic case.

The essential feature of the results (56) to (59) is that they convert loading by $\mathrm{d}\theta$ *and* $\mathrm{d}k$ *to equivalent mechanical loading by tractions* $\mathrm{s}^{\theta}\,\mathrm{d}\theta$ *and* $\mathrm{s}^{\mathrm{k}}\,\mathrm{d}k$. Local uniform fields must be added to the microstresses. These fields are isotropic in the matrix and thus do not affect its plastic deformation. Accordingly, loading by combined tractions $\mathrm{d}\bar{\boldsymbol{\sigma}}$, $\mathrm{s}^{\theta}\,\mathrm{d}\theta$, and $\mathrm{s}^{\mathrm{k}}\mathrm{d}k$ influences plastic flow in the matrix in the same way as the overall stress

$$\mathrm{d}\bar{\boldsymbol{\tau}} = \mathrm{d}\bar{\boldsymbol{\sigma}} - \mathrm{s}^{\theta}\,\mathrm{d}\theta - \mathrm{s}^{\mathrm{k}}\mathrm{d}k, \tag{60}$$

where

$$\mathrm{s}^{\theta} = [s_{\mathrm{T}}^{\theta}\ s_{\mathrm{A}}^{\theta}]^{\mathrm{T}}, \qquad \mathrm{s}^{\mathrm{k}} = [s_{\mathrm{T}}^{\mathrm{k}}\ s_{\mathrm{A}}^{\mathrm{k}}]^{\mathrm{T}}.$$

The applied stress $\mathrm{d}\bar{\boldsymbol{\tau}}$ determines the actual loading path of the composite, it must be used in evaluation of M_{ij}^{m} in (A-24) or (40), and of B_{f}, B_{m} in (42), (44), and (45).

The decomposition sequence can be changed at will, but that has no effect on the results (56) to (60). Therefore, the results are independent of the procedure used in their derivation.

Note that both $\mathbf{b}_r$ and $\mathbf{b}'_r$ depend on $B_r(r = \mathrm{f, m})$ which were derived for stress averages, and are therefore approximate when the matrix deforms plastically. However, $\mathbf{b}_r$ are exact in terms of B_r for any transverse geometry of the fibrous medium. This follows directly from the fact that the coefficients a_1 to a_3 and b_1 to b_3 in (53) depend only on phase properties and volume fractions. These coefficients are used in (55) to find $s^\theta_{\mathrm{T}}, s^\theta_{\mathrm{A}}$ which then appear in (56) and (58). On the other hand, $\mathbf{b}'_r$ depend on matrix deviatoric stresses $\mathbf{s}^*_{\mathrm{m}}$, and thus on averaged $\boldsymbol{\sigma}_{\mathrm{m}}$, through a_4 and b_4 in (55). Actually, all terms in (57) and (59) are derived from the averaging scheme which was discussed in connection with (16). Since eqns. (10) to (15) for averages were derived for the specific geometry of Fig. 1, $\mathbf{b}'_r$ are valid only for the composite cylinder model of the fibrous medium.

Again, like B_r, the $\mathbf{b}_r$ and $\mathbf{b}'_r$ depend only on phase properties. Therefore, (33) and (34) are explicit expressions for overall instantaneous compliances $\mathbf{m}$ and $\mathbf{m}'$. Another, more compact derivation of these compliances is presented in the next section.

6. Instantaneous compliances m and m′.

The procedure leading to (56)—(59) can be utilized to evaluate overall strains. For $\mathrm{d}\bar{\boldsymbol{\sigma}} = \mathbf{0}$ we can write (16) as

$$\mathrm{d}\bar{\boldsymbol{\varepsilon}} = \mathbf{m}\mathrm{d}\theta + \mathbf{m}'\mathrm{d}k = c_{\mathrm{f}}\mathrm{d}\bar{\boldsymbol{\varepsilon}}_{\mathrm{f}} + c_{\mathrm{m}}\,\mathrm{d}\bar{\boldsymbol{\varepsilon}}_{\mathrm{m}} - M[\mathrm{d}S_{\mathrm{T}}\ \mathrm{d}S_{\mathrm{A}}]^{\mathrm{T}}, \tag{61}$$

where $\mathrm{d}\bar{\boldsymbol{\varepsilon}}_{\mathrm{f}}$, and $\mathrm{d}\bar{\boldsymbol{\varepsilon}}_{\mathrm{m}}$, are found in (4), (51), and (50), respectively. Now we substitute $\mathrm{d}S_{\mathrm{T}}$ and $\mathrm{d}S_{\mathrm{A}}$ from (54) into (61); the resulting compliances are

$$\mathbf{m} = (q_\theta + \alpha_{\mathrm{m}})\begin{Bmatrix}2\\1\end{Bmatrix} - M\begin{Bmatrix}s^\theta_{\mathrm{T}}\\ s^\theta_{\mathrm{A}}\end{Bmatrix}, \tag{62}$$

$$\mathbf{m}' = q_{\mathrm{k}}\begin{Bmatrix}2\\1\end{Bmatrix} - \frac{3}{2Hk}\begin{Bmatrix}c_{\mathrm{m}}(s^*_{r\mathrm{m}} + s^*_{\phi\mathrm{m}}) + 2c_{\mathrm{f}}h(c_5 s^*_{\phi\mathrm{m}} - c_7 s^*_{r\mathrm{m}})\\ s^*_{z\mathrm{m}}\end{Bmatrix} - M\begin{Bmatrix}s^{\mathrm{k}}_{\mathrm{T}}\\ s^{\mathrm{k}}_{\mathrm{A}}\end{Bmatrix} \tag{63}$$

Here, $q_\theta = s^\theta_{\mathrm{T}}(1 - 2\nu_{\mathrm{m}})/E_{\mathrm{m}}$, $\quad q_{\mathrm{k}} = s^{\mathrm{k}}_{\mathrm{T}}(1 - 2\nu_{\mathrm{m}})/E_{\mathrm{m}}$.

These results are equivalent to those obtained from (33) and (34). The path dependence of M must be evaluated from (60), as was the case with B_{m} in (58) and (59). The latter equations are, of course, similar to (62) and (63) in many ways. The loading by $\mathrm{d}\theta$ and $\mathrm{d}k$ is again converted in a unique way to loading by equivalent mechanical tractions $\mathbf{s}^\theta\,\mathrm{d}\theta$, $\mathbf{s}^{\mathrm{k}}\,\mathrm{d}k$.

Uniform strain fields must be added to those caused by the equivalent tractions. The sequence can be altered, but it will still lead to (62) and (63), providing that the current deviatoric stress state in the matrix is kept constant in each step. Then, (62) and (63) represent total instantaneous strain rates for incremental loading by $d\theta$ and dk, respectively.

7. Overall yield surface

The matrix yield condition (A-14), when written in terms of overall stresses, represents the overall yield surface of the composite cylinder

$$f(\mathbf{s}_m) = \tfrac{1}{2}(\mathbf{s}_m - \hat{\mathbf{a}}_m)^T(\mathbf{s}_m - \hat{\mathbf{a}}_m) - k^2 = 0. \tag{A-14}$$

The deviatoric stress increment in the matrix is

$$d\mathbf{s}_m = C d\boldsymbol{\sigma}_m, \tag{64}$$

where

$$C = \tfrac{1}{3}\begin{bmatrix} 2 & -1 & -1 \\ & 2 & -1 \\ \text{SYM.} & & 2 \end{bmatrix},$$

and $d\boldsymbol{\sigma}_m$ is given by the elastic variant of (18)

$$d\boldsymbol{\sigma}_m = B_{me} d\bar{\boldsymbol{\sigma}} + \mathbf{b}_{me} d\theta + \mathbf{b}'_{me} dk. \tag{18}$$

Recall that $\mathbf{b}'_{me} \equiv \mathbf{0}$, and substitute (58) for $\mathbf{b}_{me}$ in (18). Then, (64) gives

$$\mathbf{s}_m = C B_{me}(\bar{\boldsymbol{\sigma}} - \mathbf{s}^{\theta}\,\theta), \tag{66}$$

where θ is total thermal change applied after unloading, or prior to initial yielding.

When the matrix hardens kinematically, it is necessary to consider the effect of local translation $\boldsymbol{\alpha}_m$ (A-12) on the position of the overall yield surface. This translation $\boldsymbol{\alpha}_m$ contributes the phase hardening part to macroscopic hardening. Furthermore, the interaction of the plastically deforming phases creates local residual stress fields which also affect the translation of the overall yield surface; this contribution is referred to as constraint hardening.

The general form of the overall yield surface which accounts for these effects, and also confirms with (66) is, in analogy with (A-14)

$$f(\bar{\boldsymbol{\sigma}}) = \tfrac{1}{2}(\bar{\boldsymbol{\sigma}} - \bar{\boldsymbol{\alpha}})^T Q_{me}^T Q_{me}(\bar{\boldsymbol{\sigma}} - \bar{\boldsymbol{\alpha}}) - k^2 = 0, \tag{67}$$

where

$$Q_{me} = C B_{me} \tag{68}$$

is a 3 × 2 matrix.

A comparison of (67) with (A-14) suggests the following relationship between the radii of local and overall yield surfaces

$$(\mathbf{s}_{\mathrm{m}} - \hat{\mathbf{a}}_{\mathrm{m}}) = Q_{\mathrm{me}}(\bar{\boldsymbol{\sigma}} - \bar{\boldsymbol{\alpha}}). \tag{69}$$

Hence, at any stage of loading

$$(\mathrm{d}\mathbf{s}_{\mathrm{m}} - \mathrm{d}\hat{\mathbf{a}}_{\mathrm{m}}) = Q_{\mathrm{me}}(\mathrm{d}\bar{\boldsymbol{\sigma}} - \mathrm{d}\bar{\boldsymbol{\alpha}}). \tag{70}$$

To find $\mathrm{d}\bar{\boldsymbol{\alpha}}$, it is necessary to eliminate the second row of (70), and write (70) in the form

$$(\mathrm{d}\mathbf{s}'_{\mathrm{m}} - \mathrm{d}\hat{\mathbf{a}}'_{\mathrm{m}}) = Q'_{\mathrm{me}}(\mathrm{d}\bar{\boldsymbol{\sigma}} - \mathrm{d}\bar{\boldsymbol{\alpha}}), \tag{71}$$

where

$$Q'_{\mathrm{me}} = C'B_{\mathrm{me}},$$

$$C' = \tfrac{1}{3}\begin{bmatrix} 2 & -1 & -1 \\ -1 & -1 & 2 \end{bmatrix},$$

$$\mathbf{s}'_{\mathrm{m}} = [s_r^{\mathrm{m}} \ s_z^{\mathrm{m}}] = C'\boldsymbol{\sigma}_{\mathrm{m}},$$

$$\hat{\mathbf{a}}'_{\mathrm{m}} = [\hat{a}_r^{\mathrm{m}} \ \hat{a}_z^{\mathrm{m}}] = C'\boldsymbol{\alpha}_{\mathrm{m}}.$$

The second row in (70) can be found when needed from (71), and from $\mathrm{d}s_{\mathrm{kk}}^{\mathrm{m}} = \mathrm{d}\hat{a}_{\mathrm{kk}}^{\mathrm{m}} = 0$. Now, Q'_{me} is a 2 × 2 matrix, hence (71) can be solved to give

$$\mathrm{d}\bar{\boldsymbol{\alpha}} = \mathrm{d}\bar{\boldsymbol{\sigma}} - (Q'_{\mathrm{me}})^{-1}(\mathrm{d}\mathbf{s}'_{\mathrm{m}} - \mathrm{d}\hat{\mathbf{a}}'_{\mathrm{m}}), \tag{72}$$

providing that $(Q'_{\mathrm{me}})^{-1}$ exists (c.f., Section A-3 in Appendix).

Equation (72) describes the translation of the overall yield surface. Note that $\mathrm{d}\bar{\boldsymbol{\alpha}} \neq \mathbf{0}$ when $\mathrm{d}\hat{\mathbf{a}}'_{\mathrm{m}} = \mathbf{0}$, i.e., when the matrix does not harden. The nonvanishing part of $\mathrm{d}\bar{\boldsymbol{\alpha}}$ represents the constraint hardening of the composite cylinder.

Specific forms of $\mathrm{d}\bar{\boldsymbol{\alpha}}$ in the elastic and plastic deformation ranges are as follows:

In the elastic case, i.e., if $f(\bar{\boldsymbol{\sigma}}) < 0$, or if $f(\bar{\boldsymbol{\sigma}}) = 0$, and $\mathrm{d}f < 0$, where

$$\mathrm{d}f = \left(\frac{\partial f}{\partial \bar{\boldsymbol{\sigma}}}\right)^{\mathrm{T}}(\mathrm{d}\bar{\boldsymbol{\sigma}} - \mathrm{d}\bar{\boldsymbol{\alpha}}) + \frac{\partial f}{\partial \theta}\,\mathrm{d}\theta + \frac{\partial f}{\partial k}\,\mathrm{d}k, \tag{73}$$

the composite experiences either elastic loading, neutral loading, or unloading. While $\mathrm{d}\bar{\boldsymbol{\sigma}}$ will have no effect on the yield surface (67), $\mathrm{d}\theta$ and $\mathrm{d}k < 0$ will cause (67) to change as follows:

$$f(\bar{\boldsymbol{\sigma}}) = \tfrac{1}{2}[\bar{\boldsymbol{\sigma}} - (\bar{\boldsymbol{\alpha}} + \mathrm{d}\bar{\boldsymbol{\alpha}}_{\mathrm{e}}^{\theta})]^{\mathrm{T}} Q_{\mathrm{me}}^{\mathrm{T}} Q_{\mathrm{me}}[\bar{\boldsymbol{\sigma}} - (\bar{\boldsymbol{\alpha}} + \mathrm{d}\bar{\boldsymbol{\alpha}}_{\mathrm{e}}^{\theta})]$$
$$- (k + \mathrm{d}k)^2 = 0, \tag{74}$$

where

$$d\bar{\boldsymbol{\alpha}}_e^\theta = \mathbf{s}^\theta d\theta . \tag{75}$$

This result follows directly from (58), the $-d\bar{\boldsymbol{\alpha}}_e^\theta$ is equal to the traction $-\mathbf{s}^\theta\, d\theta$ applied to the elastic composite. The dk causes no effect through (59) since $\mathbf{b}'_{me} \equiv \mathbf{0}$, i.e., no dk-dependent tractions are added to the elastic composite.

In the case of plastic loading, for $f(\bar{\boldsymbol{\sigma}}) = 0$, and $df \geqslant 0$, it is necessary to account for all traction components which affect plastic straining of the composite cylinder. These are represented by $d\bar{\boldsymbol{\tau}}$ in (60); each component will now be considered separately.

Loading by $d\bar{\boldsymbol{\sigma}} \neq \mathbf{0}$, *when* $d\theta = dk = 0$, keeps the yield condition (67) and the hardening rule (72) intact. Also,

$$d\mathbf{s}'_m = C' d\boldsymbol{\sigma}_m = Q'_m d\bar{\boldsymbol{\sigma}},$$

$$d\hat{\mathbf{a}}'_m = d\mu_m (\mathbf{s}'_m - \hat{\mathbf{a}}'_m) = [(\mathbf{s}_m - \hat{\mathbf{a}}_m)^T Q_m d\bar{\boldsymbol{\sigma}}/2k^2](\mathbf{s}'_m - \hat{\mathbf{a}}'_m).$$

Hence, (72) becomes

$$d\bar{\boldsymbol{\alpha}} = (Q'_{me})^{-1} d\hat{\mathbf{a}}'_m - [(Q'_{me})^{-1} Q'_m - I]\, d\bar{\boldsymbol{\sigma}}. \tag{76}$$

For loading by $d\theta \neq 0$, *when* $d\bar{\boldsymbol{\sigma}} = \mathbf{0}$, $dk = 0$, the hardening process can be resolved into two steps. First, the current load $\bar{\boldsymbol{\sigma}} = \text{const.}$ is temporarily removed. The composite responds elastically to $d\theta$, this causes translation $d\bar{\boldsymbol{\alpha}}_e$ given by (75). Second, the yield surface must be brought to the loading point $\bar{\boldsymbol{\sigma}}$. That is accomplished, in agreement with (58), by applying the tractions $-\mathbf{s}^\theta\, d\theta$ to the translated (by $d\bar{\boldsymbol{\alpha}}_e$) yield surface. Using (72), one obtains

$$d\bar{\boldsymbol{\alpha}}_p = -\mathbf{s}^\theta\, d\theta - (Q'_{me})^{-1}(d\mathbf{s}'_m - d\hat{\mathbf{a}}'_m), \tag{77}$$

where

$$d\mathbf{s}'_m = -Q'_m \mathbf{s}^\theta\, d\theta, \quad d\hat{\mathbf{a}}'_m = -[(\mathbf{s}_m - \hat{\mathbf{a}}_m)^T Q_m \mathbf{s}^\theta\, d\theta/2k^2](\mathbf{s}'_m - \hat{\mathbf{a}}'_m). \tag{78}$$

The total translation $d\bar{\boldsymbol{\alpha}}$ due to $d\theta \neq 0$ is obtained by adding (75) and (77). In view of (78), it follows that

$$d\bar{\boldsymbol{\alpha}} = (Q'_{me})^{-1} d\hat{\mathbf{a}}'_m + (Q'_{me})^{-1} Q'_m \mathbf{s}^\theta\, d\theta . \tag{79}$$

The yield condition (67) applies, the $d\bar{\boldsymbol{\alpha}}_e$ in (74) is now included in (79). Equations (77) to (79) can also be derived from the decomposition sequence in Section 5; this approach is used for loading by dk below.

Loading by $dk < 0$, *when* $d\bar{\boldsymbol{\sigma}} = \mathbf{0}$ *and* $d\theta = 0$ causes an isotropic contraction of the overall yield surface (67), which is accompanied by a translation of the center, so that the yield surface remains at the current loading point $\bar{\boldsymbol{\sigma}}$. To describe this hardening process, we retrace the steps leading to derivation of (59).

The fiber and matrix are separated, and tractions $\boldsymbol{\sigma}_f^0, \boldsymbol{\sigma}_m^0$ are applied to the phases to preserve the current local stress state. The yield stress change $dk < 0$ is applied in the matrix. The resulting local stress and strain increments are (c.f. (A-11) and (50))

$$d\boldsymbol{\sigma}_f = \mathbf{0}, \quad d\boldsymbol{\sigma}_m = \mathbf{0}, \quad d\boldsymbol{\varepsilon}_f = \mathbf{0}, \quad d\boldsymbol{\varepsilon}_m = \mathbf{m}'_m \, dk. \tag{80}$$

At the same time, the local yield surface contracts, and since $d\boldsymbol{\sigma}_m = \mathbf{0}$, the center must translate by $d\hat{\mathbf{a}}_m$. The translation vector can be found from the consistency equation (A-15) and the hardening rule (A-13). Since $d\mathbf{s}_m = \mathbf{0}$, one obtains

$$d\mu_m = -2k\,dk/[(\mathbf{s}_m - \hat{\mathbf{a}}_m)^T(\mathbf{s}_m - \hat{\mathbf{a}}_m)] = -dk/k \tag{81}$$

and

$$d\hat{\mathbf{a}}_m = -(\mathbf{s}_m - \hat{\mathbf{a}}_m)dk/k.$$

The translation of the overall yield surface corresponding to this first step is, from (72),

$$d\bar{\boldsymbol{\alpha}}_1 = -(Q'_{me})^{-1}(\mathbf{s}'_m - \hat{\mathbf{a}}'_m)dk/k. \tag{82}$$

Before the phases can be rejoined, it is necessary to apply local tractions (48) and (49) which now become

$$d\sigma_r^m = d\sigma_z^m = d\sigma_r^f = s_T^k\,dk, \qquad c_f d\sigma_z^f + c_m d\sigma_z^m = s_A^k\,dk. \tag{83}$$

When s_T^k, s_A^k are found from (55), the constraint equation (13) to (15), or (52), are satisfied, and the phases may be rejoined. The application of the hydrostatic stress increment $s_T^k\,dk$ in the matrix does not affect local yielding. However, the overall stress increment (83), equal to $\mathbf{s}^k dk$, must be accommodated by translation of the overall yield surface. From (71), for $d\mathbf{s}'_m = d\hat{\mathbf{a}}'_m = \mathbf{0}$, one obtains

$$d\bar{\boldsymbol{\alpha}}_2 = \mathbf{s}^k dk \tag{84}$$

for the second loading step (83).

Now, the surface tractions (83) must be removed. The corresponding translation for this third loading step is, from (72),

$$d\bar{\boldsymbol{\alpha}}_3 = -\mathbf{s}^k dk - (Q'_{me})^{-1}(d\mathbf{s}'_m - d\hat{\mathbf{a}}'_m), \tag{85}$$

where

$$d\mathbf{s}'_m = -Q'_m\,\mathbf{s}^k dk, \quad d\hat{\mathbf{a}}'_m = d\mu_m(\mathbf{s}'_m - \hat{\mathbf{a}}'_m),$$
$$d\mu_m = -(\mathbf{s}_m - \hat{\mathbf{a}}_m)^T Q_m\,\mathbf{s}^k dk/2k^2. \tag{86}$$

The total translation of the overall yield surface caused by $dk < 0$ is obtained as

$$d\bar{\boldsymbol{\alpha}} = d\bar{\boldsymbol{\alpha}}_1 + d\bar{\boldsymbol{\alpha}}_2 + d\bar{\boldsymbol{\alpha}}_3 = (Q'_{me})^{-1}d\hat{\mathbf{a}}'_m + (Q'_{me})^{-1}[Q'_m\,\mathbf{s}^k dk - (\mathbf{s}'_m - \hat{\mathbf{a}}'_m)dk/k] \tag{87}$$

The yield condition (67) applies.

Finally, the translation vectors (76), (79), and (87) can be added, and the overall hardening rule written in the familiar form

$$\mathrm{d}\bar{\boldsymbol{\alpha}} = \mathrm{d}\bar{\boldsymbol{\sigma}} - (Q'_{\mathrm{me}})^{-1}(\mathrm{ds}'_{\mathrm{m}} - \mathrm{d}\hat{\mathrm{a}}'_{\mathrm{m}}), \tag{72}$$

where

$$\mathrm{ds}'_{\mathrm{m}} = Q'_{\mathrm{m}}\mathrm{d}\bar{\boldsymbol{\tau}}, \tag{88}$$

$$\mathrm{d}\hat{\mathrm{a}}'_{\mathrm{m}} = \mathrm{d}\mu_{\mathrm{m}}(\mathrm{s}'_{\mathrm{m}} - \hat{\mathrm{a}}'_{\mathrm{m}}), \tag{89}$$

$$\mathrm{d}\mu_{\mathrm{m}} = [(\mathrm{s}_{\mathrm{m}} - \hat{\mathrm{a}}_{\mathrm{m}})^{\mathrm{T}}Q_{\mathrm{m}}\mathrm{d}\bar{\boldsymbol{\tau}} - 2k\mathrm{d}k]/2k^2.$$

Equation (67) represents the overall yield surface for the combined loading in $\bar{\boldsymbol{\sigma}}$-space.

This further clarifies the dependence of the hardening process on loading path $\mathrm{d}\bar{\boldsymbol{\tau}}$. As indicated earlier, (88) must be used as a part of loading path, to find M_{m} in (A-24), B_{m} in (45), and $\mathbf{b}_{\mathrm{m}}$, $\mathbf{b}'_{\mathrm{m}}$ in (58) and (59). We note that if $\mathrm{s}^*_{\mathrm{m}}$ is kept constant, then the steps leading to (87), or (88) and (89) are interchangeable and thus these results are not affected by the particular decomposition sequence used in their derivation.

The yield surface (67) represents a closed ellipse in the overall stress space $\bar{\boldsymbol{\sigma}}$. Therefore, each of the contributions to $\mathrm{d}\bar{\boldsymbol{\alpha}}$ will, in fact, cause a translation. Also, the composite will yield under hydrostatic stress. The yield surface closure is due to the fact that three different stress averages (1) were evaluated in the matrix. That contrasts with results obtained with other approaches, such as the self-consistent scheme [6], and the vanishing fiber diameter model [15], which provided only $\mathrm{d}\bar{\boldsymbol{\sigma}}_{\mathrm{m}}$ in (4). The overall yield surface derived from those models consisted of two parallel lines in $\bar{\boldsymbol{\sigma}}$-plane which rendered the models unsuitable for studies of thermal loading problems.

8. Concluding remarks

The results can be extended to other types of material behavior. Isotropic hardening in the matrix can be readily used, as pointed out in the Appendix. Other hardening rules can be incorporated once the matrix instantaneous compliance M_{m} in (40) has been assembled. The effect of matrix hardening is significant in dimensional stability problems, careful choice of appropriate yield surfaces and hardening rules is thus required for accurate predictions of overall response.

Space limitations prevent derivation of overall stiffnesses in (29) to (31). However, this can be accomplished by following the procedure which yielded the compliances. Numerical evaluation of the overall strains for combined loading programs, choice of material parameters for specific composite systems and temperature ranges which appear in practice will be discussed elsewhere.

Acknowledgment

This work was supported in part by grants from the U.S. Army Research Office and from the Army Materials and Mechanics Research Center.

References

1 R. Hill, J. Mech. Phys. Solids, 12 (1964) 213.
2 J.F. Mulhern, T.G. Rogers and A.J.M. Spencer, J. Inst. Math. Applics., 3 (1967) 21.
3 A.R.T. de Silva and G.A. Chadwick, J. Mech. Phys. Solids, 17 (1969) 387.
4 G.J. Dvorak and M.S.M. Rao, Int. J. Engrg. Sci., 14 (1976) 361.
5 G.J. Dvorak and M.S.M. Rao, J. Appl. Mech., 43 (1976) 619.
6 G.J. Dvorak and Y.A. Bahei-El-Din, J. Mech. Phys. Solids, 27 (1979) 51.
7 A. Phillips and M. Ricciutti, Int. J. Solids Structures, 12 (1976) 159.
8 A. Phillips and H. Moon, Acta Mech., 27 (1977) 91.
9 R. Hill, J. Mech. Phys. Solids, 11 (1963) 357.
10 N. Laws, J. Mech. Phys. Solids, 21 (1973) 9.
11 G.J. Dvorak, in: Z. Hashin and C.T. Herakovich (Eds.), Mechanics of Composite Materials: Recent Advances, Pergamon Press, 1983, pp. 73—91.
12 R. Hill, J. Mech. Phys. Solids, 12 (1964) 199.
13 R.T. Shield and H. Ziegler, ZAMP, IXa (1958) 260.
14 H. Ziegler, Q. Appl. Math., XVII (1959) 55.
15 G.J. Dvorak and Y.A. Bahei-El-Din, J. Appl. Mech., 49 (1982) 327.

Appendix

A. 1 STRESS AND STRAIN AVERAGES

The local stress components (1) and (2) in the composite cylinder element must satisfy the equilibrium equation

$$\mathrm{d}\sigma_r/\mathrm{d}r + (\sigma_r - \sigma_\phi)/r = 0.$$

Hence

$$\sigma_r + \sigma_\phi = r\mathrm{d}\sigma_r/\mathrm{d}r + 2\sigma_r.$$

The volume average of $(\sigma_r + \sigma_\phi)$ in the region $0 \leqslant r \leqslant b$ of area $A = \pi b^2$ and unit length z is

$$\frac{1}{V}\int_V (\sigma_\phi + \sigma_r)\mathrm{d}V = \frac{1}{V}\int_V (r\mathrm{d}\sigma_r/\mathrm{d}r + 2\sigma_r)\, r\mathrm{d}r\mathrm{d}\phi\mathrm{d}z$$

$$= \frac{1}{\pi b^2}\int_0^{2\pi} \mathrm{d}\phi \int_0^b \mathrm{d}(r^2\sigma_r) = 2\sigma_r|_{r=b} = 2\bar{\sigma}_r. \qquad \text{(A-1)}$$

In the fiber $0 \leqslant r \leqslant a$, $\sigma_r^{\rm f} = \sigma_\phi^{\rm f}$. Then, (A-1) can be written for volume average of local stresses as

$$2\bar{\sigma}_r = 2c_{\rm f}\sigma_r^{\rm f} + c_{\rm m}(\sigma_r^{\rm m} + \sigma_\phi^{\rm m}). \tag{A-2}$$

The volume average of $\sigma_\phi^{\rm m}$ in the region $a \leqslant r \leqslant b$, $0 \leqslant z \leqslant 1$ is

$$\frac{1}{V}\int_V \sigma_\phi \, dV = \frac{1}{V}\int_0^1 dz \int_0^{2\pi} d\phi \int_a^b \left(\sigma_r + r\frac{d\sigma_r}{dr}\right) r dr$$

$$= \frac{2\pi}{A}\left[\bar{\sigma}_r b^2 - \sigma_r^{\rm f} a^2 - \int_a^b \sigma_r r dr\right], \tag{A-3}$$

where we took

$$\sigma_r|_{r=b} = \bar{\sigma}_r, \qquad \sigma_r|_{r=a} = \sigma_r^{\rm f}.$$

The exact value of the integral of $\sigma_r r$ in (A-3) cannot be found, but it can be approximated with certain assumed distributions of σ_r.

First, in analogy with the elastic solution, we assume the distribution:

$$\sigma_r = B_1 - B_2/r^2 \quad \text{in} \quad a \leqslant r \leqslant b.$$

From stress boundary conditions at $r = b$, and $r = a$ one obtains constants B_1 and B_2 as

$$B_1 = (a^2\sigma_r^{\rm f} - b^2\bar{\sigma}_r)/(a^2 - b^2), \qquad B_2 = a^2b^2(\sigma_r^{\rm f} - \bar{\sigma}_r)/(a^2 - b^2).$$

Hence

$$\int_a^b \sigma_r r dr = \left[\frac{a^2b^2}{b^2 - a^2}\ln\left(\frac{b}{a}\right) - \frac{a^2}{2}\right]\sigma_r^{\rm f} + \left[-\frac{a^2b^2}{b^2 - a^2}\ln\left(\frac{b}{a}\right) + \frac{b^2}{2}\right]\bar{\sigma}_r.$$

From this approximation one can obtain the following expressions for the volume average of $\sigma_\phi^{\rm m}$ in (A-3)

$$\sigma_\phi^{\rm m} = q_1\bar{\sigma}_r + q_2\sigma_r^{\rm f}, \tag{A-3a}$$

where

$$q_1 = \frac{b^2}{b^2 - a^2} + 2\frac{a^2b^2}{(b^2 - a^2)^2}\ln\left(\frac{b}{a}\right),$$

$$q_2 = -\frac{a^2}{b^2 - a^2} - 2\frac{a^2b^2}{(b^2 - a^2)^2}\ln\left(\frac{b}{a}\right).$$

Also, the volume average of $\sigma_\phi^{\rm m}$ may be approximated by the area average,

$$\frac{1}{V}\int_V \sigma_\phi \mathrm{d}V = \frac{1}{b-a}\int_0^1 \mathrm{d}z \int_a^b (\sigma_r + r\mathrm{d}\sigma_r/\mathrm{d}r)\mathrm{d}r.$$

Then

$$\sigma_\phi^{\mathrm{m}} = q_1\bar{\sigma}_r + q_2\sigma_r^{\mathrm{f}}, \tag{A-3b}$$

where

$$q_1 = b/(b-a) = (1+\eta)/\eta, \quad q_2 = -a/(b-a) = -1/\eta,$$

with η given by (5).

The local strain components (1) and (2) must satisfy the compatibility equation

$$\mathrm{d}\epsilon_\phi/\mathrm{d}r + (\epsilon_\phi - \epsilon_r)/r = 0.$$

Hence

$$\epsilon_r + \epsilon_\phi = r\mathrm{d}\epsilon_\phi/\mathrm{d}r + 2\epsilon_\phi.$$

The volume average of $(\epsilon_r + \epsilon_\phi)$ in $0 \leqslant r \leqslant b$, $0 \leqslant z \leqslant 1$, is

$$\frac{1}{V}\int_V (\epsilon_r + \epsilon_\phi)\mathrm{d}V = \frac{1}{\pi b^2}\int_0^1 \mathrm{d}z \int_0^{2\pi} \mathrm{d}\phi \int_0^b \mathrm{d}(r^2\epsilon_\phi) = 2\epsilon_\phi|_{r=b}. \tag{A-4}$$

Furthermore

$$2\epsilon_\phi|_{r=b} = \frac{2}{b}u_r|_{r=b} = 2\bar{\epsilon}_r.$$

Alternatively (A-4) can be evaluated from kinematic relations

$$\epsilon_r + \epsilon_\phi = \mathrm{d}u_r/\mathrm{d}r + u_r/r.$$

Then, the volume average (A-4) is

$$\frac{1}{V}\int_V (\epsilon_r + \epsilon_\phi)\mathrm{d}V = \frac{1}{\pi b^2}\int_0^1 \mathrm{d}z \int_0^{2\pi} \mathrm{d}\phi \int_0^b \mathrm{d}(u_r r) = \frac{2}{b}u_r|_{r=b} = 2\bar{\epsilon}_r.$$

Again, in the fiber $0 \leqslant r < a$, $\epsilon_r^{\mathrm{f}} = \epsilon_\phi^{\mathrm{f}}$, and the volume average of local strains in the composite cylinder element is

$$2\bar{\epsilon}_r = 2c_{\mathrm{f}}\epsilon_r^{\mathrm{f}} + c_{\mathrm{m}}(\epsilon_r^{\mathrm{m}} + \epsilon_\phi^{\mathrm{m}}). \tag{A-5}$$

The volume average of $\epsilon_\phi^{\mathrm{m}}$ in the region $a \leqslant r \leqslant b$, $0 \leqslant z \leqslant 1$ is

$$\epsilon_\phi^m = \frac{1}{V}\int_0^1\int_0^{2\pi}\int_a^b \frac{u_r}{r} \mathrm{d}r\mathrm{d}\phi\mathrm{d}z \tag{A-6}$$

where

$$u_r = a\epsilon_r^f \text{ at } r = a, \qquad u_r = b\bar{\epsilon}_r \text{ at } r = b.$$

Again, exact evaluation of (A-6) is not possible, but several approximations are available.

First, assume u_r in the form

$$u_r = \mathrm{B}_1 r + \mathrm{B}_2/r$$

With

$$\mathrm{B}_1 = (b^2\bar{\epsilon}_r - a^2\epsilon_r^f)/(b^2 - a^2), \qquad \mathrm{B}_2 = a^2b^2(\bar{\epsilon}_r - \epsilon_r^f)/(a^2 - b^2).$$

Then

$$\epsilon_\phi^m = p_1\bar{\epsilon}_r + p_2\epsilon_r^f, \tag{A-6a}$$

where

$$p_1 = \frac{b^2}{b^2 - a^2} - 2\frac{a^2b^2}{(b^2 - a^2)^2}\ln\left(\frac{b}{a}\right),$$

$$p_2 = -\frac{a^2}{b^2 - a^2} + 2\frac{a^2b^2}{(b^2 - a^2)^2}\ln\left(\frac{b}{a}\right).$$

Second, suppose that the ,volume average of ϵ_r in (A-4) can be approximated by the area average of ϵ_r in $a \leqslant r \leqslant b, 0 \leqslant z \leqslant 1$

$$\frac{1}{V}\int_V \epsilon_r \mathrm{d}V \doteq \frac{1}{b - a}\int_0^1 \mathrm{d}z\int_a^b (\mathrm{d}u_r/\mathrm{d}r)\mathrm{d}r.$$

Then,

$$\epsilon_r^m = \bar{\epsilon}_r + (\bar{\epsilon}_r - \epsilon_r^f)a/(b - a), \tag{A-7a}$$

or, with the notation (A-6a) and (5),

$$\epsilon_\phi^m = p_1\bar{\epsilon}_r + p_2\epsilon_r^f, \tag{A-7b}$$

where $p_1 = (1 + \eta)/\eta, \qquad p_2 = -1/\eta.$

A.2 LOCAL CONSTITUTIVE EQUATIONS

The fiber is regarded as an elastic transversely isotropic solid; its thermo-mechanical response is described by

$$d\boldsymbol{\sigma}_f = L_f d\boldsymbol{\varepsilon}_f - \mathbf{l}_f d\theta, \qquad d\boldsymbol{\varepsilon}_f = M_f d\boldsymbol{\sigma}_f + \mathbf{m}_f d\theta, \tag{A-8}$$

where $d\boldsymbol{\sigma}_f$, $d\boldsymbol{\varepsilon}_f$, are given by (2), and [6]

$$L_f = \begin{bmatrix} k_f & l_f \\ l_f & n_f \end{bmatrix}, \qquad \mathbf{l}_f = \begin{bmatrix} k_f\alpha_f + l_f\beta_f \\ l_f\alpha_f + n_f\beta_f \end{bmatrix},$$

$$M_f = \frac{1}{k_f E_A^f}\begin{bmatrix} n_f & -l_f \\ -l_f & k_f \end{bmatrix}, \qquad \mathbf{m}_f = \begin{bmatrix} \alpha_f \\ \beta_f \end{bmatrix} = \begin{bmatrix} 2\alpha_T^f \\ \alpha_A^f \end{bmatrix} \tag{A-9}$$

The Hill's [12] elastic moduli k, l, n are related to the axial and transverse Young's moduli E_A, E_T, Poisson's ratios ν_A, ν_T, and to the transverse shear modulus G_T by

$$k_r = [-E_T G_T/(E_T - 4G_T + 4\nu_A^2 G_T E_T/E_A)]_r;$$

$$l_r = [2k\nu_A]_r, \qquad n_r = [E_A + l^2/k]_r; r = f, m. \tag{A-10}$$

The α_A, α_T are linear thermal expansion coefficients in the axial (z), and transverse (r) directions, respectively.

The matrix is an elastic-plastic solid of the Mises type, with either kinematic or isotropic strain hardening, and with a variable yield stress. The constitutive relations are

$$d\boldsymbol{\sigma}_m = L_m d\boldsymbol{\varepsilon}_m - \mathbf{l}_m d\theta - \mathbf{l}'_m dk, \qquad d\boldsymbol{\varepsilon}_m = M_m d\boldsymbol{\sigma}_m + \mathbf{m}_m d\theta + \mathbf{m}'_m dk, \tag{A-11}$$

where dk indicates a change in the matrix yield stress, and $d\boldsymbol{\sigma}_m$, $d\boldsymbol{\varepsilon}_m$ are given by (1). When the matrix is elastic, both $\mathbf{l}'_m$ and $\mathbf{m}'_m$ shall vanish.

In the case of *kinematic hardening* we adopt the Ziegler—Shield hardening rule [13, 14]

$$d\boldsymbol{\alpha}_m = d\mu_m(\boldsymbol{\sigma}_m - \boldsymbol{\alpha}_m), \tag{A-12}$$

or, in the deviatoric stress plane $\mathbf{s}_m$:

$$d\hat{\mathbf{a}}_m = d\mu_m(\mathbf{s}_m - \hat{\mathbf{a}}_m), \tag{A-13}$$

where

$$\hat{a}^m_{ij} = \alpha^m_{ij} - \tfrac{1}{3}\delta_{ij}\alpha^m_{kk}.$$

The Mises yield condition is

$$f(\mathbf{s}_m) = \tfrac{1}{2}(\mathbf{s}_m - \hat{\mathbf{a}}_m)^T(\mathbf{s}_m - \hat{\mathbf{a}}_m) - k^2 = 0, \tag{A-14}$$

and the consistency equation is

$$df = (\partial f/\partial \mathbf{s}_m)^T(d\mathbf{s}_m - d\hat{\mathbf{a}}_m) - 2kdk = (\mathbf{s}_m - \hat{\mathbf{a}}_m)^T(d\mathbf{s}_m - d\hat{\mathbf{a}}_m) - 2kdk = 0. \tag{A-15}$$

From (A-13) and (A-15)

$$d\mu_m = [(\mathbf{s}_m - \hat{\mathbf{a}}_m)^T d\mathbf{s}_m - 2kdk]/2k^2. \tag{A-16}$$

The total strain increment consists of elastic and plastic parts

$$d\boldsymbol{\varepsilon}_m = d\boldsymbol{\varepsilon}^e_m + d\boldsymbol{\varepsilon}^p_m, \tag{A-17}$$

where

$$d\boldsymbol{\varepsilon}^p_m = \mathbf{0} \text{ if } f(\mathbf{s}_m) < 0, \quad \text{or if} \quad f(\mathbf{s}_m) = 0 \quad \text{and} \quad df < 0, \tag{A-18a}$$

$$d\boldsymbol{\varepsilon}^p_m = d\lambda_m(\partial f/\partial \mathbf{s}_m) = d\lambda_m(\mathbf{s}_m - \hat{\mathbf{a}}_m) \quad \text{if} \quad f(\mathbf{s}_m) = 0 \quad \text{and} \quad df \geqslant 0. \tag{A-18b}$$

To obtain $d\lambda_m$, we assume that the magnitude of vector $c\,d\boldsymbol{\varepsilon}^p_m$ is equal to the total translation of the center of the yield surface in the direction of its exterior normal at $\mathbf{s}_m$, which gives

$$(\partial f/\partial \mathbf{s}_m)^T d\boldsymbol{\sigma}_m + (\partial f/\partial k)dk = c(\partial f/\partial \mathbf{s}_m)^T d\boldsymbol{\varepsilon}^p_m. \tag{A-19}$$

The magnitude of c can be evaluated from a hardening rule in simple tension and compression, which we choose as

$$d\epsilon^p = (d\sigma - \sqrt{3}\,dk)/H, \tag{A-20}$$

where H is the instantaneous hardening parameter. When (A-19) is written for the case of simple tension and compared with (A-20), one obtains

$$c = 2H/3. \tag{A-21}$$

Now, (A-18) and (A-21) can be substituted in (A-19), and the differentiation carried out. The result is

$$(\mathbf{s}_{\mathrm{m}} - \hat{\mathbf{a}}_{\mathrm{m}})^{\mathrm{T}} \mathrm{d}\boldsymbol{\sigma}_{\mathrm{m}} - 2k\mathrm{d}k = \frac{2H}{3}(\mathbf{s}_{\mathrm{m}} - \hat{\mathbf{a}}_{\mathrm{m}})^{\mathrm{T}} \mathrm{d}\boldsymbol{\varepsilon}_{\mathrm{m}}^{\mathrm{p}}$$

$$= \frac{2H}{3}\mathrm{d}\lambda_{\mathrm{m}}(\mathbf{s}_{\mathrm{m}} - \hat{\mathbf{a}}_{\mathrm{m}})^{\mathrm{T}}(\mathbf{s}_{\mathrm{m}} - \hat{\mathbf{a}}_{\mathrm{m}})$$

$$= \frac{4Hk^2}{3}\mathrm{d}\lambda_{\mathrm{m}}. \qquad \text{(A-22)}$$

Accordingly, the total strain increment $\mathrm{d}\boldsymbol{\varepsilon}_{\mathrm{m}}$ in (A-17) follows from (A-18), with $\mathrm{d}\lambda_{\mathrm{m}}$ taken from (A-22),

$$\mathrm{d}\boldsymbol{\varepsilon}_{\mathrm{m}} = M_{\mathrm{me}}\mathrm{d}\boldsymbol{\sigma}_{\mathrm{m}} + \mathbf{m}_{\mathrm{m}}\mathrm{d}\theta + \frac{3}{4Hk^2}[(\mathbf{s}_{\mathrm{m}} - \hat{\mathbf{a}}_{\mathrm{m}})(\mathbf{s}_{\mathrm{m}} - \hat{\mathbf{a}}_{\mathrm{m}})^{\mathrm{T}}\mathrm{d}\boldsymbol{\sigma}_{\mathrm{m}}$$

$$-2k(\mathbf{s}_{\mathrm{m}} - \hat{\mathbf{a}}_{\mathrm{m}})\mathrm{d}k], \qquad \text{(A-23)}$$

where M_{me} is the elastic compliance matrix. This can be compared with (A-11), and the instantaneous elastic-plastic compliances can be evaluated as a sum of their respective parts

$$M_{\mathrm{m}} = \frac{1}{E_{\mathrm{m}}}\begin{bmatrix} 1 & -\nu_{\mathrm{m}} & -\nu_{\mathrm{m}} \\ & 1 & -\nu_{\mathrm{m}} \\ \text{SYM.} & & 1 \end{bmatrix} + \frac{3}{4Hk^2}\begin{bmatrix} (s_r^*)^2 & s_r^* s_\phi^* & s_r^* s_z^* \\ & (s_\phi^*)^2 & s_\phi^* s_z^* \\ \text{SYM.} & & (s_z^*)^2 \end{bmatrix}_{\mathrm{m}}$$

$$= M_{\mathrm{me}} + M_{\mathrm{mp}}. \qquad \text{(A-24)}$$

$$\mathbf{m}_{\mathrm{m}} = [\alpha_{\mathrm{m}}\ \alpha_{\mathrm{m}}\ \alpha_{\mathrm{m}}]^{\mathrm{T}} = \mathbf{m}_{\mathrm{me}}, \qquad \mathbf{m}_{\mathrm{mp}} = 0 \qquad \text{(A-25)}$$

$$\mathbf{m}'_{\mathrm{m}} = -\frac{3}{2Hk}[s_r^*\quad s_\phi^*\quad s_z^*]_{\mathrm{m}}^{\mathrm{T}} = \mathbf{m}'_{\mathrm{mp}}, \qquad \mathbf{m}'_{\mathrm{me}} = 0, \qquad \text{(A-26)}$$

where α_{m} is the linear thermal expansion coefficient, and

$$\mathbf{s}_{\mathrm{m}}^* = \mathbf{s}_{\mathrm{m}} - \hat{\mathbf{a}}_{\mathrm{m}}.$$

Of course, $\mathbf{m}_{\mathrm{m}}$ above was known from the outset, omitted in the derivations, and listed for completeness.

Evaluation of the instantaneous stiffness matrices L_{m}, $\mathbf{l}'_{\mathrm{m}}$, in (A-11) can be accomplished by inversion of (A-23).

From (A-22), with notation (A-27),

$$(\mathbf{s}_{\mathrm{m}}^*)^{\mathrm{T}}\mathrm{d}\boldsymbol{\sigma}_{\mathrm{m}} = \frac{4Hk^2}{3}\mathrm{d}\lambda_{\mathrm{m}} + 2k\mathrm{d}k. \qquad \text{(A-28)}$$

Since the elastic part of deviatoric matrix strain $\mathrm{de}_m^e = \mathrm{ds}_m/2G_m$, we can write

$$\begin{aligned}(\mathrm{s}_m^*)^T \mathrm{d}\boldsymbol{\sigma}_m &= (\mathrm{s}_m^*)^T \mathrm{ds}_m = (\mathrm{s}_m^*)^T 2G_m \mathrm{de}_m^e \\ &= 2G_m(\mathrm{s}_m^*)^T(\mathrm{de}_m - \mathrm{de}_m^p) \\ &= 2G_m(\mathrm{s}_m^*)^T \mathrm{de}_m - 4G_m k^2 \mathrm{d}\lambda_m, \end{aligned} \tag{A-29}$$

using the relationship $\frac{1}{2}(\mathrm{s}_m^*)^T \mathrm{s}_m^* = k^2$.

From (A-28) and (A-29)

$$\mathrm{d}\lambda_m = \frac{1}{S}[(\mathrm{s}_m^*)^T \mathrm{d}\boldsymbol{\varepsilon}_m - Q\mathrm{d}k], \tag{A-30}$$

where $S = 2k^2(1 + H/(3G_m))$, $Q = k/G_m$, and the equality $(\mathrm{s}_m^*)^T \mathrm{de}_m = (\mathrm{s}_m^*)^T \mathrm{d}\boldsymbol{\varepsilon}_m$ was employed.

Now

$$\mathrm{d}\boldsymbol{\sigma}_m = \mathrm{ds}_m + \tfrac{1}{3}\mathrm{d}\sigma_{kk}^m[1\ 1\ 1]^T, \tag{A-31}$$

$$\mathrm{ds}_m = 2G_m(\mathrm{de}_m - \mathrm{d}\lambda_m \mathrm{s}_m^*), \tag{A-32}$$

$$\mathrm{d}\sigma_{kk}^m = 3K_m(\mathrm{d}\epsilon_{kk}^m - 3\alpha_m \mathrm{d}\theta), \tag{A-33}$$

and (A-30), (A-32), (A-33) can be substituted into (A-31). If the result is compared with (A-11), then the instantaneous stiffness matrices are

$$L_m = \begin{bmatrix} 2G_m + \lambda_m & \lambda_m & \lambda_m \\ & 2G_m + \lambda_m & \lambda_m \\ \text{SYM.} & & 2G_m + \lambda_m \end{bmatrix} - \frac{2G_m}{S}\begin{bmatrix} (s_r^*)^2 & s_r^* s_\phi^* & s_r^* s_z^* \\ & (s_\phi^*)^2 & s_\phi^* s_z^* \\ \text{SYM.} & & (s_z^*)^2 \end{bmatrix}_m \tag{A-34}$$

$$\mathrm{l}_m = 3K_m \alpha_m [1 \quad 1 \quad 1]^T \tag{A-35}$$

$$\mathrm{l}_m' = -(2G_m Q/S)[s_r^* \quad s_\phi^* \quad s_z^*]_m^T. \tag{A-36}$$

$\lambda_m = K_m - 2G_m/3$ is a Lame's elastic constant of the matrix. It can be shown that $M_m = L_m^{-1}$, $\mathrm{l}_m = L_m \mathbf{m}_m$, and $\mathrm{l}_m' = L_m \mathbf{m}_m'$.

In the case of *isotropic hardening* of the matrix, it is advantageous to introduce the definitions of effective stresses and strains

$$\hat{\sigma} = (\tfrac{3}{2}\mathrm{s}^T\mathrm{s})^{1/2}, \qquad \mathrm{d}\hat{\epsilon}^p = (\tfrac{2}{3}\mathrm{de}_p^T\,\mathrm{de}_p)^{1/2}, \tag{A-37}$$

and the familiar relationship

$$d\hat{\epsilon}^{p}_{m} = \tfrac{2}{3}\, d\lambda_m \hat{\sigma}_m . \tag{A-38}$$

The dependence of k on plastic strain is taken in the form†

$$k = k_0(\hat{\epsilon}^{p}_{m}) + k_1 . \tag{A-39}$$

In agreement with the concept of dk change discussed in Section 3, (A-39) gives $dk = dk_1$. Again, the uniaxial hardening rule (A-20) is chosen, and that leads directly to

$$d\hat{\epsilon}^{p}_{m} = \frac{1}{H}(d\hat{\sigma}_m - \sqrt{3}\, dk_1). \tag{A-40}$$

From (A-37) we have

$$\hat{\sigma}_m = \sqrt{3}k, \qquad \hat{\sigma}_m\, d\hat{\sigma}_m = \tfrac{3}{2} \mathbf{s}^{T}_{m}\, d\boldsymbol{\sigma}_m , \tag{A-41}$$

which, together with (A-38) and (A-40) gives

$$d\lambda_m = \frac{3}{4Hk^2}[\mathbf{s}^{T}_{m}\, d\boldsymbol{\sigma}_m - 2k\, dk_1]. \tag{A-42}$$

It is readily seen that this result coincides with $d\lambda_m$ in (A-22) providing that $\mathbf{s}_m$ and dk_1 are used to replace $\mathbf{s}^{*}_{m}$ and dk there. It follows that (A-24)–(A-26) yield the instantaneous compliances of an isotropically hardening matrix if $\mathbf{s}^{*}_{m}$ is replaced by $\mathbf{s}_m$ in these equations. The same substitution in (A-34)–(A-36) leads to expressions for instantaneous local stiffnesses. And replacing dk by dk_1 in (16)–(25) leads to the governing equations in Section 3.

A.3 INVERSE OF Q'_{me} IN (72).

A sufficient condition that Q'_{me} be invertible is that $E_m \neq (l_f + 2k_f)(1 - 2\nu_m)$.

Proof:

The real quadratic form $\bar{\boldsymbol{\sigma}}^{T} Q'^{T}_{me} Q'_{me} \bar{\boldsymbol{\sigma}}$ can be written as

$$\bar{\boldsymbol{\sigma}}^{T} Q'^{T}_{me} Q'_{me} \bar{\boldsymbol{\sigma}} = \mathbf{s}'^{T}_{m} \mathbf{s}'_{m} = (s^{m}_{r})^2 + (s^{m}_{z})^2 \geqslant 0; \tag{A-43}$$

(A-43) is equal to or greater than zero for all real values of $\bar{\boldsymbol{\sigma}}$ and is zero only for $s^{m}_{r} = s^{m}_{z} = s^{m}_{\phi} = 0$, i.e., when there is a hydrostatic stress field in matrix,

$$\sigma^{m}_{r} = \sigma^{m}_{\phi} = \sigma^{m}_{z} = \sigma . \tag{A-44}$$

The values of $\bar{\boldsymbol{\sigma}}$ which would create a hydrostatic matrix stress field can be found by substituting (A-44) into equilibrium and constraint equations

†In the kinematic hardening case we took $k = k_1$ and $k_0 \equiv 0$.

(10)—(15). One finds:

from (A-44), (10), and (11)

$$\bar{\sigma}_r = \sigma_r^{\mathrm{f}} = \sigma_\phi^{\mathrm{f}} = \sigma, \tag{A-45a}$$

from (A-44), (A-45a), and (15)

$$\sigma_z^{\mathrm{f}} = \left(\frac{E_{\mathrm{A}}^{\mathrm{f}}(1-2\nu_{\mathrm{m}})}{E_{\mathrm{m}}} + \frac{l_{\mathrm{f}}}{k_{\mathrm{f}}}\right)\sigma, \tag{A-45b}$$

$$\bar{\sigma}_z = c_{\mathrm{m}}\sigma_z^{\mathrm{m}} + c_{\mathrm{f}}\sigma_z^{\mathrm{f}},$$

and from (A-44), (A-45a,b) and (13), (14)

$$\sigma[E_{\mathrm{m}} - (l_{\mathrm{f}} + 2k_{\mathrm{f}})(1-2\nu_{\mathrm{m}})] = 0. \tag{A-45c}$$

(A-45) suggests that if E_{m} is not equal to $(l_{\mathrm{f}} + 2k_{\mathrm{f}})(1-2\nu_{\mathrm{m}})$ then $\bar{\sigma}_r = \bar{\sigma}_z = \sigma = 0$ is the only possible answer. Therefore, (A-43) is zero only for $\bar{\boldsymbol{\sigma}} = \mathbf{0}$, hence $\bar{\boldsymbol{\sigma}}^{\mathrm{T}} Q_{\mathrm{me}}^{\prime\mathrm{T}} Q_{\mathrm{me}}^{\prime} \bar{\boldsymbol{\sigma}}$ is positive definite. The determinant $|Q_{\mathrm{me}}^{\prime\mathrm{T}} Q_{\mathrm{me}}^{\prime}|$ is always positive, $|Q_{\mathrm{me}}^{\prime}|$ is nonzero, and $(Q_{\mathrm{me}}^{\prime})^{-1}$ always exists, as asserted.

The Prediction of Large Permanent Deformations in Rigid-Plastic Impulsively Loaded Frames

P.D. GRIFFIN and J.B. MARTIN

Department of Civil Engineering, University of Cape Town (South Africa)

Abstract

This paper considers the behaviour of metal beams and frames subjected to large impulsive loading. The elastic response is ignored, and the material is considered as rigid-plastic. Earlier work of the authors is extended to provide an analytical technique which models the material as homogeneous viscous in order to account for strain rate effects, and which uses a direct numerical method of analysis to model non-modal behaviour of the structure. The technique is shown to give very good predictions even when displacements are very large (in the order of the dimensions of the frame).

1. Introduction

The analysis of ductile metal structures subjected to very large impulsive loads can be carried out very successfully on the assumption that elastic behaviour can be ignored. The most commonly used constitutive equations used under these circumstances are rigid-viscoplastic; a particular example is the relation

$$\begin{aligned} \frac{\dot{\epsilon}}{\dot{\epsilon}_0} &= \left(\frac{\sigma}{\sigma_0} - 1\right)^n \quad \text{for} \quad |\sigma| > \sigma_0, \\ \dot{\epsilon} &= 0 \quad \text{for} \quad |\sigma| < \sigma_0. \end{aligned} \tag{1}$$

In this equation $\dot{\epsilon}$, σ denote strain rate and stress respectively, and $\dot{\epsilon}_0$, σ_0 are constants with the dimensions of strain rate and stress. The index n is typically taken to have a value in the range 4—5.

Because eqn. (1) is nonhomogeneous, and because of the rigid regions which occur in the structure when it is used, rigid viscoplastic dynamic analysis is relatively complex when viewed as a computational problem. The problem can be simplified considerably, however, by making use of the concept, put forward by Symonds [1], of replacing equation (1) by a homogeneous viscous relation of the form

$$\dot{\epsilon}/\dot{\epsilon}_0 = (\sigma/\sigma_0')^{n'} \tag{2}$$

where n', σ_0' are chosen to match eqn. (2) to eqn. (1) for the problem under consideration. The homogeneity of eqn. (2) has a number of helpful consequences; one of these is the existence of mode solutions (i.e. solutions in which the velocity field can be written as a product of independent functions of space and time) for geometrically linear problems. Based on the scheme put forward for rigid-plastic problems by Martin and Symonds [2], approximate solutions using the mode concepts have been generated for both geometrically linear and geometrically non-linear problems (Symonds [1], Symonds and Chon [3, 4], Symonds and Raphanel [5]).

One of the difficulties inherent in the use of a matched homogeneous viscous relation, however, is that n' in eqn. (2) is usually large, in the range 9—15. This leads to difficulties in both spatial and temporal integration which require particular attention in numerical schemes. In beam and frame problems, for example, conventional finite element methods would make use of cubic interpolation functions for the transverse velocity. If this is used along with a lump mass idealisation, the inconsistency arises, particularly significant for large n', in that the variation in bending moment between nodes should be linear from the point of view of dynamics, but may be highly nonlinear as a result of the interpolation function.

This suggests that force method rather than displacement method formulations may be appropriate for problems of this class. In two earlier papers, Griffin and Martin [6, 7] have considered the application of the mode approximation technique. In the first paper, a systematic method of determining the mode shape was introduced, and adapted for geometrically linear problems to the use of the force method for numerical implementation. A complete solution was also generated by integrating forward the difference between the mode approximation and the actual solution, based on a mixed variational formulation. In the second paper the mode approximation was extended to geometrically nonlinear problems, following the instantaneous mode technique of Symonds and Chon [3]. The structure is assumed to be geometrically linear over a short interval Δt; the geometrically nonlinear approximation is then a series of geometrically linear approximations with the geometry being updated at the beginning of each time step.

These methods lead to very efficient computational schemes which provide excellent agreement with experimental results when the deformations of the structure are essentially modal. The mode approximation does not work very efficiently, however, when the actual solution does not converge rapidly onto a mode type response, where the shape of the velocity field does not change significantly with time. In such cases direct forward integration of the equations of motion is necessary. In this paper we shall formulate a procedure for this integration based on the experience gained in the mode approximation calculations: the procedure is of the force method type, and uses an implicit time integration scheme. While the direct integration procedure is much less efficient than the mode approximation in those cases where the use of the mode approximation is appropriate, it is very much more efficient than a general purpose finite element program in which elastic effects are included.

Specifically, we consider beam and polygonal frame structures which lie in one plane and which are supported at their ends, and which are subjected to large impulsive loads in their own plane. The problem is discretised by identifying nodes along the centre line of the structure.

It will be assumed that the displacements, and hence velocities at the constrained nodes or supports are identically zero. Rotations, or rotation rates, will only be included in the description of the displacements and displacement rates if they are constrained. Furthermore, anticipating the force method formulation, we designate three independent constrained node displacement components as those required to prevent rigid body motion of the structure, and we do not include these components in the description of the displacements and displacement rates. By this process we define a statically determinate "released" structure. The remaining displacement, velocity and acceleration components are grouped into the vectors $\mathbf{u}(t)$, $\dot{\mathbf{u}}(t)$ and $\ddot{\mathbf{u}}(t)$ respectively, where t denotes time.

Mass is lumped at the nodes, and a diagonal mass matrix $[G]$ is defined in such a way that the kinetic energy of the structure is given by

$$K = \tfrac{1}{2}\dot{\mathbf{u}}^{\mathrm{T}}[G]\dot{\mathbf{u}}, \tag{3}$$

at any instant. The mass terms corresponding to constrained velocity components of $\dot{\mathbf{u}}$ can be arbitrarily defined; this includes the rotatory inertia associated with constrained (support) rotation rates. No other rotatory inertia terms appear in $[G]$.

At time $t = 0$ an impulse is applied to each node, represented by the vector $\mathbf{I}$. The impulsive load imparts an initial velocity to each node, given by

$$\mathbf{I} = [G]\dot{\mathbf{u}}^0 = [G]\dot{\mathbf{u}}(0). \tag{4}$$

Our problem is to determine the resulting motion of the structure, with initial displacements $\mathbf{u}(0) = 0$ and initial velocities $\dot{\mathbf{u}}(0)$ by eqn. (4).

The constitutive equations, which relate bending moment M, axial force N to curvature rate $\dot{\kappa}$, centre line extension rate $\dot{\epsilon}$, will be formulated for a sandwich beam, using eqn. (2). The sandwich beam comprises two flanges separated by a core which carries shear force without shear deformation, as shown in Fig. 1. Following Symonds and Chon [3], the constitutive equation is written in the form

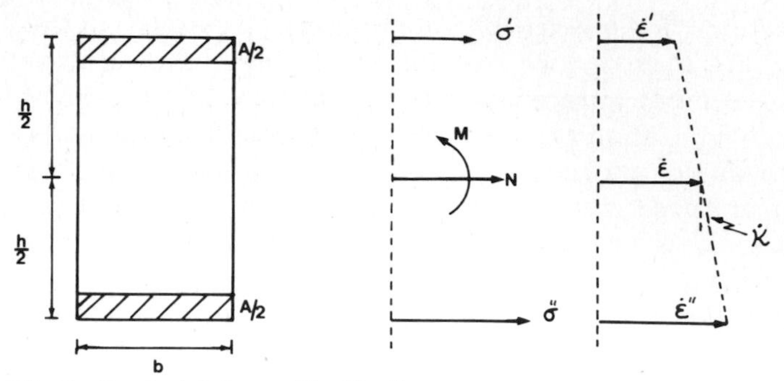

Fig. 1. Sandwich beam idealization.

$$N_0'\dot{\epsilon} = \frac{\partial\psi}{\partial(N/N_0)}, \qquad M_0'\dot{\kappa} = \frac{\partial\psi}{\partial(M/M_0)} \tag{5a}$$

$$\psi = \frac{N_0'\dot{\epsilon}_0}{2(n'+1)}\left[\left|\frac{N}{N_0'}+\frac{M}{M_0'}\right|^{n'+1}+\left|\frac{N}{N_0'}-\frac{M}{M_0'}\right|^{n'+1}\right] \tag{5b}$$

where

$$N_0' = A\sigma_0', \qquad M_0' = \tfrac{1}{2}Ah\sigma_0'. \tag{5c}$$

In the matching process we put

$$\sigma_0' = \mu\sigma_0, \qquad n' = \nu n, \tag{6}$$

and the values of N_0, M_0 for the rigid-viscoplastic beam are

$$N_0 = A\sigma_0 = \frac{1}{\mu}A\sigma_0', \quad M_0 = \frac{1}{2}Ah\sigma_0 = \frac{1}{2}Ah\sigma_0'. \tag{7}$$

The values of N_0, M_0 are taken from the actual section properties as the limiting axial force and the limiting moment for a perfectly plastic material with yield stress σ_0.

2. Initial moments and axial forces in the structure

As a sub-problem of the general problem of integrating the equations of motion, the moments and axial forces at time $t = 0$ must be determined. This can be treated as a static problem. We have a *statically determinate* structure (the supports being the three node displacement components which prevent rigid body motion) with the node velocity components $\dot{\mathbf{u}}^0 = \dot{\mathbf{u}}(0)$ completely prescribed. Note that $\dot{\mathbf{u}}(0)$ contains both velocity components defined by the impulsive load and velocity components constrained to be zero. In addition, the geometry of the structure is defined by the initial displacements $\mathbf{u}(0) = 0$.

Let the forces conjugate to the velocity vector $\dot{\mathbf{u}}(t)$ be denoted by $\mathbf{X}(t)$;

these forces are defined as loads which when acting *statically* on the structure lead to node velocities $\dot{\mathbf{u}}$.

Using the principle of virtual work, we can readily compute the node velocities $\dot{\mathbf{u}}$ in terms of the nodal forces **X**. First, we formulate the nodal moments, represented by the vector **M**, in terms of the loads **X**;

$$\mathbf{M} = [m]\mathbf{X}. \tag{8}$$

Each row of the influence matrix $[m]$ is the set of nodal moments due to a unit value of some component of **X**. Moments are distributed linearly across each element: if a, b are adjacent nodes separated by distance l_e, and M_a, M_b are the nodal moments, the bending moment at a distance s from node a is given by

$$M(s) = M_a(1 - s/l_e) + M_b(s/l_e). \tag{9}$$

Using these relations, we can define the bending moment m_j along each element resulting from a unit value of the jth component X_j of the load vector **X**.

The axial forces are constant along each element, and are represented by an element axial force vector **N**, given by

$$\mathbf{N} = [n]\mathbf{X}. \tag{10}$$

Each row of the influence matrix $[n]$ is the set of element axial forces due to a unit value of some component of **X**. From this, we can define the axial force n_j in each element resulting from a unit value of the jth component X_j of the load vector **X**.

Using the constitutive equations (5) we can thus write the curvature rate $\dot{\kappa}$ and the strain rate $\dot{\epsilon}$ at each point on the structure in terms of **X**. With the curvature rates $\dot{\kappa}$, axial strain rates $\dot{\epsilon}$ and velocities $\dot{\mathbf{u}}$ as the kinematic system, and a unit value of the jth component of **X**, together with its associated m_j, n_j, as the static system, the principle of virtual velocities gives the jth component $\dot{u}_j$ of $\dot{\mathbf{u}}$ as

$$\dot{u}_j = \sum_{\text{elements}} \left\{ \int_{l_e} m_j \dot{\kappa} \mathrm{d}s + n_j \dot{\epsilon} l_e \right\}, \tag{11}$$

where l_e is the length of an element. This process is repeated for each component of $\dot{\mathbf{u}}$, giving finally

$$\dot{\mathbf{u}} = \mathbf{F}(\mathbf{X}). \tag{12}$$

It is a straightforward computational problem to determine $\dot{\mathbf{u}}$ given **X**; we require, however, **X** given $\dot{\mathbf{u}}$. This is a nonlinear problem, and a Newton—Raphson iterative procedure is used to determine the solution. Equation (12) is written as

$$\dot{\mathbf{u}} - \mathbf{F}(\mathbf{X}) = 0, \tag{13}$$

and a matrix of partial derivatives of $\mathbf{F}$ with respect to $\mathbf{X}$ is defined

$$[A(\mathbf{X})] = \partial \mathbf{F}/\partial \mathbf{X}. \tag{14}$$

In the iterative scheme the kth trial value of $\mathbf{X}$ is denoted by $\mathbf{X}^k$. An improved trial value, $\mathbf{X}^{k+1}$, is then given by

$$\mathbf{X}^{k+1} = \mathbf{X}^k - [A^k]\{\mathbf{F}(\mathbf{X}^k) - \dot{\mathbf{u}}\}. \tag{15}$$

This process is repeated until an estimate of $\mathbf{X}$ of acceptable accuracy is obtained. At this point, the moment and axial force vectors $\mathbf{M}$, $\mathbf{N}$ may be evaluated.

This procedure is applied to the determination of the initial moments and axial forces, given the initial velocities and the initial geometry. Note, however, that it might be applied at any instant, provided that the velocities $\dot{\mathbf{u}}(t)$ and the configuration, described by $\mathbf{u}(t)$, are given. We shall make use of this in the next section, but for instants after $t = 0$ the iteration scheme will be broadened to include forward integration.

3. An implicit time integration scheme

The forward integration of the equations of motion of impulsively loaded homogeneous viscous structures is not trivial, owing to the high degree of nonlinearity of the constitutive equations. Explicit forward integration schemes, although simple to formulate and implement, were found in general to be inadequate as they resulted in an unstable solution unless very small time steps were taken. In this section we present an implicit integration scheme in which equilibrium iterations are performed at each time step in order to improve the accuracy of the solution.

Let subscripts t, $t+1$ denote the instants t, $t + \Delta t$ respectively and let superscript i denote the ith iteration in the algorithm which will be outlined below. At time t velocities $\dot{\mathbf{u}}_t$ and displacements $\mathbf{u}_t$ are known, as are the nodal forces $\mathbf{X}_t$. The nodal forces at time $t = 0$ are calculated by the procedure set out in the previous section; thereafter $\mathbf{X}_t$ is calculated in the forward integration algorithm.

For the equation of motion, with the assumption that no external forces are applied to the structure at $t > 0$,

$$[G]\ddot{\mathbf{u}}_t + \mathbf{X}_t = 0, \tag{16a}$$

or

$$\ddot{\mathbf{u}}_t = -[G]^{-1}\mathbf{X}_t. \tag{16b}$$

Rewriting eqn. (13) at time $t + 1$, we have

$$\dot{\mathbf{u}}_{t+1} = \mathbf{F}(\mathbf{X}_{t+1}) = \mathbf{F}_{t+1}. \tag{17}$$

It is implicitly assumed that the function $\mathbf{F}$, evaluated according to eqns. (11) and (12), refers to the geometry of the structure at time $t + 1$. Thus $\mathbf{F}_{t+1}$ can be found only when $\mathbf{u}_{t+1}$ (or an estimate of $\mathbf{u}_{t+1}$) is available. Nonlinear geometrical effects are thus taken into account; because we are working with a *viscous* material, and are computing velocities in an instantaneously defined configuration, no further complications arise from the inclusion of large displacements.

Increments in $\ddot{\mathbf{u}}$, $\dot{\mathbf{u}}$, $\mathbf{u}$ and $\mathbf{X}$ are defined by the equations

$$\ddot{\mathbf{u}}_{t+1} = \ddot{\mathbf{u}}_t + \Delta\ddot{\mathbf{u}}, \qquad \dot{\mathbf{u}}_{t+1} = \dot{\mathbf{u}}_t + \Delta\dot{\mathbf{u}},$$
$$\mathbf{u}_{t+1} = \mathbf{u}_t + \Delta\mathbf{u}, \qquad \mathbf{X}_{t+1} = \mathbf{X}_t + \Delta\mathbf{X}. \tag{18}$$

Substituting $\ddot{\mathbf{u}}_{t+1}$, $\mathbf{X}_{t+1}$ from eqns. (18) into the equations of motion (16a) at time $t + 1$, we have

$$[G](\ddot{\mathbf{u}}_t + \Delta\ddot{\mathbf{u}}) + (\mathbf{X}_t + \Delta\mathbf{X}) = 0. \tag{19a}$$

Hence

$$[G]\Delta\ddot{\mathbf{u}} + \Delta\mathbf{X} = -([G]\ddot{\mathbf{u}}_t + \mathbf{X}_t). \tag{19b}$$

Substituting also into eqn. (17), we may put

$$\dot{\mathbf{u}}_t + \Delta\dot{\mathbf{u}} = \mathbf{F}_t + [A_t]\Delta\mathbf{X}, \tag{20a}$$

where

$$[A_t] = [A(\mathbf{X}_t)] = \partial\mathbf{F}/\partial\mathbf{X}|_t \tag{20b}$$

and is given by the last evaluation of eqn. (14) in the iterative procedure to determine $\mathbf{X}_t$, described in the previous section. Note that as the constitutive relation used is homogeneous, the partial derivatives of $\mathbf{F}$ with respect to $\mathbf{X}$ may be formed explicitly. Integration is then carried out over the length of an element, and the contribution of each element is summed over the structure. Hence

$$\Delta\dot{\mathbf{u}} = [A_t]\Delta\mathbf{X} + (\mathbf{F}_t - \dot{\mathbf{u}}_t). \tag{20c}$$

Using the trapezoidal rule, we put

$$\dot{\mathbf{u}}_{t+1} = \dot{\mathbf{u}}_t + \tfrac{1}{2}\Delta t(\ddot{\mathbf{u}}_t + \ddot{\mathbf{u}}_{t+1}), \tag{21}$$

and hence, from the first of eqns. (18),

$$\Delta\ddot{\mathbf{u}} = \ddot{\mathbf{u}}_{t+1} - \ddot{\mathbf{u}}_t = \frac{2}{\Delta t}\Delta\dot{\mathbf{u}} - 2\ddot{\mathbf{u}}_t. \tag{22}$$

Substituting eqn. (22) into eqn. (19b), we thus have

$$\frac{2}{\Delta t}[G]\Delta\dot{\mathbf{u}} + \Delta\mathbf{X} = -(-[G]\ddot{\mathbf{u}}_t + \mathbf{X}_t).$$

Finally, substituting for $\Delta\dot{\mathbf{u}}$ from eqn. (20c), and rearranging, we have

$$\{[A_t] + \tfrac{1}{2}\Delta t[G]^{-1}\}\Delta\mathbf{X} = -(\mathbf{F}_t - \dot{\mathbf{u}}_t) + \tfrac{1}{2}\Delta t(\ddot{\mathbf{u}}_t - [G]^{-1}\mathbf{X}_t). \tag{24}$$

Equation (24) is solved for $\Delta\mathbf{X}$, and $\Delta\dot{\mathbf{u}}$ follows from eqn. (20c). Equations (18) then give $\dot{\mathbf{u}}_{t+1}$, $\mathbf{X}_{t+1}$, and $\mathbf{u}_{t+1}$ is found by a further application of the trapezoidal rule,

$$\mathbf{u}_{t+1} = \mathbf{u}_t + \tfrac{1}{2}\Delta t(\dot{\mathbf{u}}_t + \dot{\mathbf{u}}_{t+1}). \tag{25}$$

This procedure will be numerically stable, but will introduce errors which will propagate as the solution advances in time. In particular, eqn. (20a) does not include the effects of change in geometry, and hence the equation of motion at time $t + 1$ will not be exactly satisfied. In order to improve estimates of $\Delta\dot{\mathbf{u}}$, $\Delta\mathbf{X}$, and to incorporate the error in the equation of motion at the previous time step, an iteration scheme is introduced. Letting superscript $(i + 1)$ denote the $(i + 1)$th iteration, we write eqns. (16a) and (13) as

$$[G]\ddot{\mathbf{u}}_{t+1}^{i+1} + \mathbf{X}_{t+1}^{i+1} = 0, \tag{26}$$

and

$$\dot{\mathbf{u}}_{t+1}^{i+1} = \mathbf{F}(\mathbf{X}_{t+1}^{i+1}) = \mathbf{F}_{t+1}^{i+1}. \tag{27}$$

Redefining the increments of eqn. (18) as residuals, we have

$$\ddot{\mathbf{u}}_{t+1}^{i+1} = \ddot{\mathbf{u}}_{t+1}^{i} + \Delta\ddot{\mathbf{u}}_{t+1}^{i}, \tag{28a}$$

$$\dot{\mathbf{u}}_{t+1}^{i+1} = \dot{\mathbf{u}}_{t+1}^{i} + \Delta\dot{\mathbf{u}}_{t+1}^{i}, \tag{28b}$$

$$\mathbf{u}_{t+1}^{i+1} = \mathbf{u}_{t+1}^{i} + \Delta\mathbf{u}_{t+1}^{i}, \tag{28c}$$

$$\mathbf{X}_{t+1}^{i+1} = \mathbf{X}_{t+1}^{i} + \Delta\mathbf{X}_{t+1}^{i}. \tag{28d}$$

From the trapezoidal rule at the ith and $(i + 1)$th iterations, we may write

$$\dot{\mathbf{u}}_{t+1}^{i} = \dot{\mathbf{u}}_t + \tfrac{1}{2}\Delta t(\ddot{\mathbf{u}}_t + \ddot{\mathbf{u}}_{t+1}^{i}), \tag{29a}$$

and

$$\dot{\mathbf{u}}_{t+1}^{i+1} = \dot{\mathbf{u}}_t + \tfrac{1}{2}\Delta t(\ddot{\mathbf{u}}_t + \ddot{\mathbf{u}}_{t+1}^{i+1}) \tag{29b}$$

and hence

$$\Delta\dot{\mathbf{u}}_{t+1}^{i} = \tfrac{1}{2}\Delta t\Delta\ddot{\mathbf{u}}_{t+1}^{i}. \tag{29c}$$

From the first of eqns. (28) and eqn. (29c), we find

$$\ddot{\mathbf{u}}_{t+1}^{i+1} = \ddot{\mathbf{u}}_{t+1}^{i} + \frac{2}{\Delta t}\Delta\dot{\mathbf{u}}_{t}^{i}. \tag{30}$$

Substituting eqns. (30) and (28d) into eqn. (26), and rearranging, we have

$$\frac{2}{\Delta t}[G]\Delta\dot{\mathbf{u}}_{t+1}^{i} + \Delta\mathbf{X}_{t+1}^{i} = -([G]\ddot{\mathbf{u}}_{t+1}^{i} + \mathbf{X}_{t+1}^{i}). \tag{31}$$

From eqn. (20a), we write

$$\dot{\mathbf{u}}_{t+1}^{i} + \Delta\dot{\mathbf{u}}_{t+1}^{i} = \mathbf{F}_{t+1}^{i} + [A_{t+1}^{i}]\Delta\mathbf{X}_{t+1}^{i}, \tag{32a}$$

where

$$[A^i_{t+1}] \;=\; \partial \mathbf{F}/\partial \mathbf{X}|\mathbf{x}^i_{t+1}. \tag{32b}$$

The matrix $[A^i_{t+1}]$ is re-evaluated at the beginning of each equilibrium iteration by taking the partial derivatives of the current value of $\mathbf{F}$, given for the $(i+1)$th iteration by eqn. (27), with respect to the current value of the body forces, $\mathbf{X}^i_{t+1}$ and for the configuration denoted by $\mathbf{u}^i_{t+1}$. In order to find the final equation for $\mathbf{X}^i_{t+1}$, we substitute eqn. (32a) into eqn. (31), and rearrange;

$$\{[A^i_{t+1}] + \tfrac{1}{2}\Delta t[G]^{-1}\}\Delta\mathbf{X}^i_{t+1} \;=\; -(\mathbf{F}^i_{t+1} - \dot{\mathbf{u}}^i_{t+1}) - \tfrac{1}{2}\Delta t\,(\ddot{\mathbf{u}}^i_{t+1} + [G]^{-1}\mathbf{X}^i_{t+1}). \tag{33}$$

Equation (33) provides $\Delta\mathbf{X}^i_{t+1}$, and $\Delta\dot{\mathbf{u}}^i_{t+1}$ is then obtained from eqn. (31). By the same process which led to eqn. (29c), we have

$$\Delta\mathbf{u}^i_{t+1} \;=\; \tfrac{1}{2}\Delta t\Delta\dot{\mathbf{u}}^i_{t+1}. \tag{34}$$

We may thus find revised estimates $\dot{\mathbf{u}}^{i+1}_{t+1}$, $\mathbf{u}^{i+1}_{t+1}$, $\mathbf{X}^{i+1}_{t+1}$ from eqns. (28). The iterative procedure is repeated until the residual quantities $\Delta\dot{\mathbf{u}}$, $\Delta\mathbf{u}$, $\Delta\mathbf{X}$ are acceptably small.

Once the solution quantities at time $t+1$ have been computed to the required tolerance, the solution proceeds to the next time step. The algorithm has been found to be an efficient procedure for the homogeneous viscous structures under consideration. Much larger time steps than can be used in an explicit scheme are possible, and even including the iteration within the time step, this leads to a much less costly computational scheme.

In the following section we shall discuss briefly the matching strategy used to choose σ_0' and n', and this discussion will be followed by examples of the application of the numerical scheme described here.

4. The matching procedure

The use of a homogeneous viscous relation for rigid-plastic dynamic analysis is based on the supposition that the rigid-visco-plastic relation

$$\begin{aligned} \frac{\dot{\epsilon}}{\epsilon_0} &= \left(\frac{\sigma}{\sigma_0} - 1\right)^n \quad \text{for} \quad \sigma \geqslant \sigma_0, \\ \frac{\dot{\epsilon}}{\epsilon_0} &= 0 \qquad\qquad\quad \text{for } \sigma \leqslant \sigma_0, \end{aligned} \tag{35}$$

can be adequately approximated in any particular problem by a relation of the form

$$\dot{\epsilon}/\epsilon_0 \;=\; (\sigma/\sigma_0')^{n'}, \tag{36a}$$

with

$$\sigma_0' = \mu\sigma_0, \tag{36b}$$

$$n' = \nu n. \tag{36c}$$

A strategy for choosing μ and ν is thus an essential part of the application of the homogeneous viscous material in dynamic problems.

Symonds [1] suggested that the factors μ and ν should be chosen such that eqns. (35) and (36a) have a common intercept and slope at a value of strain rate which is the largest occurring in structure at time $t = 0$. If this largest value is denoted by $\dot{\epsilon}_{\max}$ this strategy gives

$$\nu = \frac{1 + (\dot{\epsilon}_{\max}/\dot{\epsilon}_0)^{1/n}}{(\dot{\epsilon}_{\max}/\dot{\epsilon}_0)^{1/n}}\ , \tag{37a}$$

$$\mu = \frac{1 + (\dot{\epsilon}_{\max}/\dot{\epsilon}_0)^{1/n}}{(\dot{\epsilon}_{\max}/\dot{\epsilon}_0)^{1/\nu n}}\ . \tag{37b}$$

This matching is shown diagrammatically in Fig. 2; the rigid-visco-plastic relation is given by curve 1, and curve 2 depicts the homogeneous viscous relation matched by the procedure outlined above.

In general, this strategy appears to be effective in simple problems; difficulties occur under two circumstances, however, when a generalisation is attempted. The first is when it is difficult to estimate, or interpret, the maximum initial strain rate in the structure. The second, with which we are concerned here, is that the value of n' is typically in the range 10—15.

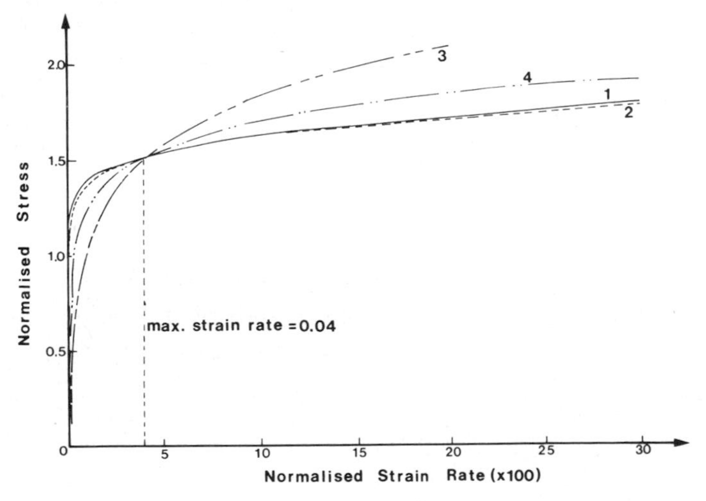

Fig. 2. The matching procedure. Curve 1: rigid-viscoplastic curve; $n = 5$. Curve 2: homogeneous viscous curve matched on shape and intercept at $\dot{\epsilon}/\dot{\epsilon}_0 = 0.04$; $\mu = 1.904$, $n' = 14.52$. Curve 3: homogeneous viscous curve matched on intercept alone at $\dot{\epsilon}/\dot{\epsilon}_0 = 0.04$; $\mu = 2.904$, $n = n' = 5$. Curve 4: homogeneous viscous curve matched on intercept alone at $\dot{\epsilon}/\dot{\epsilon}_0 = 0.04$ with increased n'; $\mu = 2.204$, $n' = 8.75$.

This results in considerable numerical problems, particularly in direct solution techniques.

In analyses given by Griffin and Martin [6, 7] using the mode technique, and for some problems using the direct analysis technique set out in this paper, the full matching procedure of eqns. (37) was found to be unnecessary. Satisfactory results were obtained by setting $n' = n$, and choosing μ so that the curves intersect at the value of $\dot{\epsilon}_{max}$. This has the obvious numerical advantage of keeping the value of n' low, thereby eliminating potential numerical problems. The scheme is illustrated by curve 3 in Fig. 2.

This scheme does not always lead to a satisfactory solution, however; much depends on the magnitude of $\dot{\epsilon}_{max}$ in relation to $\dot{\epsilon}_0$. A compromise is to choose the value of n', with $n' > n$, as large as possible, with the choice being dictated by the ability of the solution procedure to carry through the analysis without computational difficulties. As before, μ is then chosen so that the rigid-viscoplastic relation and the homogeneous relation intersect at $\dot{\epsilon}_{max}$. This is shown diagrammatically by curve 4 in Fig. 2.

Whilst this compromise matching procedure has disadvantages in that a prior estimate of the largest n' which can be tolerated must be made, it seems a reasonable approach in the context of beams and frames. The results of analysis performed using the approach have shown that the higher the value of the n' the better the correlation with test data. The errors introduced by low values of n' are not consistent, and thus upper or lower bounds cannot be established.

The best choice of the strain rate magnitude on which the matching is based is also an open question. Symonds [8] has also suggested that matching can be based on an average strain rate. Another possible approach in numerical analysis is to rematch at the beginning of each time step. While the compromise procedure given in this section has given the best results in our study, further work is clearly needed to give firm guidelines on the matching strategy in any particular case.

5. Illustrative examples

A computer program which performs the numerical procedures outlined in this paper has been used successfully to analyse a variety of beam and frame structures. To illustrate the application of the program, the results of analyses of three types of structures are presented. They are:

(a) a cantilever struck transversely at its tip;

(b) steel and aluminum rectangular, fixed end, portal frames subjected to a uniform sideways impulse applied along the length of one column; and

(c) aluminum rectangular, fixed end, portal frames subjected to a uniform transverse impulse applied to half the length of the beam.

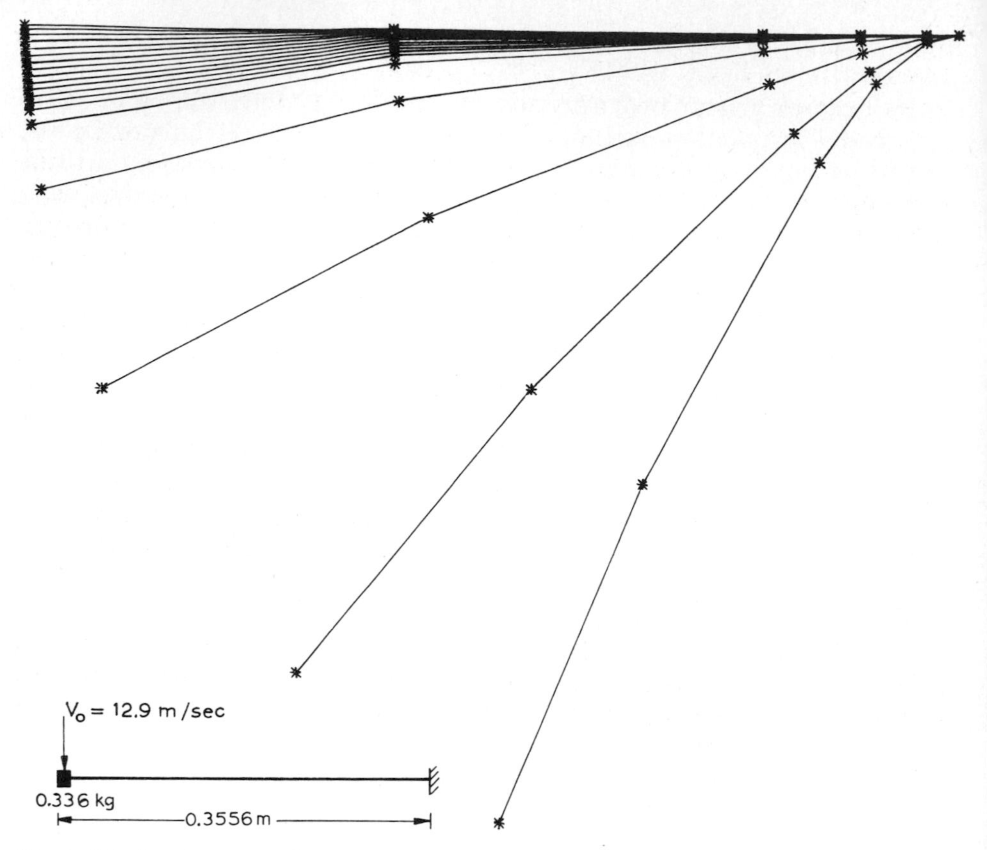

Fig. 3. Displaced shape of cantilever beam at successive time intervals; mass density $\rho = 7850\,\text{kg/m}^3$, $\sigma_0 = 200\,\text{MPa}$, section depth $h = 4.5\,\text{mm}$, section breadth $b = 16.3\,\text{mm}$, $\dot{\epsilon} = 40/\text{s}$, $n = 5$, $n' = 11.0$, $\mu = 1.99$.

The problem of the cantilever beam struck transversely at its tip has received considerable attention both experimentally and analytically (for example, Bodner and Symonds [9], Ting [10], Lee and Martin [11]). A particular beam E4 from the tests by Bodner and Symonds [9] was analysed. Their experiments gave a tip rotation of 53°, and they estimated the total time of deformation t_f to be 0.052 s. Deflections were not presented in their results. Since rotations are not calculated in the analysis procedure outlined in this paper, the only parameter which may be used to compare the results obtained here to test data and most previous analytical solutions is t_f. Nevertheless, the cantilever is an important standard problem, and indicates the capabilities of the program. From a small displacement rigid-plastic analysis which included strain rate sensitivity Bodner and Symonds [9] estimated t_f to be 0.064 s. Ting [10], using a rigid-viscoplastic material model with small displacement assumptions calculated t_f to be 0.065 s. Lee and Martin [11], using a rigid-viscoplastic material model and the "piecewise stationary mode" technique, and Symonds [1] using the mode approximation technique with a matched viscous constitutive relation, estimated t_f to be 0.064 s and 0.066 s respectively. Both analyses neglected geometric effects. A more comprehensive comparison can be made with the results obtained by Griffin and Martin [7] who used a matched viscous stress—strain rate relation together with the instantaneous mode technique, a large displacement method of analysis based on the mode approximation technique. They obtained a tip transverse displacement of 0.32 m, and estimated the total time of deformation to be 0.065 s. The present direct method of analysis used a homogeneous viscous constitutive relation matched on both shape and intercept at the maximum initial curvature rate, which resulted in the power $n' = 11.04$, and a stress matching factor $\mu = 1.99$. The transverse displacement at the tip was 0.303 m, and t_f was 0.065 s. Computer plots of the displaced shape at successive time intervals for this analysis are given in Fig. 3, together with a physical description of the cantilever. The deformed shape at successive time intervals, shown in Fig. 3, compares excellently with the sequence of photographs of the deforming cantilever given by Bodner and Symonds [9].

Figure 4 shows a comparison between the results obtained by the present technique and test results by Wegener [12] for rectangular steel portal frames subjected to uniform sideways impulse along the length of one column. Matching was performed at the maximum initial curvature

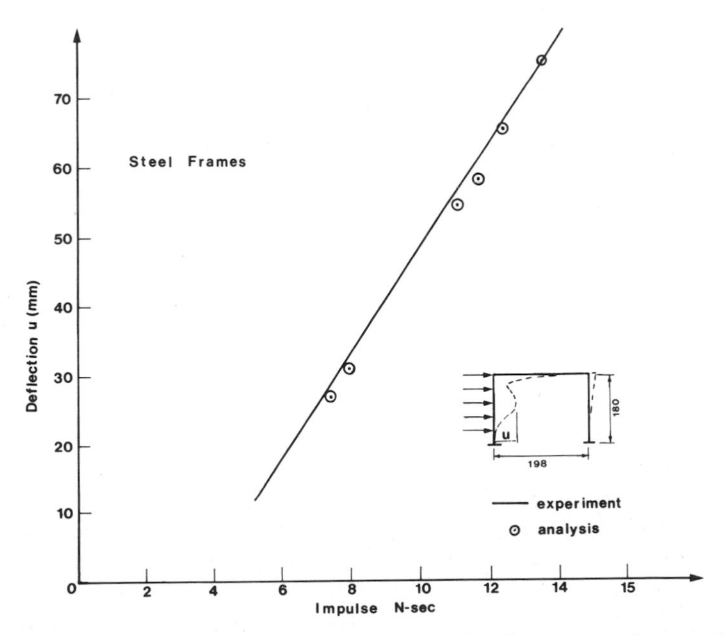

Fig. 4. Plot of deflection versus impulse for rectangular steel frames subjected to uniform sideways impulse.

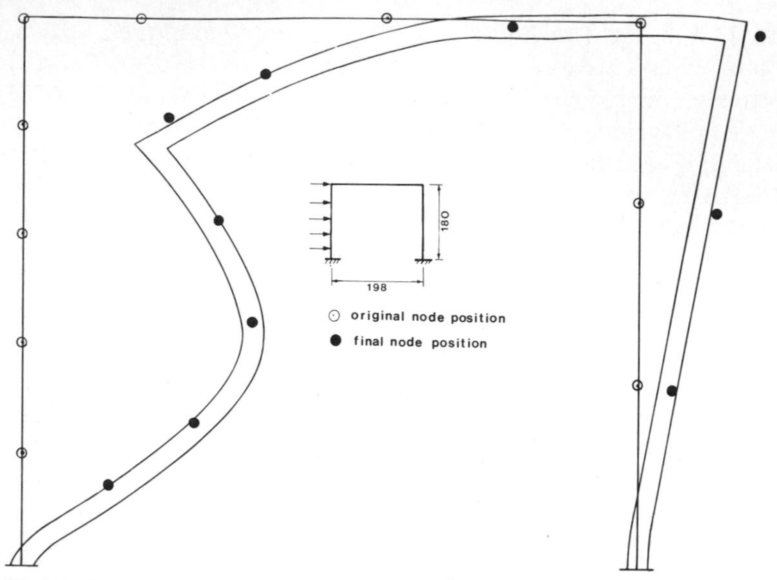

Fig. 5. Deformed shape of rectangular steel portal frame subjected to a uniform sideways impulse of 13.47 N s; mass density $\rho = 7850\ \text{kg/m}^3$, $\sigma_0 = 350$ MPa, section depth $h = 6$ mm, section breadth $b = 20$ mm, $\dot{\epsilon}_0 = 40/\text{s}$, $n = n' = 5$, $\mu = 2.23$.

rate on intercept alone, with $n = n' = 5$. The stress factor μ was found to range between 2.20 for the highest impulse and 2.37 for the lowest impulse. As shown in Fig. 4, the analyses agree excellently with experimental values, even with the crude matching scheme used. In Fig. 5 the deformed shape of a typical frame is shown, together with the original and deformed nodal position, and a physical description of the frame.

A similar series of analyses was performed on rectangular aluminum frames, and the results were compared with the test data obtained by Hashmi and Al-Hassani [13]. The frame is shown in Fig. 6, with its geometric and material properties. Curve 1 in Fig. 7 shows results of the tests performed by Hashmi and Al-Hassani. The results of analyses using a homogeneous viscous relation which is matched on intercept alone at the maximum initial curvature rate, with $n = n' = 4$, is given by curve 3, Fig. 7. Considerable numerical difficulties were encountered, however, when attempting analyses using a constitutive relation matched on intercept *and* shape. Such a matching procedure required that n' exceed 12. For n' greater than 8, the initial moments and axial forces required by the direct analysis procedure could not be calculated using the numerical technique outlined in Section 2. Results were obtained by using a value of n' as high as would permit a solution; this was found for these examples to lie in the range between 6.0 and 8.0. These results are shown by curve 2 in Fig. 7. In Fig. 8, computer plots of the displaced shape at successive time intervals for a typical side loaded aluminum frame are given.

Similar numerical difficulties were encountered in the analyses of

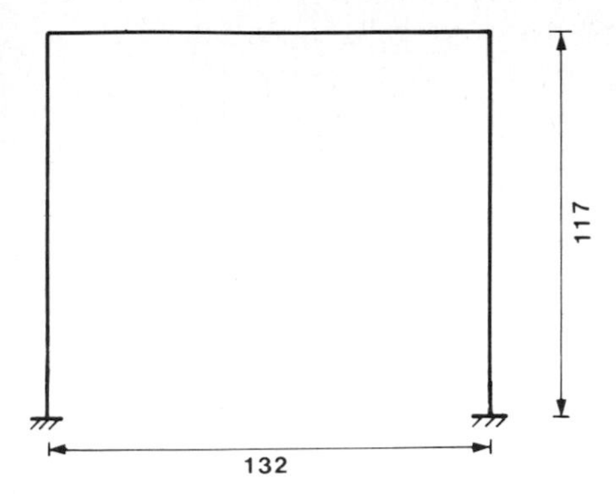

Fig. 6. Aluminum rectangular portal frame; mass density $\rho = 2670\,\text{kg/m}^3$, $\sigma_0 = 83\,\text{MPa}$, section depth $h = 0.91$ mm, section breadth $b = 12.7$ mm, $\dot{\epsilon}_0 = 6500/\text{s}$, $n = 4$.

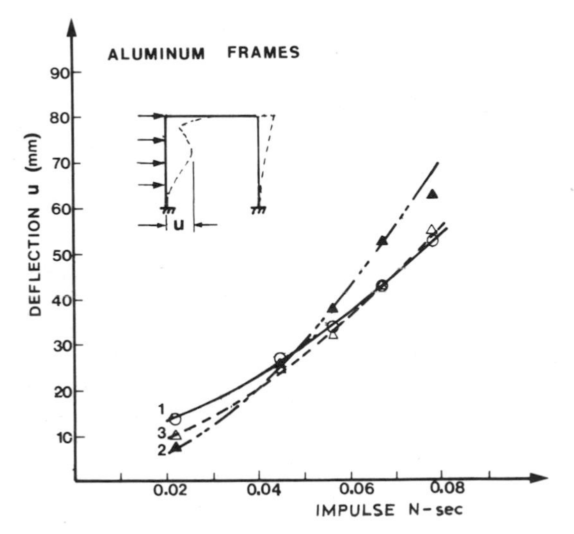

Fig. 7. Plot of deflection versus impulse for rectangular aluminum portal frames subjected to uniform sideways impulse. Curve 1: test results by Hashmi and Al-Hassani [13]. Curve 2: analyses using homogeneous viscous relation matched on intercept alone but with n' between 6 and 8. Curve 3: analyses using homogeneous viscous relation matched on intercept alone with $n = n' = 4$.

aluminum rectangular frames of the type shown in Fig. 6, subjected to a uniform impulse over half of the beam length. In Fig. 9, the test data of Hashmi and Al-Hassani [13] are shown by curve 1. Curve 2 shows the results obtained by the present analyses when matching was performed on intercept alone at the maximum curvature rate, with $n = n' = 4$. Much better correlation with experimental results was obtained when a higher n' was used. In this series of analyses, solutions were achieved for n' between 8 and 10 and are shown by curve 3 in Fig. 9. Plots of the displaced shape

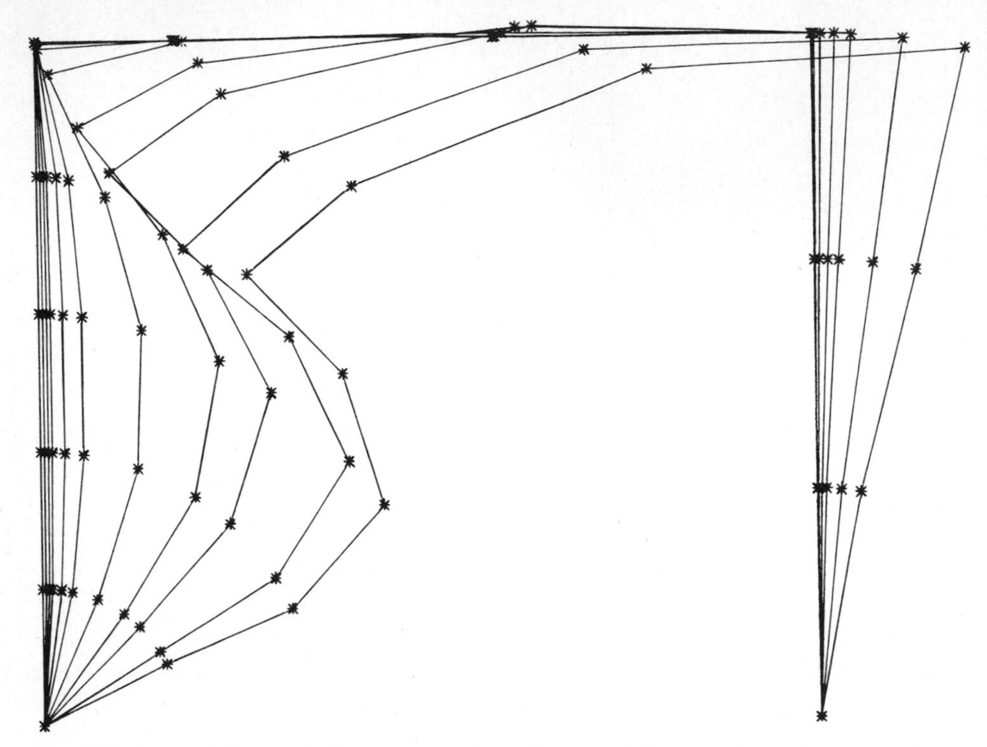

Fig. 8. Deformed shape of aluminum rectangular portal frame subjected to a uniform sideways impulse of 0.0757 N s, at successive time intervals.

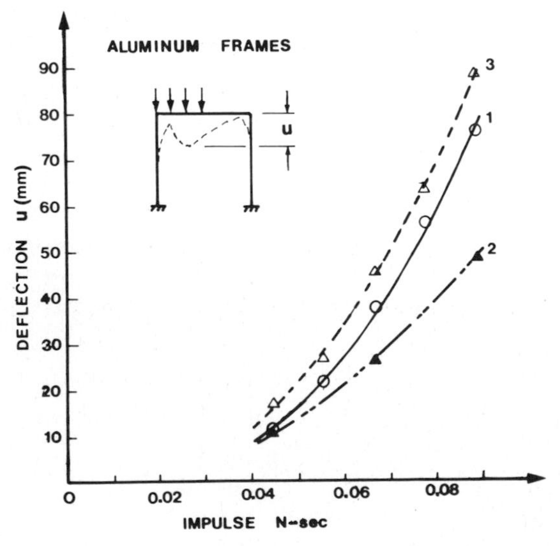

Fig. 9. Plot of deflection versus impulse for aluminum rectangular portal frames subjected to a uniform impulse over half the length of the beam. Curve 1: test results by Hashmi and Al-Hassani [13]. Curve 2: analyses using homogeneous viscous relation matched on intercept alone with $n = n' = 4$. Curve 3: analyses using homogeneous viscous relation matched on intercept alone but with n' between 8 and 10.

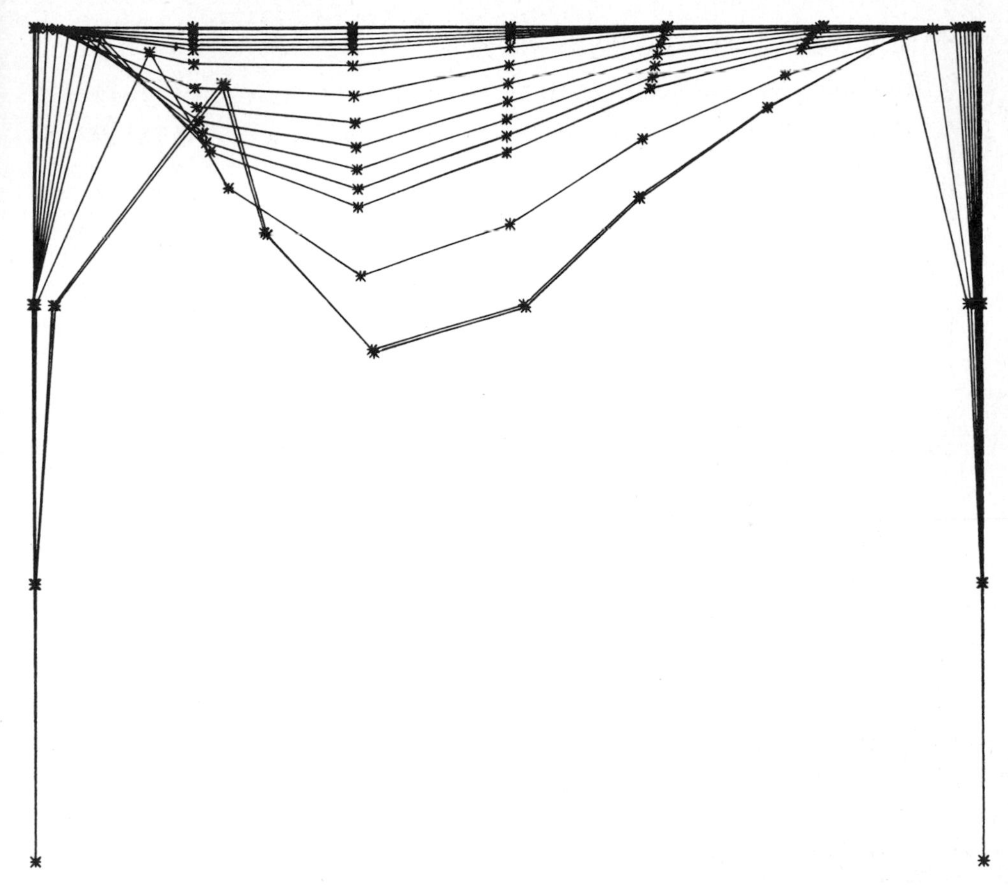

Fig. 10. Deformed shape of aluminum rectangular portal frame subjected to a uniform impulse of 0.066 N s along half the length of the beam, at successive time intervals.

at successive time intervals for the analysis of a typical frame in this series of analyses is shown in Fig. 10.

All analyses were performed on the University of Cape Town UNIVAC 1100/81 computer. For the analysis of the cantilever with the full matching scheme described above, the CPU time was 2 min 43 s. For the analyses of the second series of aluminum frames described above, the CPU time ranged between 10 min 54 s for the lowest impulse ($n' = 10$) and 32 min 13 s for the highest impulse ($n' = 8$). The CPU time increases substantially with increasing n', as for higher n' more iterations are required to obtain a solution. For example, for the frame in this series of analyses which was subjected to an impulse of 0.0667 N s, the CPU time with $n' = 8$ was 21 min 15 s, and with $n' = 8.8$ was 35 min 36 s.

6. Conclusions

The numerical procedure outlined in this paper is a reasonably efficient method for analysing ductile metal beam and frame structures subjected to large impulses. Good agreement with experimental results can be obtained if the homogeneous viscous relation is suitably matched to the rigid-viscoplastic constitutive equation. It is also clear that, if a solution cannot be obtained with matching on both intercept and slope at the maximum initial strain rate, a compromise is possible with n' chosen as large as possible for numerical stability; the larger the value of n' the better the correlation with experimental results.

The direct integration procedure presented in this paper is far more costly than solutions obtained using the mode approximation technique, and show conclusively that where the mode approximation technique is appropriate it should be used. Where the deformations are not modal, however, the use of the direct integration technique cannot be avoided.

As remarked earlier, the question of matching strategy is still an open question. Solutions are sensitive to the choice of μ and ν, and unambiguous methods for the choice of these factors should form the subject of further research.

Acknowledgement

The support of the Council for Scientific and Industrial Research, South Africa, is acknowledged.

References

1 P.S. Symonds, Approximation techniques for impulsively loaded structures of rate sensitive plastic behaviour. SIAM J. Appl. Math., 25 (1973) 462—473.
2 J.B. Martin and P.S. Symonds, Mode approximations for impulsively loaded rigid-plastic structures. J. Eng. Mech. Div. ASCE, 92 (1966) 43—66.
3 P.S. Symonds and C.T. Chon, Approximation techniques for impulsive loading of structures of time-dependent plastic behaviour with finite deflections. In: J. Harding (Ed.), Proc. Conf. Mechanical Properties of Materials at High Rates of Strain, Inst. Phys. Conf. Ser. No. 21, London and Bristol, 1974, pp. 299—316.
4 P.S. Symonds and C.T. Chon, Large viscoplastic deflections of impulsively loaded plane frames. Int. J. Solids Struct., 15 (1979) 15—31.
5 P.S. Symonds and J.L. Raphanel, Large deflections of impulsively loaded plane frames — extensions of the mode approximation technique. In: J. Harding (Ed.), Proc. Conf. Mechanical Properties of Materials at High Rates of Strain, Inst. Phys. Conf. Ser. No. 47, London and Bristol, 1979, pp. 277—282.
6 P.D. Griffin and J.B. Martin, Finite element analysis of dynamically loaded homogeneous viscous beams. J. Struct. Mech., 10 (1982) 93—115.
7 P.D. Griffin and J.B. Martin, Geometrically nonlinear mode approximations for impulsively loaded homogeneous viscous beams and frames. Int. J. Mech. Sci., 25 (1983) 15—26.

8 P.S. Symonds, Approximation techniques for impulsively loaded plastic structures. Course on Dynamics of Plastic Structures, Centre Internationale Des Sciences Mecanique, Udine, Italy, 1979.

9 S.R. Bodner and P.S. Symonds, Experimental and theoretical investigation of the plastic deformation of cantilever beams subjected to impulsive loading. J. Appl. Mech., 29 (1962) 719—728.

10 T.C.T. Ting, The plastic deformation of a cantilever beam with strain-rate sensitivity under impulsive loading. J. Appl. Mech., 31 (1964) 38—42.

11 L.S.S. Lee and J.B. Martin, Approximate solutions of impulsively loaded structures of a rate sensitive material. ZAMP, 21 (1970) 1011—1032.

12 R.B. Wegener, Impulsive Loading Tests on Portal Frames. Tech. Rep. 17, Non-linear Structural Mechanics Research Unit, University of Cape Town, April 1982.

13 S.J. Hashmi and S.T.S. Al-Hassani, Large deflection response of square frames to distributed impulsive loading. Int. J. Mech. Sci., 17 (1975) 513—523.

Dynamic Analysis of Elastic-Plastic Structures by Mode-Superposition and Equivalent Plastic Loads

W.K. HO and T.H. LIN

Mechanics and Structures Department, University of California, Los Angeles (U.S.A.)

Abstract

A method is shown for the dynamic analysis of elastic-plastic structures by mode superposition and treating plastic strain as additional applied loads. This method reduces the elastic-plastic analysis to elastic analysis of the same structure with additional forces. In the elastic analysis, the modes of vibration and stiffness matrix are elastic and remain the same in all time steps. This eliminates the calculation of a new stiffness matrix for each time step and hence greatly reduces the amount of numerical computation. Dynamic elastic-plastic responses of an unsymmetrical portal frame were calculated as an illustration of this method. The elastic modes and stiffness matrix were calculated by the finite element method.

Introduction

The equation of motion of an elastic body may be written as

$$[M]\{\ddot{U}\} + [K]\{U\} = \{Q\} \tag{1}$$

where $[M]$ and $[K]$ are mass and stiffness matrices, $\{Q\}$ the external load vector, $\{U\}$, $\{\ddot{U}\}$ are the displacement and acceleration vectors of a finite element system. The equation is generally numerically solved by two methods: direct integration and mode superposition [1]. In direct integration, the equation is integrated step by step. Instead of trying to satisfy the equations at all instants of time, this procedure satisfies the equations at discrete time intervals Δt apart. An arbitrary variation of displacements, velocity and acceleration within each time interval Δt is assumed. The amount of calculation to cover a given length of time increases as Δt decreases. Errors in direct integration often arise when the time step is too large. In the mode superposition method, the equations of motion are decoupled. The decoupled differential equations are readily solved by Duhamel's integral [2]. Superposition of the solutions of all modes gives the total response of the system. The main difference between direct integration and mode superposition is that in the former, the same time step is used for all modes and all modes are included in the calculation while in the latter, we can consider only those modes which contribute to the primary response of the structure. Consider the case, where we are only interested in integrating the first p of the modes. In mode superposition we can use a time step of $T_p/10$ where T_p is the period of the pth mode. However, in direct integration if we use the same time step, the response in the higher modes is integrated with this same time step and this time step may be large compared to the natural period, introduce a serious error and thus render the integration of the lower modes worthless. As indicated by Bathe and Wilson in their excellent book "Numerical Methods in Finite Element Analysis" [1], finite element analysis approximates the lowest frequencies best, and has little or no accuracy in approximating the higher frequencies and mode shapes. The finite element mesh is often chosen in such a way that all important exact frequencies and mode shapes are well approximated, and thus the solution needs only to include the response in these modes. This can be obtained by using mode superposition analysis by considering only the important modes of the finite element system. Hence, the mode superposition method has an inherent advantage over direct integration of excluding the non-essential high frequency modes. This enables the use of larger time steps. Much work has been published in this extension of the above two methods to dynamic nonlinear analysis of structures. Wu and Witmer [3] have indicated that for such structures, the stiffness matrix and the normal modes vary with time. They concluded that the mode superposition method is impractical. Clough and Penzien [4] stated in their well-known book, "Dynamics of Structures" that probably the most powerful technique for nonlinear analysis is the step-by-step integration procedure. In this procedure, the nonlinear nature of the structure is accounted for by using new properties appropriate to the current deformed state at the beginning of each time increment. The displacement and velocity at the end of one computational interval are used as the initial conditions for the next interval. Thus this process is continued step by step from the initial instant to any length of time approximating the nonlinear behavior as a sequence of successively changing linear systems.

In the calculation of the strain field of an elastic-plastic body, the plastic strain gradient has been shown [5] to have the same effect as the applied body force. Using this approach for dynamic analysis, the plastic strain is accounted for as an additional force applied to each element. The stiffness matrix in the calculation is elastic and the normal modes in the numerical integration remain the same. Since the stiffness matrix needs not be calculated for each time increment, the amount of numerical calculation in mode superposition is greatly reduced. In 1981, Liu and Lin [6] used this approach to analyze the dynamic responses of elastic-plastic simply supported beams and plates. In these structures the

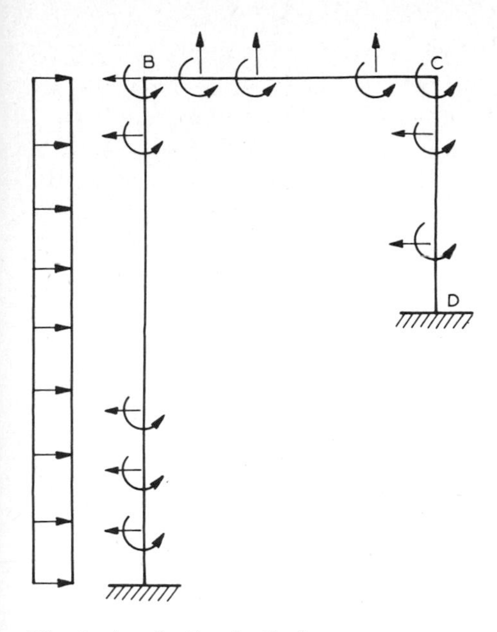

Fig. 1. An elastic-plastic frame.

dynamic elastic solutions are available and hence the calculation of eigenvectors is not required. The dynamic elastic solutions of structures are generally not available and these eigenvectors are to be calculated by the method of finite elements.

The present paper aims to show how the mode superposition method can be applied to such structures. The structure chosen for this study is an unsymmetric rigid frame as shown in Fig. 1. The frame is divided into a number of beam elements; Fig. 2. The element stiffness and consistent mass matrices can be written as:

$$[K] = \frac{EI}{l^3}\begin{bmatrix} 12 & 6l & -12 & 6l \\ 6l & 4l^2 & -6l & 2l^2 \\ -12 & -6l & 12 & -6l \\ 6l & 2l^2 & -6l & 4l^2 \end{bmatrix} \tag{2}$$

$$[M] = \frac{\rho l}{420}\begin{bmatrix} 150 & 22l & 54 & -13l \\ 22l & 4l^2 & 13l & -3l \\ 54 & 13l & 156 & -22l \\ -13l & -3l^2 & -22l & 4l^2 \end{bmatrix} \tag{3}$$

These beam element matrices are assembled into the global mass and stiffness matrices.

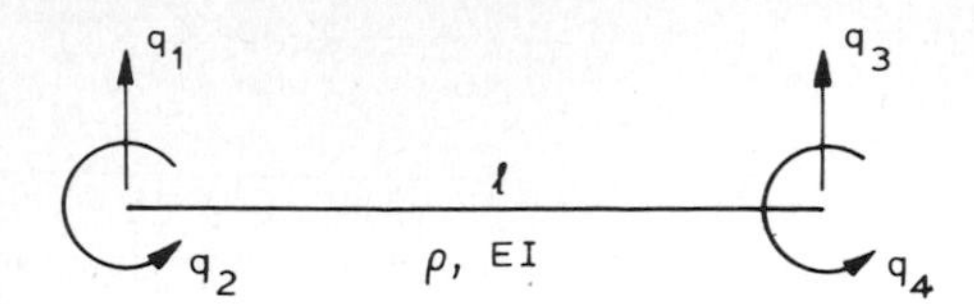

Fig. 2. Standard notation for a beam element used in the direct stiffness method. q_i are nodal displacements, $i = 1, 2, 3, 4$; l is element length; ρ is element mass per unit length; and EI is elastic flexural rigidity.

Dynamic influence coefficient

The eigenvalues and eigenvectors are found from the equation

$$[K] - \omega^2 [M] \{U\} = 0. \tag{4}$$

Let $[\phi]$ be the modal matrix, which is normalized to give

$$[\phi]^{\mathrm{T}} [M] [\phi] = [I] \tag{5}$$

where $[\phi] = [\phi_1; \phi_2; \ldots; \phi_i; \ldots; \phi_n]$ and ϕ_i is the eigenvector of the ith mode, with

$$\{U\} = [\phi] \{q\}$$

where $\{q\}$ is the normal coordinates. Let $[\Omega^2]$ be the diagonal matrix with ω_i^2 as the elements

$$[\phi]^{\mathrm{T}} [K] [\phi] = [\Omega^2]. \tag{6}$$

Consider a generalized force of a unit step function applied at the jth generalized coordinate. $Q_j = H(t)$ where $H(t)$ is the Heaviside step function

$$[M] [\phi] \{\ddot{q}\} + [K] [\phi] \{q\} = H(t) \begin{Bmatrix} 0 \\ 0 \\ \cdot \\ \cdot \\ \cdot \\ 1 \\ 0 \\ 0 \end{Bmatrix} \tag{7}$$

$$[I]\{\ddot{q}\} + [\Omega^2]\{q\} = H(t)\begin{Bmatrix} \phi_j^1 \\ \phi_j^2 \\ \cdot \\ \cdot \\ \cdot \\ \phi_j^n \end{Bmatrix} \tag{8}$$

where the superscript of ϕ denotes the eigenvector number. Written in index notation,

$$\ddot{q}_j^i + \Omega_i^2 q_j^i = H(t)\phi_j^i\,, \qquad i = 1, 2 \ldots n, \tag{9}$$

$$q_j^i(t) = \frac{\phi_j^i}{\omega_i^2}(1 - \cos\omega_i t) \tag{10}$$

where j denotes the global degree of freedom at which the unit step function of load is applied, and i denotes the ith mode shape.

$$\{U_j\} = [\phi]\{q_j^i\}. \tag{11}$$

Let the ith component of the vector $\{U_j\}$ be denoted by u_j^i

$$u_j^i(t) = \sum_{m=1}^{N} \frac{\phi_i^m \phi_j^m}{\omega_m^2}(1 - \cos\omega_m t) \tag{12}$$

Thus $u_j^i(t)$ is the displacement at the ith degree of freedom at time t due to a unit step force applied at the jth degree of freedom at time $t = 0$. Let time t be expressed as $k\Delta t$. $u_j^i(t)$ is written as $G(i, j, k)$, which is referred to as the dynamic influence coefficient. The displacement in the ith degree of freedom at $t = p\Delta t$ is denoted by $u(i, p)$ due to a unit step load applied at the jth degree of freedom at a preceding time instant $t = k\Delta t$ $(k < p)$.

An arbitrary force function at any node j at time $t = m\Delta t$ can be approximated by a sequence of finite rectangular steps as shown in Fig.3. The displacement at node i at $t = m\Delta t$ due to this force Q^j is

$$u(i, m) = \sum_{k=0}^{m} G(i, j, m-k)\Delta Q(j, k)$$

and due to forces at all nodes is then

$$u(i, m) = \sum_{k=0}^{m}\sum_{j=1}^{N} G(i, j, m-k)\Delta Q(j, k) \tag{13}$$

where N is the number of nodal forces considered. This gives the elastic response of the structure.

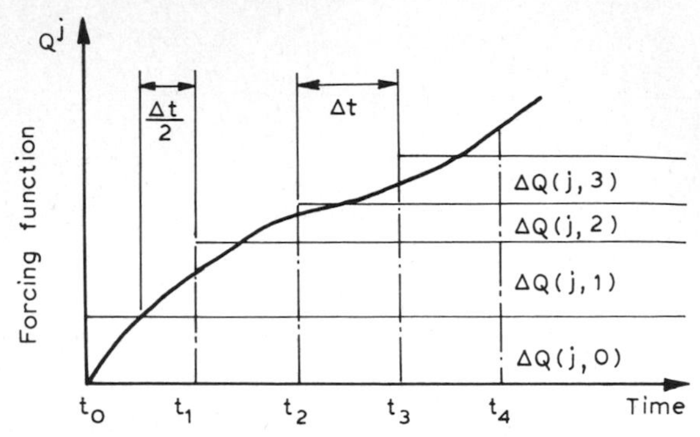

Fig. 3. Approximation of forcing function.

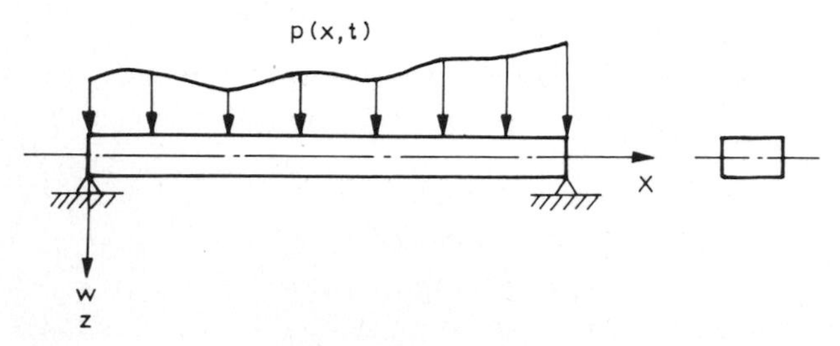

Fig. 4. Beam of uniform cross-section subjected to dynamic loading.

Equivalent load caused by plastic strain

The fiber stress in a beam is

$$\sigma = E(e - e^{\mathrm{p}})$$

where E is Young's modulus, e, the total strain and e^p the plastic strain. Consider a beam of uniform cross-section as shown in Fig. 4.

The equation of motion for a length dx of the beam is given by

$$\frac{\partial^2 M}{\partial x^2} + p(x, t) = \rho \frac{\partial^2 w}{\partial t^2} \tag{14}$$

where M is the bending moment, ρ is the mass per unit length of the beam, w the lateral displacement. With Euler--Bernoulli's assumption of planes remaining plane in bending,

$$e = -z \frac{\partial^2 w}{\partial x^2}, \qquad M = -EI \frac{\partial^2 w}{\partial x^2} - \int E e^{\mathrm{p}} z \,\mathrm{d}A. \tag{15}$$

From (14) and (15), we have

$$EI \frac{\partial^4 w}{\partial x^4} + \rho \frac{\partial^2 w}{\partial t^2} = p(x, t) + \bar{p}(x, t) \tag{16}$$

where

$$\bar{p}(x, t) = -\frac{\partial^2}{\partial x^2} \int_A E e^{\mathrm{p}} z \mathrm{d}A. \tag{17}$$

$\bar{p}(x, t)$ can be considered as an additional applied load causing the displacement w.

Equation (16) corresponds to eqn. (1) with the first term corresponding to $[K]\{U\}$, the second term to $[M]\{\ddot{U}\}$ and the two terms on the right side corresponding to $\{Q\}$. The force $\{Q\}$ in (1) and (13) may be written as the sum of the actual applied force $\{Q\}$ and the equivalent force $\{\bar{Q}\}$ due to plastic strain. This is similar to the initial strain formulation in the finite element method [7]. Equation 13 is then written as

$$u(i, m) = \sum_{k=1}^{m} \sum_{j=1}^{N} G(i, j, m-k)[\Delta Q(j, k) + \Delta \bar{Q}(j, k)] \tag{18}$$

$\Delta \bar{Q}(j, m)$ depends on $u(i, m)$ and $u(i, m)$ depends on $\Delta \bar{Q}(j, m)$, hence an iteration is used so that eqn. 18 is satisfied at each step.

Numerical results

The stress—strain relationship is taken to be of the bilinear form as shown in Fig. 5. Using the method described in the previous sections, a frame of this bilinear stress—strain relation was analyzed. The dimensions of the frame are shown in Fig. 6 and the material properties were taken

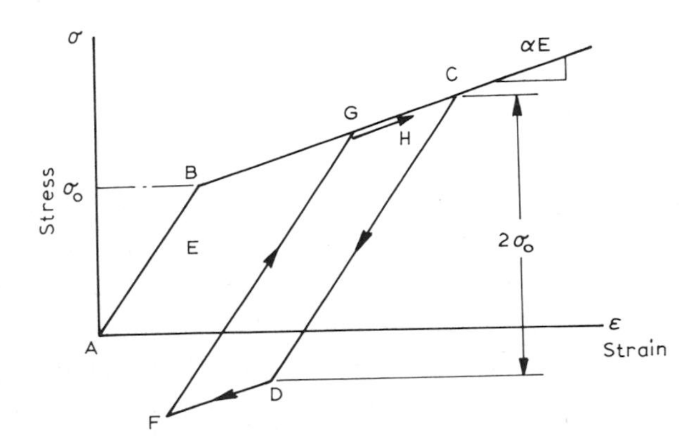

Fig. 5. Bilinear stress—strain curve.

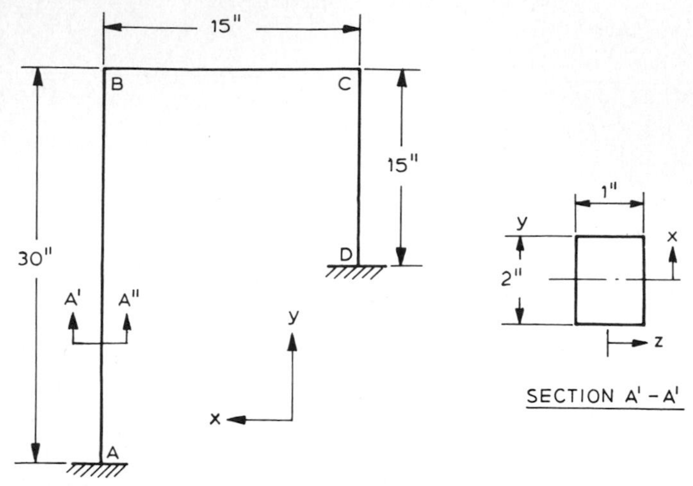

Fig. 6. Frame dimensions and properties.

as follows: Modulus of elasticity $E = 30 \times 10^6$ psi; Yield stress $\sigma_0 =$ 50,000 psi; Poisson's ratio $\nu = 0.30$; Mass density $\rho = 0.733 \times 10^{-3}$ lb sec^2/in^4.

This frame was subject to a uniformly distributed step load applied on the left column AB. Ten modes were used. Ten lowest frequencies were calculated for this elastic structure by dividing the frame into 20 elements, 60 elements and 75 elements. It was found that the agreements between the 60 and 75 element were very good. The fundamental period of the frame was found to be 5.98×10^{-3} s. A time increment Δt of 1.5×10^{-4} s was used. The step forcing function is expressed in terms of the static yield load p_0 at which yielding starts in the column. This yielding load is about 85% of the static collapse load. The lateral deflection at B is expressed in terms of the static deflection δ at B under load p.

To calculate the plastic strains at different points in a cross-section of the beam element, the cross-section was divided into ten grids on each side of the centroidal axis. Then the integral $\int \Delta e^p z dA$ was evaluated to calculate $\Delta \bar{p}$ in eqn. 16 or $\Delta \bar{Q}$ in eqn. 18.

The dynamic responses at the upper left corner of the frame under different uniform step loads for an ideally plastic frame were calculated and given in Fig. 7. The plastic effect is small until the step load reaches $0.70\, p_0$. At this load the plastic effect becomes significant. Plastic hinges occur near the two fixed ends. As the load is further increased, another plastic hinge occurs also at the upper right corner C. Similar dynamic responses at the upper left corner of the frame with strain-hardening $\alpha = 1/2$ (Fig. 5) were calculated and plotted in Fig. 8 with the ideally plastic and purely elastic cases.

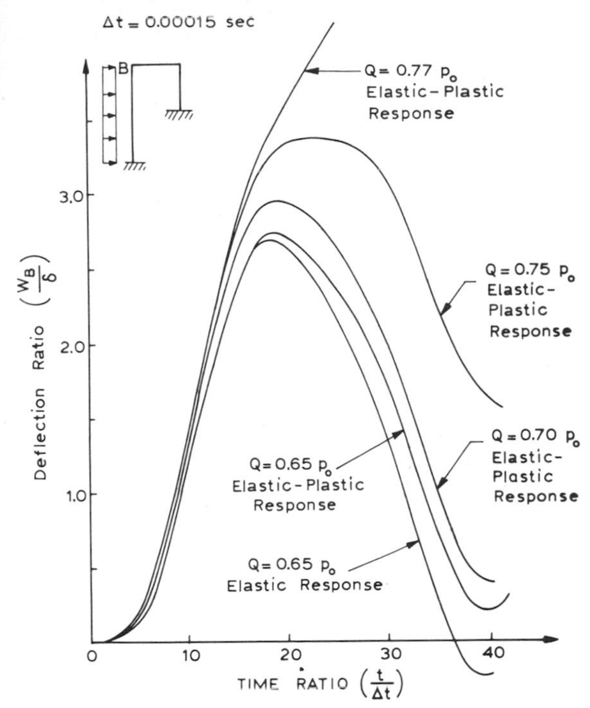

Fig. 7. Response history of an elastic perfectly-plastic frame.

Conclusion

This paper gives a method to calculate the dynamic elastic-plastic response of structures by mode superposition. Plastic strain is considered as an additional applied force. The stiffness matrix and the natural modes in mode superposition calculation are elastic and hence remain the same at all time steps. This eliminates the calculation of a new stiffness matrix at each time step and hence reduces greatly the numerical calculations. This method has the inherent advantage of excluding non-essential high-frequency modes over direct integration and will in many cases predict responses more accurately than direct integration. A dynamic elastic-plastic analysis of a frame is given as an illustration.

References

1 K. Bathe and E.L. Wilson, Numerical Methods in Finite Element Analysis. Prentice-Hall, 1976, pp. 308—344.

2 L. Meirovitch, Analytical Methods in Vibrations. MacMillan Company, 1967, p. 15.

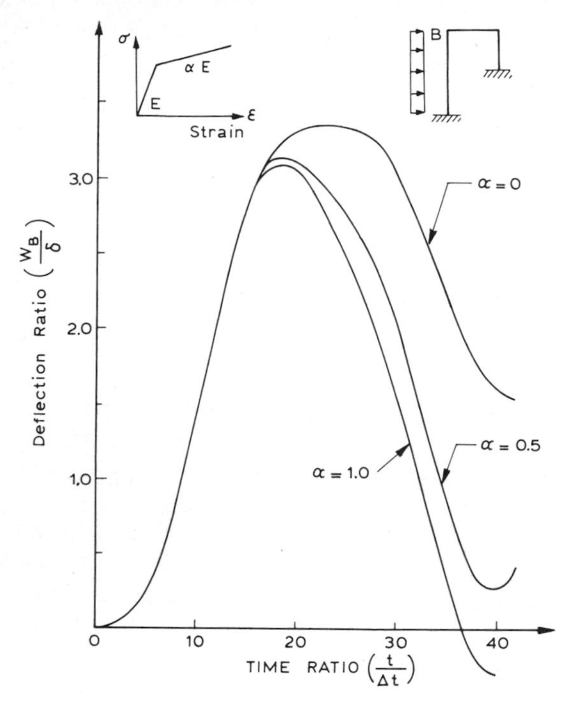

Fig. 8. Elastic-plastic responses of various strain hardening coefficients α.

3 R. Wu and E.A. Witmer, Finite-Element Analysis of Large Transient Elastic-Plastic Deformations of Simple Structures, with Application to the Engine Rotor Fragment Containment/Deflection Problem. NASA CR-120886, 1972.
4 R.W. Clough and J. Penzien, Dynamics of Structures. McGraw Hill Book Co., 1975, pp. 118—128.
5 T.H. Lin, Theory of Inelastic Structures. John Wiley & Sons, 1968, pp. 43—55, 108.
6 S.C. Liu and T.H. Lin, Elastic-plastic dynamic analysis of structures using known elastic solutions. Earthquake Engr. and Structural Dynamic, 7 (1979) 147—159.
7 O.C. Zienkiewiez and Y.K. Cheung, The Finite Element Method in Structural and Continuum Mechanics. McGraw Hill Book Co., 1968, p. 77, 108.

Simple Examples of Complex Phenomena in Plasticity

PHILIP G. HODGE, Jr.

Department of Aerospace Engineering and Mechanics, University of Minnesota, Minneapolis, MN 55455 (U.S.A.)

Abstract

In 1951 D.C. Drucker presented a classic example of a simple three-bar system in which a single monotonically-increasing external load produced an actual plastic stress reversal with one bar yielding first in tension and then in compression. This example is reviewed and built upon. Other simple examples illustrate general loading, unloading, and reloading of elastic/plastic structures, shakedown, and lack of uniqueness in contained plastic deformation.

1. Introduction

There are many complications in applications of plasticity theory, particularly in comparison with the linear theory of elasticity. If these complications are encountered for the first time in a "real" problem with large amounts of analysis and computation even for an elastic material, the peculiar features of plastic behavior may be masked, and the probability of undetected errors is greatly increased. For this reason, there is an obvious advantage to discussing plasticity first in terms of simple structures, the simpler the better. Of course, if a problem is too simple, then some essential features of plasticity may not appear. However, one of the surprising features of plasticity theory is that so many of its important features can be illustrated with problems which are easily comprehended by students in a beginning class in statics.

In the present paper we shall illustrate several of the features of plasticity theory with two very simple three-bar statically indeterminate trusses, as shown in Figs. 1 and 8. Specifically, in Sec. 2 we will use the truss of Fig. 1 to show the difference between hardening and perfectly-plastic materials, and the effect of including or neglecting elastic strain components. Section 3 extends these results to unloading and reloading. We then examine some undesirable effects which may occur when loads are applied repeatedly. The next two sections illustrate types of behavior which are drastically different from those encountered in elasticity. In Sec. 5 we look at the truss in Fig. 8 and show that under a monotonically increasing single load a bar may yield first in tension and then in compression. Thus proportional "loading" does not by any means guarantee proportional "stressing". Section 6 returns to the truss in Fig. 1 and shows that a boundary-value problem that is "well-posed" in the theory of elasticity may have a non-unique solution in plasticity. The final section of the paper will quote analytic and numerical solutions to more complex problems where the same types of responses are observed.

2. Loading of different plasticity models

The defining equations for any problem in continuum or structural mechanics come from three sources: statics, kinematics, and constitutive. For the truss in Fig. 1, the two statics equations are obtained from horizontal and vertical equilibrium of point D:

$$(F_1 - F_3)/\sqrt{2} = H \qquad (F_1 + F_3)/\sqrt{2} + F_2 = V \tag{1a, b}$$

where F_i is the tensile force in bar i. The kinematics expresses the elongation e_i of each bar in terms of the displacements u and v of point D:

$$e_1 = (u + v)/\sqrt{2} \qquad e_2 = v \qquad e_3 = (-u + v)/\sqrt{2} \tag{2}$$

Obviously, we are using a strictly linear theory in writing eqns. (1) and (2).

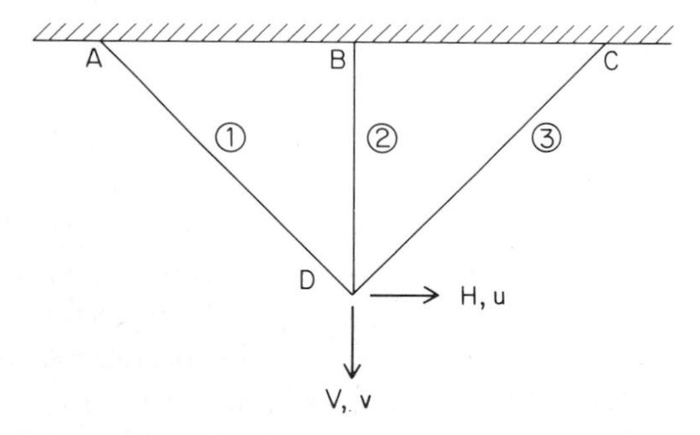

Fig. 1. Three-bar truss.

The force and elongation in a bar may be related to its stress and strain by

$$F_i = A_i \sigma_i \qquad e_i = L_i \epsilon_i \tag{3a}$$

where A_i and L_i are the area and length, respectively. Moduli and yield stresses are defined in Fig. 2, and corresponding bar quantities are

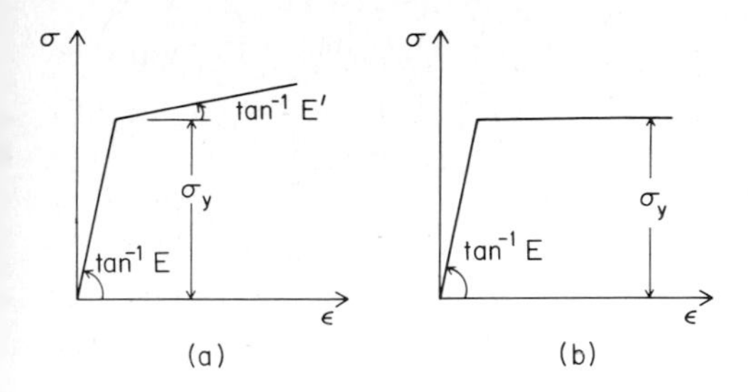

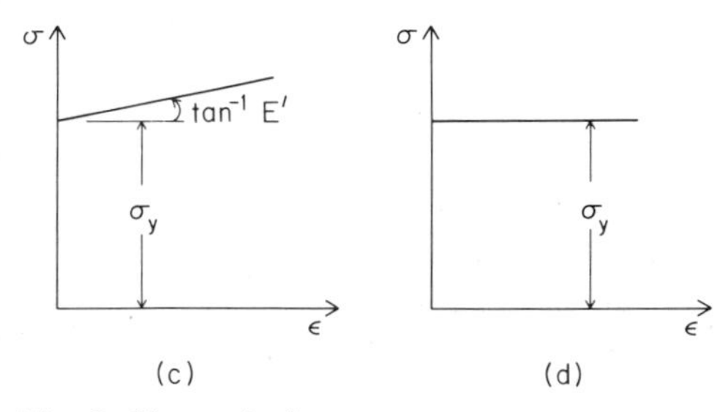

Fig. 2. Stress-strain curves.
(a) Elastic/strain-hardening (*E/SH*);
(b) Elastic/perfectly-plastic (*E/PP*);
(c) Rigid/strain-hardening (*R/SH*);
(d) Rigid/perfectly-plastic (*R/PP*).

$$Y_i = A_i\sigma_{yi} \qquad k_i = A_iE_i/L_i \qquad k_i' = A_iE_i'/L_i \qquad r_i = k_i'/k_i \tag{3b}$$

Therefore, the constitutive behavior of bar i can be summarized by

$$E/ \qquad 0 < F_i < Y_i \qquad F_i = k_ie_i \tag{4a}$$

$$E/SH \qquad Y_i < F_i \qquad F_i = k_i'e_i + Y_i(1-r_i) \tag{4b}$$

$$E/PP \qquad e_i > Y_i/k_i \qquad F_i = Y_i \tag{4c}$$

$$R/ \qquad 0 < F_i < Y_i \qquad e_i = 0 \tag{4d}$$

$$R/SH \qquad F_i > Y_i \qquad F_i = k_i'e_i + Y_i \tag{4e}$$

$$R/PP \qquad e_i > 0 \qquad F_i = Y_i \tag{4f}$$

for the various piecewise linear models.

Let the three bars of the truss in Fig. 1 all have the same moduli, area, and yield stress. Then the bar stiffnesses are

$$k_1 = k_3 = k/\sqrt{2}\,; \qquad k_2 = k\,; \qquad k = AE/L \tag{5}$$

We consider a deformation-controlled loading program in which $u = 0$ and v is slowly increased. It then follows from eqns. (2) that

$$e_1 = e_3 = v/\sqrt{2} \qquad e_2 = v \tag{6}$$

Regardless of which of eqns. (4) apply, we see that $F_1 = F_3$, whence $H = 0$ and eqn. (1b) can be written

$$V = \sqrt{2}F_1 + F_2 \tag{7}$$

We first examine the *E/SH* material. For v sufficiently small all bars will be elastic so that (4a) applies. Thus

$$\text{Stage 1:} \quad F_1 = F_3 = kv/2; \quad F_2 = kv; \quad V = kv(1 + 1/\sqrt{2}) \tag{8}$$

This stage will reach its limit when bar 2 yields at $kv = Y$.

In the next stage bar 2 is plastic and is governed by eqn. (4b) but (4a) still applies to bars 1 and 3. Thus eqns. (8) are replaced by

$$\begin{aligned} \text{Stage 2:} \quad F_1 &= F_3 = kv/2; \qquad F_2 = k'v + (1-r)Y \\ V &= kv(1/\sqrt{2} + r) + Y(1-r) \end{aligned} \tag{9}$$

When $kv = 2Y$ bars 1 and 3 reach yield and the stage ends. As v is still further increased, all bars will be plastic and the solution from now on will be

$$\begin{aligned} \text{Stage 3:} \quad F_1 &= F_3 = k'v/2 + Y(1-r); \qquad F_2 = k'v + Y(1-r) \\ V &= k'v(1/\sqrt{2} + 1) + (\sqrt{2} + 1)Y(1-r) \end{aligned} \tag{10}$$

The solid curve in Fig. 3 shows the resulting load-deflection curve defined by eqns. (8—10).

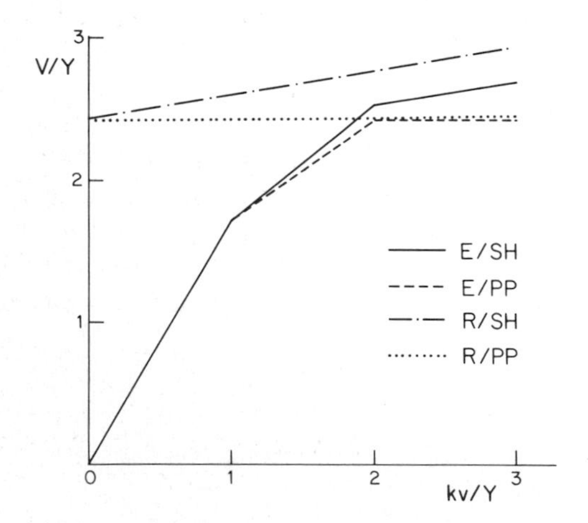

Fig. 3. Load—deformation curves.

Solutions for the other models in Fig. 2 are obtained by a similar analysis or by taking appropriate limiting values for k and k' in eqns. (8—10). Figure 3 shows the resulting load-deformation curves. It is drawn with dimensionless scales with an assumed ratio $r = 0.1$. Let us define the "yield-point load" of the truss as that value $V = V_0$ for which the truss becomes very much more flexible. For the *E/SH* model it is given by the last eqn. (9) with $kv = 2Y$, and for the other three models it has the precise value $Y(\sqrt{2}+1)$. Therefore, if only V_0 is required, even the simplest *R/PP* model gives a very good approximation to it.

For values of the load less than the yield-point load, the effect of strain hardening is very small, and it appears quite reasonable to use the *E/PP* model as an approximation of the more accurate *E/SH* one. On the other hand, if the applied load is to be much larger than V_0, then the effect of elastic strains becomes increasingly less important, and the *R/SH* model gives a good approximation.

3. Unloading and reloading

Equations (4) are no longer adequate to describe the constitutive behavior if the stress in a bar is allowed to both increase and decrease. Instead, we must formulate them in terms of rates or increments. To this end, at any given instant we characterize the behavior of a bar as either elastic (*E*), strain-hardening (*SH*), or perfectly-plastic (*PP*). Then the rate equations corresponding to Fig. 2 may be written

$$E: \quad \dot{\sigma} = E\dot{\epsilon} \tag{11a}$$

$$SH: \quad \dot{\sigma} = E'\dot{\epsilon} \tag{11b}$$

$$PP: \quad \dot{\sigma} = 0 \tag{11c}$$

Equations (11) must be supplemented by rules to tell which set of equations are applicable at any given time. For the *PP* material the magnitude of the stress can never exceed the initial yield stress σ_y. If $\sigma = \sigma_y$ and ϵ is increasing or if $\sigma = -\sigma_y$ and ϵ is decreasing, then the bar is plastic; otherwise it is elastic.

For the SH material, there exist two numbers σ' and σ'' which represent the current yield stresses at any time. If these are known, we can write

$$\sigma'' \leqslant \sigma \leqslant \sigma' \tag{11d}$$

$$IF \quad (\sigma = \sigma' \; AND \; \dot{\epsilon} > 0) \quad OR \quad (\sigma = \sigma'' \; AND \; \dot{\epsilon} < 0) \tag{11e}$$

THEN (PLASTIC) *ELSE* (ELASTIC)

Equations (11d, e) apply to a *PP* material if we assign the constant values

$$PP: \quad \sigma' = \sigma_y \qquad \sigma'' = -\sigma_y \tag{11f}$$

For the SH material it is still necessary to formulate a rule for the variation of the current yield stresses. We begin by stating that whenever a bar is elastic its yield stresses do not change. Next, we require that during plastic behavior the "active" yield stress remains equal to the bar stress:

$$SH: \; IF \text{ (PLASTIC TENSION) } THEN \; \sigma' = \sigma$$
$$IF \text{ (PLASTIC COMPRESSION) } THEN \; \sigma'' = \sigma \tag{11g}$$

Various models have been suggested for describing the change in the "passive" yield stress, i.e., the change in σ'' when $\sigma = \sigma'$. We consider here two common models known as "isotropic" (*IH*) and "kinematic" (*KH*) hardening. For an *IH* material the two yield stresses are always equal in magnitude and opposite in sign, whereas for a *KH* material the "elastic range" between the two yield stresses maintains a constant value:

$$IH: \; \sigma' + \sigma'' = 0 \tag{11h}$$

$$KH: \; \sigma' - \sigma'' = 2\sigma_y \tag{11i}$$

Figure 4 shows typical load—unload—reload curves for the three different materials.

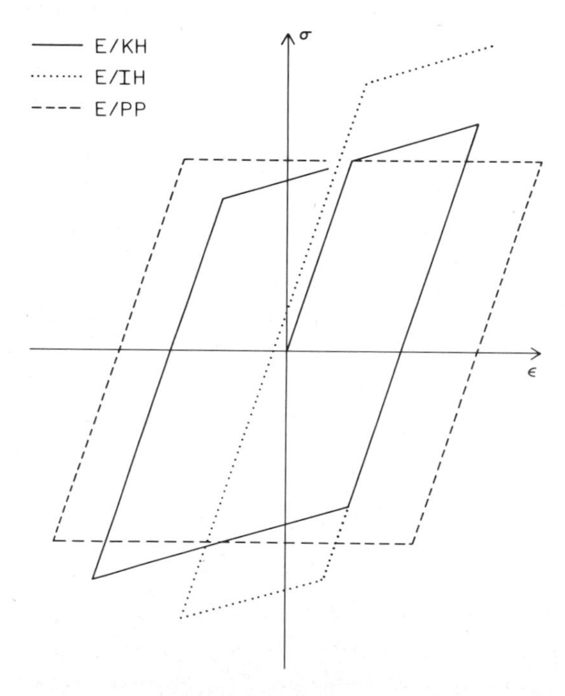

Fig. 4. Stress—strain curves for unloading and reloading.

We consider again the truss in Fig. 1. The controlled displacement is vertical and is first increased to $8Y/k$, then decreased to $-6Y/k$, and finally increased to 0. Combining the rate form of the kinematic equations with eqns. (11a—c) we obtain the constitutive equations in the form

$$E: \quad \dot{F}_1 = k\dot{v}/2 \qquad \dot{F}_2 = k\dot{v} \tag{12a}$$

$$SH\colon\ \dot{F}_1 = k'\dot{v}/2 \qquad \dot{F}_2 = k'\dot{v} \tag{12b}$$

$$PP\colon\ \dot{F}_1 = 0 \qquad \dot{F}_2 = 0 \tag{12c}$$

For the remainder of eqns. (11) one merely needs to replace σ, ϵ, and σ_y by, respectively, F_i, e_i, and Y_i.

The solution through Stage 3 is the same as in Sec. 2, and its limit, Stage 3L, is obtained by setting $kv/Y = 8$ in Stage 3:

$$\text{Stage 3L:}\quad kv/Y = 8; \qquad F_1/Y = 1 + 3r; \qquad F_2/Y = 1 + 7r$$
$$V/Y = (\sqrt{2} + 1) + r(3\sqrt{2} + 7) \tag{13}$$

In stage 4 we decrease v which will cause all bars to revert to elastic behavior until bar 2 yields in compression. For KH the total change in F_2 during stage 4 will be $\Delta F_2 = -2Y$. Therefore, using (12a) for all three bars we may write the total change in solution during stage 4 as

$$\Delta F_2/Y = \Delta vk/Y = -2; \qquad \Delta F_1/Y = -1 \tag{14}$$

Adding these increments to the values at Stage 3L, we obtain

$$\text{Stage 4L } (KH)\colon\ kv/Y = 6;\ \ F_1/Y = 3r;\ \ F_2/Y = -1 + 7r \tag{15}$$

In Stage 5 bar 2 is plastic so we use (12a) for bar 1 and (12b) for bar 2. This stage will end when bar 1 becomes plastic:

$$\text{Stage 5L(KH):}\ kv/Y = 4;\ \ F_1/y = -1 + 3r;\ F_2/Y = -1 + 5r \tag{16}$$

For the rest of the unloading both bars are plastic and take eqn. (12b). This stage ends when we again reverse v and start to reload. During reloading we enter stages 7, 8, and 9 corresponding to all bars elastic, bar 2 plastic, and all bars plastic respectively. The various solutions are

$$\text{Stage 6L(KH):}\ kv/Y = -6;\ F_1/Y = -1 - 2r;\ F_2/Y = -1 - 5r \tag{17a}$$

$$\text{Stage 7L(KH):}\ kv/Y = -4;\ F_1/Y = -2r; \qquad F_2/Y = 1 - 5r \tag{17b}$$

$$\text{Stage 8L(KH):}\ kv/Y = -2;\ F_1/Y = 1 - 2r; \quad F_2/Y = 1 - 3r \tag{17c}$$

$$\text{Stage 9L(KH):}\ kv/Y = 0; \quad F_1/Y = 1 - r; \quad F_2/Y = 1 - r \tag{17d}$$

The load in each stage is given by eqn. (7).

For an IH material the solution will be the same up through Stage 3L, and the truss will have the same rates in Stage 4. However, we must now keep track of the current yield force Y' for each bar. During Stage 2, Y'_2 will increase and remain equal to F_2, and during Stage 3 both yield forces will increase with the bar forces. Thus at Stage 3L the yield forces will be

$$\text{Stage 3L(IH):} \qquad Y'_1 = 1 + 3r; \qquad Y'_2 = 1 + 7r \tag{18}$$

and these values will continue to hold in Stage 4. Stage 4L will occur when $F_1 = -Y'_1$, hence

Stage 4L(IH): $F_1/Y = -4r;\ F_2/Y = -(1 + 7r);\ kv/Y = 6 - 14r$ (19)

The solution will be qualitatively similar to the *KH* material, except that the magnitude of both yield forces will always grow when a bar is plastic in either tension or compression. As a result, if $r > 0.0674$, the yield force in bar 1 will increase to the point that $v = 0$ during reloading while bar 1 is still elastic. Hence the residual bar forces when $v = 0$ will be

Stage 8L(IH):
$$F_1/Y = 2 - 8r + 6r^2$$
$$F_2/Y = 1 + 23r - 52r^2 + 28r^3 \qquad (20)$$

The results for a *PP* material can be obtained from either of the above *SH* solutions by setting $r = 0$, or can easily be found directly.

For all solutions the external load V is found by substituting the listed bar forces in eqn. (10). Figure 5 shows the resulting load—displacement curves for the three different materials. Results for the rigid/plastic materials will, of course, be very similar.

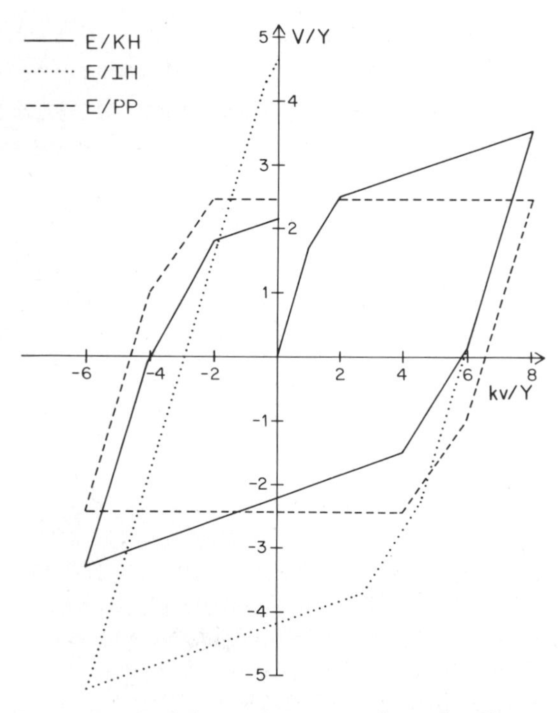

Fig. 5. Load—deformation curves for unloading and reloading.

4. Shakedown

We consider first an *E/PP* material which is subjected to a prescribed loading program which continues indefinitely, usually as a repeated cycle.

If the loads are always sufficiently small, the structure will remain everywhere elastic, and is of no concern in the present context. At the other extreme, if the loads ever exceed the yield-point load of the structure, there will be no equilibrium solution. The structure will collapse catastrophically, and the rest of the loading program will be meaningless. In the present section we are concerned with loading programs between these two extremes, i.e., the loading program includes loads which exceed the elastic limit but which nowhere exceed the yield-point load.

We illustrate various loading programs with regard to the truss in Fig. 1. In every case we take all three bars to have the same cross-sectional area and the same material properties.

Consider first a program in which $H = 0$, always, and V oscillates between 0 and 2.2 Y. During the initial increase of V eqn. (8) and eqn. (9) with $r = 0$ apply but stage 2 ends with the limiting value of V:

Stage 2L:
$$F_1 = F_3 = 0.6\sqrt{2}\,Y; \quad F_2 = Y; \qquad kv/Y = 1.2\sqrt{2}; \quad V = 2.2\,Y \tag{21}$$

When V is decreased from 2.2 Y, the changes in both bars will be elastic. Therefore, it follows from (12a) and (7) that

$$\Delta F_1 = 0.5\,k\Delta v \qquad \Delta F_2 = k\Delta v \qquad \Delta V = (1 + 1/\sqrt{2})k\Delta v \tag{22}$$

Setting $\Delta V = -2.2\,Y$ to reduce V to zero, leads to the values

Stage 3L:
$$F_1 = F_3 = (1.7\sqrt{2} - 2.2)Y; \quad F_2 = (2.2\sqrt{2} - 3.4)Y; \qquad kv/Y = 3.4\sqrt{2} - 4.4 \tag{23}$$

Since both forces are within the elastic range, this solution is the valid one for Stage 3L. Further, when V is again increased to 2.2 Y, eqns. (22) will hold for the entire reloading process, and the solution at stage 4L will be exactly the same as that at Stage 2L as given by eqns. (21). Clearly further cycles will simply alternate between eqns. (21) and (23) with both bars continuing to behave elastically.

The above behavior can be summarized as follows. There is a limited amount of plastic flow at the beginning of the loading process, but after this has taken place the rest of the cycle is completely elastic. A process with this property is called a "shakedown" process.

As an example of a process which does not shake down, suppose that after the initial loading to 2.2 Y the load V is alternated between $-2.2\,Y$ and $+2.2\,Y$. Stages 1 and 2 will be the same as before, and the change in Stage 3 will still be given by eqns. (22). However this solution is valid only until $\Delta F_2 = -2\,Y$. Further unloading takes place with bar 2 plastic, hence $\Delta F_2 = 0$, and stage 4 ends when V reaches its final value of $-2.2\,Y$.

The reloading process is similar, consisting of Stage 5 with all bars elastic and Stage 6 with bar 2 plastic in tension. The results are summarized in Table 1a. We observe that the solution for Stage 6L is exactly the

TABLE 1a
Alternating Load: *E/PP*

Cycle	Stage	V/Y	kv/Y	F_2/Y	$\Delta W_p k/Y^2$	$\Sigma W_p k/Y^2$
0	1L	1.707	1.000	1.000	0	0
	2L	2.200	1.697	1.000	0.697	0.697
1	3L	−1.214	−0.303	−1.000	0	0.697
	4L	−2.200	−1.697	−1.000	1.394	2.091
	5L	1.214	0.303	1.000	0	2.091
	6L	2.200	1.697	1.000	1.394	3.485

same as the one at Stage 2L, so that this cycle of 4 stages will be exactly repeated in each cycle of V from $+2.2\,Y$ to $-2.2\,Y$ and back to $+2.2\,Y$ as given in Table 1a. In particular, the value of the displacement will never exceed $1.697\,Y/k$ which is less than double the maximum elastic displacement.

At first glance it might appear that this loading program was a safe one for the truss since bar 1 never yields and the displacement is always bounded. However, from a materials point of view, the process of plastic displacement is not a reversible one. Although the gross displacement of bar 2 is the same after a complete cycle, the tensile plastic behavior in Stage 6 does not "undo" the compressive plastic elongation in Stage 4, but rather superimposes a tensile plastic elongation on it. To obtain some insight as to why this behavior is undesirable, we introduce the concept of "plastic work".

The differential internal work done on a bar is given quite generally by $dW = F\,de$. For an *E/PP* bar at yield there is no change in the bar force and hence no change in the elastic elongation, so that all work done is plastic. Further, since the force has its constant yield value, and since the elongation change is positive in tension and negative in compression, we can write the increment of plastic work in the form $\Delta W_p = Y|\Delta e|$. In particular, since bar 2 is the only one to yield in the present example, we may use the middle eqn. (2) to write

$$\Delta W_p = \begin{cases} Y|\Delta v| & \text{when bar 2 plastic} \\ 0 & \text{when bar 2 elastic} \end{cases} \tag{24}$$

Equation (24) shows that plastic work is done during each even-numbered stage and is always positive. The last column of Table 1a shows the cumulative plastic work through Stage 6. Clearly a further increment of $1.394\,Y^2/k$ must be added during each additional half cycle. Since the capacity of any real material to absorb plastic work is limited, the truss will become unserviceable after a relatively small number of cycles. The solid curve in Fig. 6 shows the accumulation of plastic work during the first few cycles. A loading program in which this phenomen occurs is called "alternating collapse".

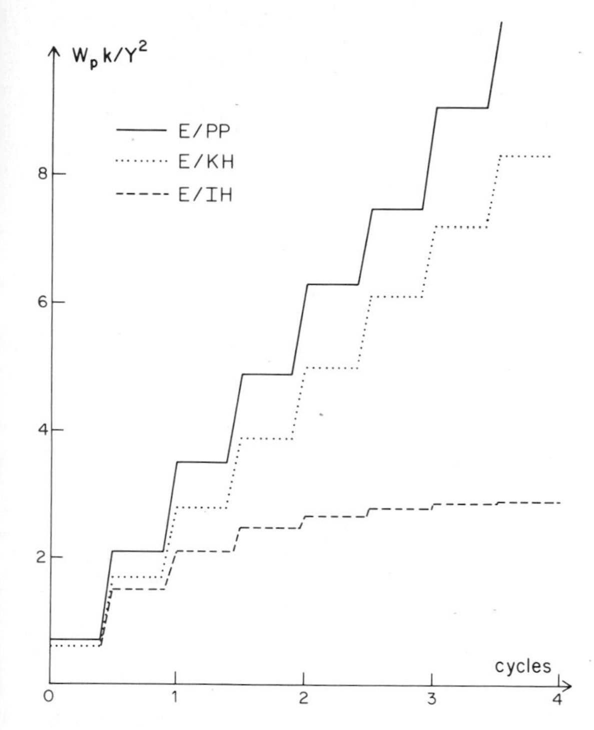

Fig. 6. Plastic work for load cycle with alternating collapse.

In Sec. 2 we concluded that it was reasonable to neglect hardening at loads less than the yield-point load, at least for a single loading. To what extent is that conclusion valid in relation to shakedown? It turns out that the answer is quite different depending upon the type of hardening. We consider first an *E/KH* material and obtain the results in Table 1b. Since conditions are exactly the same at Stages 6L and 2L, further cycles will simply repeat the last four lines of Table 1b.

When a bar is plastic, the incremental work can be written

$$\mathrm{d}W = F\,\mathrm{d}e = F\dot{e}\,\mathrm{d}t = F\dot{F}/k'\,\mathrm{d}t = \mathrm{d}(F^2/2k') \tag{25}$$

However, to find the plastic work differential we must subtract the elastic differential $\mathrm{d}W = \mathrm{d}(F^2/2k)$. If F_0 and ΔF represent the beginning value and increment, respectively, of F during a plastic stage, we may write the plastic work increment for bar 2:

$$\Delta W_\mathrm{p} = (1-r)\Delta v(F_0 + 0.5\,\Delta F) \tag{26}$$

The last two columns of Table 1b show the increments and cumulative values of the plastic work, and the dashed curve in Fig. 6 shows the cumulative plastic work over the first few cycles for $r = 0.1$.

Comparing the results for the *E/PP* and *E/KH* models we see that the former is a reasonable approximation to the latter. In particular, both

Table 1b
Alternating Load: E/KH

Cycle	Stage	V/Y	kv/Y	F_2/Y	$\Delta W_p k/Y^2$	$\Sigma W_p k/Y^2$
0	1L	1.707	1.000	1.000	0	0
	2L	2.200	1.611	1.061	0.566	0.566
1	3L	−1.214	−0.389	−0.939	0	0.566
	4L	−2.200	−1.611	−1.061	1.099	1.666
	5L	1.214	0.389	0.939	0	1.666
	6L	2.200	1.611	1.061	1.099	2.765

models predict that the plastic work will increase without bound as the cycles continue.

The *E/IH* material behaves quite differently. The increments are still governed by the same equations, but the passive yield force comes from (11h) rather than from (11i). The resulting solution is exactly the same through Stage 3, but Stage 3L occurs when F_2 is equal to the negative of its value at Stage 2L, rather than when $\Delta F = -2\,Y$. The net result is that the truss spends more of its unloading time in the elastic Stage 3 and less in the partially-plastic Stage 4. The same thing happens during the reloading phase. Therefore, as shown in Table 1c, the solution at Stage 6L is different from that at Stage 2L, so that the next cycle of unloading–reloading will be different. In particular, we note that the value of F_2 at the end of each cycle has increased. Since this value is the current yield stress of bar 2, a larger proportion of each succeeding cycle is spent in a fully elastic state.

Results for the first few cycles are shown in Table 1c and by the dotted curve in Fig. 6. Although bar 2 is plastic during part of every cycle, it is apparent that the process is converging to an entirely elastic one, and that the total amount of plastic work is finite.

Let us return to the *E/PP* material and consider a different loading cycle. We begin in the same way by increasing V to 2.2 Y, but we then superpose an alternating load at 45°. Therefore, Stages 1 and 2 are the same as in the previous example, but beginning with Stage 3 the load increments are defined by $\Delta H = -\Delta V = \Delta\lambda/\sqrt{2}$, where λ is cycled between 0 and 1.1 Y. Therefore, eqns. (1) can be written

$$\Delta F_1 - \Delta F_3 = \Delta\lambda; \qquad \Delta F_1 + \Delta F_3 + \sqrt{2}\,\Delta F_2 = -\Delta\lambda \tag{27}$$

As λ is increased from 0 to 1.1 all bars are first elastic and then bar 1 yields in tension. As λ is decreased to 0 all bars are first elastic and then bar 2 yields in tension. In each stage the increments of all bar forces and displacements are linear in $\Delta\lambda$ with coefficients given in Table 2. Table 3 shows the resulting solution at the end of each stage through 6L.

The bar forces are all exactly the same at Stage 6L as they were at Stage 2L. Therefore, they will repeat the same sequence of values during

TABLE 1c
Alternating Load: E/IH

Cycle	Stage	V/y	kv/Y	F_2/Y	$\Sigma W_p/kY^2$
0	1L	1.707	1.000	1.000	0
	2L	2.200	1.611	1.061	0.566
1	3L	−1.423	−0.511	−1.061	0.566
	4L	−2.200	−1.474	−1.157	1.528
	5L	1.752	0.840	1.157	1.528
	6L	2.200	1.396	1.213	2.121
2	7L	−1.941	−1.030	−1.213	2.121
	8L	−2.200	−1.351	−1.245	2.475
	9L	2.051	1.139	1.245	2.475
	10L	2.200	1.324	1.263	2.684
3	12L	−2.200	−1.309	−1.274	2.806
	14L	2.200	1.301	1.280	2.877
4	16L	−2.200	−1.296	−1.284	2.918
	18L	2.200	1.293	1.286	2.941
5	20L	−2.200	−1.291	−1.287	2.955
	22L	2.200	1.290	1.288	2.963
6	24L	−2.200	−1.289	−1.288	2.967
	26L	2.200	1.289	1.288	2.970

TABLE 2
Summary of incremental solutions for diagonal load

	All bars elastic (Stages 3, 5)	bar 1 plastic (Stage 4)	bar 2 plastic (Stage 6)
$\Delta F_1/\Delta\lambda$	$1-1/\sqrt{2}$	0	0
$\Delta F_2/\Delta\lambda$	$1-\sqrt{2}$	0	0
$\Delta F_3/\Delta\lambda$	$-1/\sqrt{2}$	−1	−1
$k\Delta u/\Delta\lambda$	1	2	1
$k\Delta v/\Delta\lambda$	$1-\sqrt{2}$	0	−1

each additional cycle of load. However, the vertical and horizontal displacements have each increased by $(0.3+0.2\sqrt{2})Y/k = 0.583\ Y/k$ during the cycle. Clearly they will increase by the same amount during each succeeding cycle. Therefore, although at any given instant the truss has at least two bars elastic and can support the given load, the deformations grow indefinitely with time and the truss eventually becomes unserviceable. This behavior has been termed "incremental collapse". A history of the displacements during the first few cycles is shown in Fig. 7.

Physically, we can observe that if bars 1 and 2 were both in plastic tension at the same time the truss would be a mechanism capable of a rotation about the end of bar 3 with $\dot{u} = \dot{v}$. Of course this does not occur,

TABLE 3
Diagonal Load: *E/PP*

Cycle	Stage	λ/Y	F_1/Y	F_2/Y	F_3/Y	ku/Y	kv/Y
0	1L	—	0.500	1.000	0.500	0	1.000
	2L	0	0.849	1.000	0.849	0	1.697
1	3L	0.517	1.000	0.786	0.483	0.517	1.483
	4L	1.100	1.000	0.786	−0.100	1.683	1.483
	5L	0.583	0.849	1.000	0.266	1.116	1.697
	6L	0	0.849	1.000	0.849	0.583	2.280
one	6L—2L	0	0	0	0	0.583	0.583

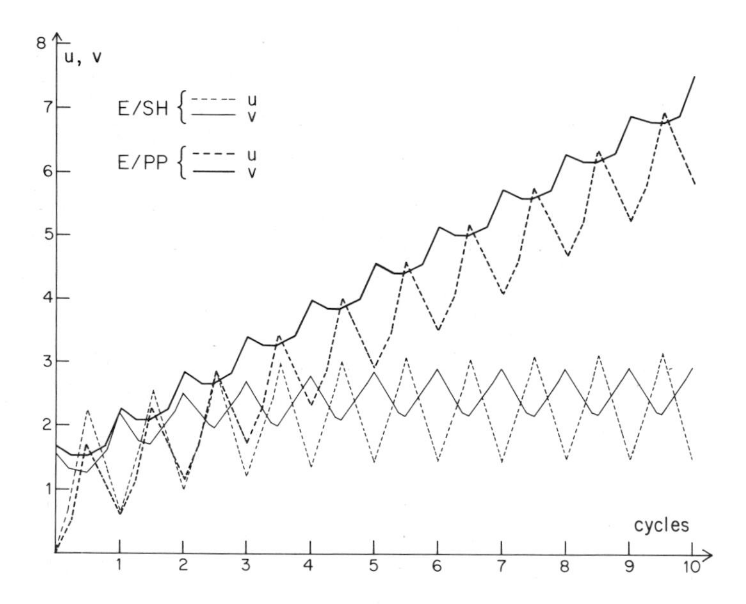

Fig. 7. Deformations for load cycle with incremental collapse.

but in the course of a complete cycle bars 1 and 2 are plastic at different times. The net effect at the end of a complete cycle is a limited mechanism motion $\Delta u = \Delta v = 0.583\ Y/k$, as shown by the detailed analysis.

For this loading program the two strainhardening models are the same. Stage 2L is again given by the appropriate line from Table 1b or 1c. The qualitative description of each cycle is the same as for the *E/PP* truss. As λ is increased all bars are elastic at the start but bar 1 becomes plastic before λ reaches 1.1 Y; as λ is decreased all bars are first elastic but bar 2 becomes plastic before the cycle ends. However each time bar 1 or bar 2 yields its yield force is increased so that during the next cycle the truss

TABLE 4
Diagonal Load: *E/SH*

Cycle	Stage	λ/Y	F_1/Y	F_2/Y	F_3/Y	ku/Y	kv/Y
0	1L	—	0.500	1.000	0.500	0	1.000
	2L	0	0.805	1.061	0.804	0	1.611
1	3L	0.470	1.000	0.786	0.335	0.665	1.335
	4L	1.100	1.072	0.684	−0.483	2.202	1.234
	5L	0.457	0.805	1.061	0.159	1.292	1.611
	6L	0	0.765	1.118	0.765	0.646	2.177
2	7L	0.740	1.072	0.684	0.025	1.692	1.743
	8L	1.100	1.113	0.626	−0.443	2.571	1.685
	9L	0.261	0.765	1.118	0.396	1.385	2.177
	10L	0	0.742	1.150	0.742	1.015	2.500
3	11L	0.894	1.113	0.626	−0.152	2.280	1.976
	12L	1.100	1.136	0.593	−0.419	2.782	1.943
	13L	0.149	0.642	1.150	0.531	1.437	2.500
	14L	0	0.729	1.169	0.729	1.226	2.685

will remain fully elastic until nearer the end of the cycle. Table 4 gives the complete results through Stage 14L, corresponding to three complete cycles.

Of particular interest are the total displacements added during each cycle. Using the results of Table 4 it is easy to show that the ratio of the increase in either u or v to its increase in the previous cycle is the constant value $\eta = 0.571636$. Thus the net displacement can be written as a geometric series which converges to the final values

$$u = 1.508\ Y/k \qquad v = 2.932\ Y/k \tag{28}$$

Therefore, at least in this simple example, the introduction of strain-hardening leads to finite values of the total displacements as compared with the infinite values predicted by the *E/PP* model.

5. Reverse stressing under monotonic loading

In the previous sections we have seen various examples where the force ratios between bars will change as one or more bars becomes plastic, even though the external load is uniformly increasing. Here we present a simple example in which the ratios not only change magnitude, but actually change sign. The truss in Fig. 8a consists of three numbered vertical bars made of an *E/PP* material and joined to a perfectly rigid horizontal bar. The three vertical bars have equal areas A, lengths L and moduli E, but are assumed to have different yield stresses given by

$$Y_1 = Y; \qquad Y_2 = 20\,Y; \qquad Y_3 = 12\,Y \tag{29}$$

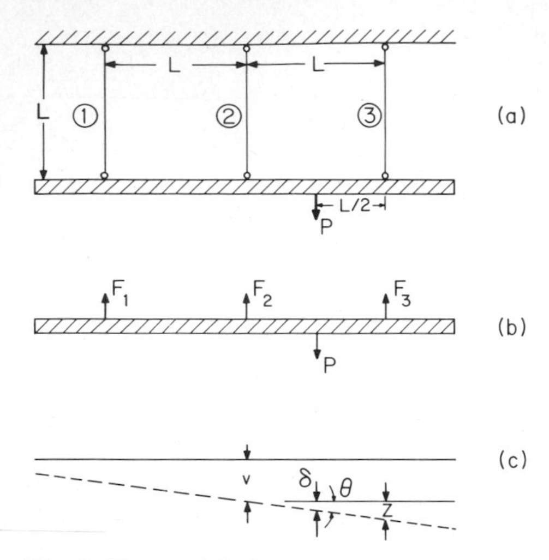

Fig. 8. Truss with three vertical bars.
(a) Loaded truss;
(b) Statics;
(c) Kinematics.

A vertical load P is applied as indicated. It is increased just to its yield-point value, and then decreased to zero. There are no horizontal loads, and we assume zero horizontal motion.

Static equations are obtained by considering vertical and moment equilibrium of the horizontal bar (Fig. 8b):

$$F_1 + F_2 + F_3 = P; \qquad F_3 - F_1 = 0.5P \tag{30}$$

There are two degrees of kinematic freedom which we take to be the elongation v of bar 2 and the difference z between the elongations of bars 3 and 2, Fig. 8c. Clearly $z = L\theta$ where θ is the clockwise rotation of the horizontal bar. In terms of these variables the bar elongation rates are

$$\dot{e}_1 = \dot{v} - \dot{z}; \qquad \dot{e}_2 = \dot{v}; \qquad \dot{e}_3 = \dot{v} + \dot{z} \tag{31}$$

The constitutive behavior is given by the rate form of eqns. (4a) and (4c). Combining these with eqns. (31) we obtain

$$E/: \quad \dot{F}_1 = k(\dot{v} - \dot{z}); \qquad \dot{F}_2 = k\dot{v}; \qquad \dot{F}_3 = k(\dot{v} + \dot{z}) \tag{32a}$$

$$/PP: \quad F_i = \pm Y_i; \qquad F_i\dot{e}_i \geqslant 0 \tag{32b}$$

where $k = AE/L$.

As P is first increased from 0 all three bars are elastic. Equations (30) and (32a) in integrated form lead easily to the solution which includes $F_1 = P/12$, $F_3 = 7P/12$. Although bar 3 is the most highly stressed, bar 1 will be the first to yield since its yield force is so much smaller. Thus Stage 1 ends with the solution

$$\text{Stage 1L:} \quad F_1 = Y; \qquad F_2 = 4\,Y; \qquad F_3 = 7\,Y$$
$$v = 4\,Y/k; \qquad z = 3\,Y/k; \qquad P = 12\,Y \tag{33}$$

In Stage 2, F_1 maintains the known value Y and bars 2 and 3 are elastic. The stage will end when bar 3 reaches its yield force 12 Y:

$$\text{Stage 2L:} \quad F_1 = Y; \qquad F_2 = 9\,Y; \qquad F_3 = 12\,Y$$
$$v = 9\,Y/k; \qquad z = 3\,Y/k; \qquad P = 22\,Y \tag{34}$$

If bars 1 and 3 both remain plastic, the truss will be at its yield-point load. A formal solution of the equilibrium equations and the middle eqn. (32a) regains the values in (34) for the bar forces, load, and v, with z still undetermined. However, with $\dot{v} = 0$, eqns. (31) show that $\dot{e}_1$ and $\dot{e}_3$ cannot both be positive, hence the inequalities in (32b) cannot both be satisfied. The same result is physically obvious from Fig. 8c. If bar 2 remains rigid, a yield mechanism would rotate the horizontal bar about the end of bar 2 which would result in a shortening of bar 1.

The above arguments show that in Stage 3 bar 3 will be plastic, but bar 1 will return to an elastic state. Thus we set $F_2 = 12\,Y$, use the first two (32a), and integrate the rates using eqns. (34) as initial conditions. We find that the force in bar 1 is now decreasing, and it will reach its yield value in compression while bar 2 is still elastic:

$$\text{Stage 3L:} \quad F_1 = -Y; \quad F_2 = 15\,Y; \quad F_3 = 12\,Y; \quad P = 26\,Y; \quad kv/Y = 15 \tag{35a}$$
$$kz/Y = 11 \tag{35b}$$

If the load is maintained at 26 Y with bars 1 and 3 remaining plastic, the solution includes eqns. (35a), but z is undetermined with any $\dot{z} > 0$ satisfying both relevant inequalities in (32b). Therefore eqns. (35) represent the beginning of the yield-point solution.

Instead of maintaining P at 26 Y, let us immediately reduce it. In most problems a reversal of the only load will cause all plastic bars to become elastic. However, if we superimpose the elastic solution with a negative load increment on eqns. (35), the force in bar 1 will immediately exceed its compressive yield strength. Therefore, even though bar 1 has just reached yield under an increasing load, it will remain at yield when we decrease the load. The resulting solution is easily found to be

$$\text{Stage 4:} \quad F_1 = -Y; \qquad F_2 = 0.5P + 2\,Y; \qquad F_3 = 0.5P - Y$$
$$kv = 2\,Y + 0.5P \qquad kz = 11\,Y \tag{36}$$

When the truss is fully unloaded it has a set of residual forces and permanent displacements given by eqns. (36) with $P = 0$.

The solution through Stage 3 will be qualitatively similar for a strain hardening material. However, Stage 3L will be determined by the current

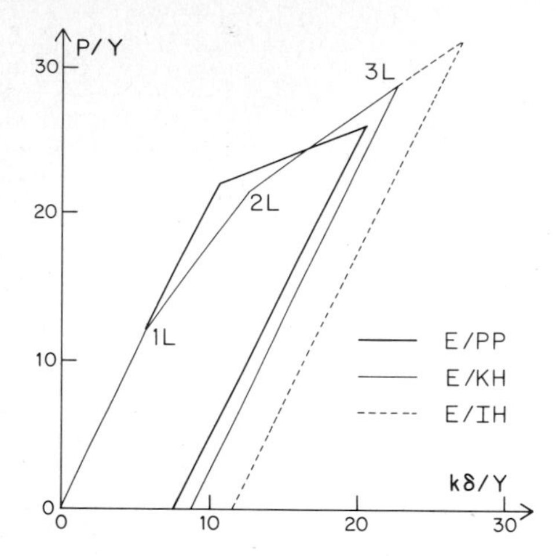

Fig. 9. Load—deflection curves.

compressive yield stress of bar 1 which will be different for the *PP*, *KH*, and *IH* materials.

Figure 9 shows the relation between the load P and the deflection

$$\delta = v - z/2 \tag{37}$$

of the point of load application. The unloading portion shows unloading from Stage 3L. Part of the difference between the curves is caused by the fact that they show unloading from different values of the load.

In the presence of strainhardening there is no maximum allowable load. Therefore, instead of reducing the load from Stage 3L, let us form Stage 4 by continuing to increase it. Bar 1 will be in plastic compression, bar 3 in plastic tension, and bar 2 will be still elastic with increasing force. The stage will end when F_2 reaches its maximum value of 20 Y. Table 5 shows the resulting solutions for both types of hardening at the various limit stages, assuming $r = 0.1$.

In Stage 5, bar 2 will be plastic. If all bars were plastic, the bar force increments would be in the same proportion as in Stage 1 when all bars were elastic. In particular, ΔF_1 would be positive which would mean that it would no longer be in plastic compression. Therefore, in Stage 5 bar 1 will again be elastic, while bars 2 and 3 are in plastic tension. Stage 5L occurs when bar 1 reaches its current tensile yield stress. As shown in Table 5 and Fig. 10, the load and displacement at Stage 5L are quite different for the different types of hardening.

However, this difference is really qualitative but not quantitative. To see this, let us continue to increase the load to a value of 80 Y. During this Stage 6 all bars are plastic in tension. Table 5 shows that at the final load all bar forces and displacements have only very small differences, and the two curves are virtually indistinguishable in Fig. 10.

TABLE 5

Three-bar truss of Fig. 8

Stage	1L	2L	3L		4L		5L		6L		5L*	
Bar States[a]	*EEE*	*TEE*	*EET*		*CET*		*ETT*		*TTT*		*CEE*	
Hardening			K	I	K	I	K	I	K	I	K	I
F_1/Y	1	1.3	− 0.7	− 1.3	− 1.5	− 1.7	0.5	1.7	2.6	2.7	− 2.7	− 2.8
F_2/Y	4	8.1	15.8	18.2	20.0	20.0	26.2	30.3	34.8	34.6	5.3	5.6
F_3/Y	7	12.0	13.8	14.3	15.4	15.0	27.6	35.3	42.6	42.7	− 2.7	− 2.8
kv/Y	4	8.1	15.8	18.1	20.0	20.0	103.3	123.3	167.7	165.5	5.3	5.6
kz/Y	3	3.9	13.7	16.7	26.2	22.0	99.9	122.0	150.5	153.7	22.9	18.7
P/Y	12	21.4	28.9	31.2	33.9	33.3	54.3	67.3	80.0	80.0	0	0
$k\delta/Y$	5.5	12.6	22.7	26.6	34.1	31.0	125.1	184.3	242.9	242.4	16.7	14.9

[a] E = Elastic, T = Plastic tension, C = Plastic Compression

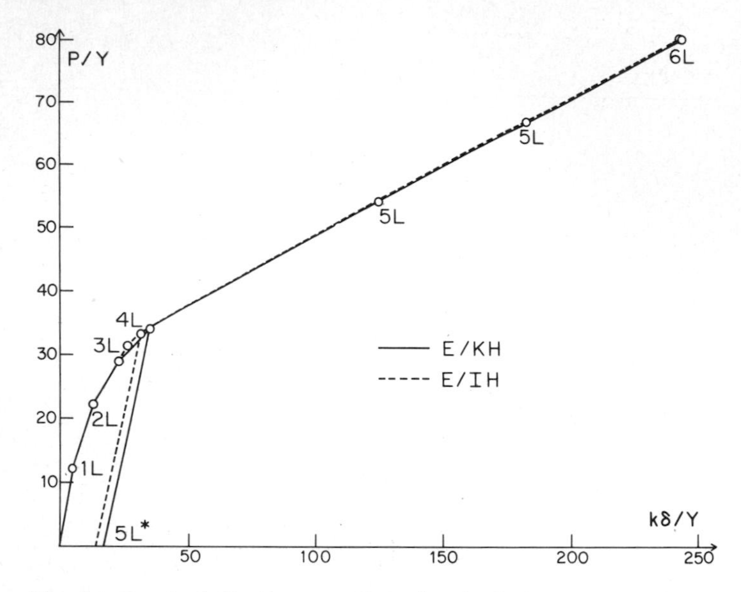

Fig. 10. Load—deflection curves for hardening truss.

Let us review the behavior of bar 1 of this simple strain-hardening truss. As the single load P is monotonically increased, bar 1 is first elastic, then yields in plastic tension, then unloads and yields in plastic compression, then reloads, and finally yields for a second time in plastic tension.

To conclude this section we consider one more loading program in which P is increased until bar 2 just reaches yield and is then decreased to zero. Stage 4L is the same as before, but a new Stage 5* will result from the unloading. Since fully elastic unloading would decrease F_1 which is already at compressive yield, Stage 5* will find bar 1 plastic and bars 2 and 3 elastic. But the original Stage 5 also had bar 1 in plastic compression. In other words, once the truss is at Stage 4L, bar 1 will continue to yield in plastic compression whether the load is increased or decreased!

6. Non-uniqueness

We return to the truss of Fig. 1, with the following modification: the bars have unequal yield forces with $Y_1 = Y_3 = Y$, $Y_2 = 3\,Y$. We consider a load-controlled program with $H = 0$ and V increasing to the yield-point load. However, we do not make any assumptions of symmetry. The bars are all *E/PP*.

For convenience we shall repeat the defining equations in a form specialized to the particular loading program. The equilibrium equations are

$$F_1 + F_3 + \sqrt{2}\,F_2 = \sqrt{2}\,V; \qquad F_1 - F_3 = 0 \qquad \text{(38a, b)}$$

It turns out that bar 2 is always elastic and bars 1 and 3 are either elastic or in plastic tension. Therefore, we may combine the kinematic and constitutive equations to obtain

$$F_2 = kv \tag{38c}$$

$$\mathit{EITHER} \quad F_1 = (k/2)(v+u) \quad \mathit{OR} \quad F_1 = Y \tag{38d}$$

$$\mathit{EITHER} \quad F_3 = (k/2)(v-u) \quad \mathit{OR} \quad F_3 = Y \tag{38e}$$

together with inequality conditions which may be written

$$F_1 \leqslant Y \qquad F_2 \leqslant Y \qquad |\dot{u}| \leqslant \dot{v} \tag{38f}$$

Equations (38) provide a total of five equations to determine F_1, F_2, F_3, u, and v.

Stage 1, of course, is elastic, so we use the first branch in eqns. (38d, e) to obtain a unique solution as in Sec. 2. However, with the stronger bar 2, bars 1 and 3 yield first, so that Stage 1L is

$$u = 0; \ F_1 = F_3 = Y; \ F_2 = kv = 2Y; \ V = V_1 \equiv (2+\sqrt{2})Y \tag{39}$$

In Stage 2 bars 1 and 3 are both plastic so the second branch is used in (38d, e). However, this means that (38b) is satisfied identically. We can still use (38a, c) to obtain the unique values

$$F_1 = F_3 = Y; \quad F_2 = kv = \Delta V + 2Y; \quad \Delta V = V - V_1 \tag{40a}$$

but we cannot determine the value of the horizontal displacement u, although (38f) does provide the bounds

$$|u| < \Delta V/k \tag{40b}$$

In particular, Stage 2L at the yield-point load is obtained when $\Delta V = Y$ in eqns. (40).

Figure 11 shows the unique solution at Stage 1L and several solutions for Stage 2L. For example, Fig. 11b shows the largest allowable value of u, $u = Y/k$. In this solution bar 3 rotates as a rigid body, bar 2 elongates elastically, and bar 1 elongates plastically. Notice that any value of u larger than Y/k would require a shortening of bar 3 which is not permissible when it is yielding in tension. On the other hand any position of point D between the limiting ones in Figs. 11a and b requires a lengthening of both bars 1 and 3 which is consistent with their both being at tensile yield.

Mathematically, the non-uniqueness was caused by the fact that two bars yielded simultaneously and reduced one of the equilibrium equations to an identity. In reality, of course, infinitesimal differences between the two bars would make it extremely unlikely that they would both yield at exactly the same instant. To examine this facet of the problem, let us make a perturbation of the problem by taking $Y_3 = Y + X$ where X is positive but otherwise arbitrary. Stages 1 and 1L will be the same as

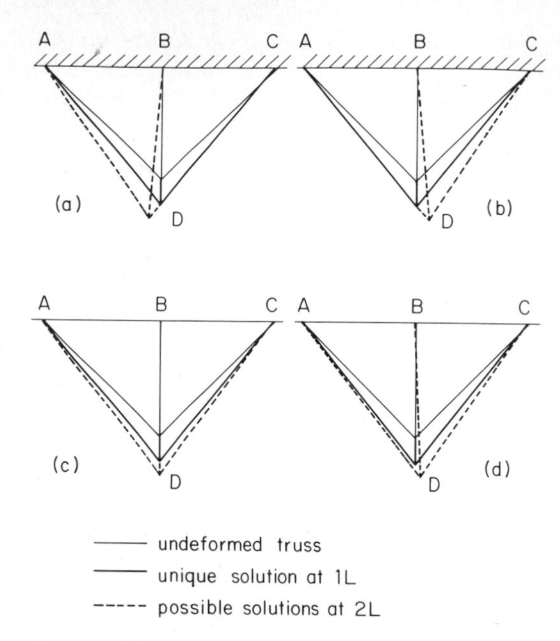

Fig. 11. Non-unique truss configurations.
(a) minimum horizontal displacement;
(b) maximum horizontal displacement;
(c) symmetric solution;
(d) generic solution.

before, although only bar 1 will yield at Stage 1L. In Stage 2 we now use the first branch of (38e) along with the second branch of (38d). The resulting unique solution consists of (40a) together with the value $u = \Delta V/k$.

Figure 11b shows the resulting configuration. This solution is completely independent of X, provided only that X is positive. In particular, X may be arbitrarily small so that there is no practical difference between the two yield forces.

However, it is clear that if we leave $Y_3 = Y$ and set $Y_1 = Y + X$, the solution in Stage 2 will consist of eqns. (40a) plus $u = -Y/k$ as pictured in Fig. 11a. This solution, too, is valid for arbitrarily small X. Therefore, although the non-unique solution of the original problem can be made unique by an arbitrarily small perturbation, two different perturbations give two very different unique solutions.

Let us now consider the effect of strainhardening. In doing so we shall allow for different rates of hardening in bars 1 and 3 by defining

$$k'_1 = kr_1 \qquad k'_3 = kr_3 \tag{41}$$

Stages 1 and 1L are the same as before, and the complete unique solution for Stages 2 and 2L is easily obtained. If the strainhardening tends to zero the bar forces and vertical displacement will tend to the

E/PP values. The unique value of u at Stage 2L is given by

$$u = \frac{r_1 - r_3}{r_1 + r_3} \frac{Y}{k} \tag{42}$$

If the strainhardening in the two bars is the same, $r_1 = r_3$, we obtain the symmetric solution $u = 0$ as pictured in Fig. 11c. On the other hand, if $r_1 = 0$, and only bar 3 hardens, we get $u = -Y/k$ as in Fig. 11a. Likewise, if only bar 1 hardens the solution is Fig. 11b. Clearly unequal non-zero hardening in the two bars can produce any value between these two extremes; for example $r_1 = 0.1$, $r_3 = 0.05$ gives the result $u = Y/3k$ as shown in Fig. 11d.

7. Conclusions

In the preceding sections we have used two very simple trusses to illustrate several important concepts in plasticity. The ideas presented are not new, of course, nor are most of the applications. The truss in Fig. 1 has been used many times starting at least as early as 1948 [1, 2, 3]* to illustrate plastic and other inelastic behavior. It has also been used to illustrate some singular features in plastic design [4]. The truss in Sec. 5 (Fig. 8) was first introduced by Drucker [5], and much of the present development was taken from that reference.

Countless texts on plasticity are available [6—10] (to name just a few). The basic constitutive equations presented in Secs. 1 and 2 can be generalized to two and three dimensions and to various structural problems. For the particular case of a simply-supported circular plate generalizations of the four stress-strain curves shown in Fig. 2 have been applied [11]. Figure 12, taken from Ref. [11] shows the relation between the pressure and the displacement of the plate center. The results are certainly qualitatively similar to those for the three-bar truss as shown in Fig. 3.

Shakedown was first introduced by Melan [12]. Many results have been presented by Symonds and by Neal [13, 14, 10]. Important theoretical work has been done by Koiter [15, 16]. Recently, an entire book by Gokhfeld and Cherniavsky [17] has been devoted to the subject.

The idea that plastic stress reversal can occur even under a single monotonic loading has many important implications. For example, the somewhat simpler theories known as "plastic-deformation" theories which directly relate stress to strain rather than relating their rates is obviously inappropriate when this phenomenon occurs.

Figure 13 shows an eleven-bar truss which does not look particularly unusual. The top four bars all have the same cross-section and are made

---

*Numbers in square brackets refer to references collected at the end of the paper.

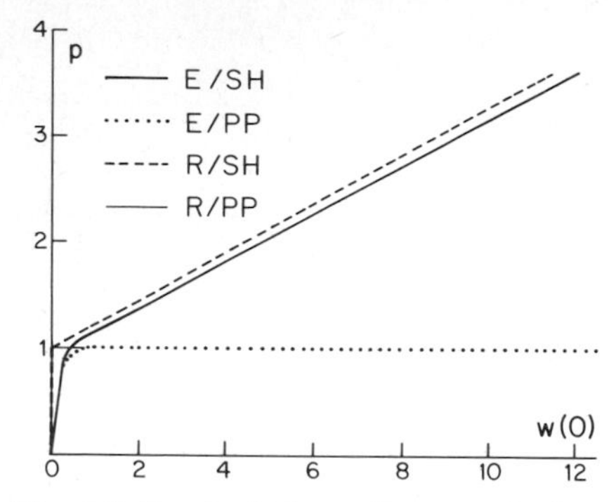

Fig. 12. Load—deformation curves for circular plate.

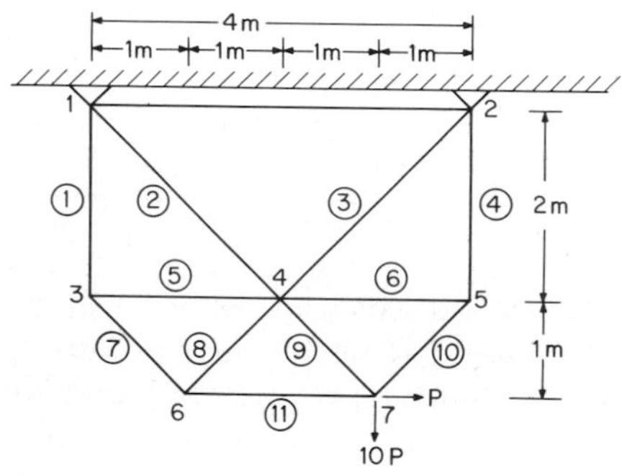

Fig. 13. Eleven-bar truss.

of an E/KH material with $A = 100\,\text{mm}^2$, $E = 80\,\text{GPa}$, $E' = 8\,\text{GPa}$, and unequal yield forces

$$Y_1 = 10\,\text{MPa}; \quad Y_2 = Y_3 = 120\,\text{MPa}; \quad Y_4 = 50\,\text{MPa} \tag{43}$$

The bottom bars all have $A = 1000\,\text{mm}^2$, $E = 240\,\text{GPa}$, and have yield stress high enough to prevent yielding. As the load P is increased, bar 1 first yields in tension, then bar 4 reaches its yield force 50. Bar 1 must now unload, and it then yields in compression. Next, bar 2 yields in tension, and bar 1 again unloads, goes into tension, and eventually yields again in tension. Table 6 shows the bar forces, load, and displacement of the point of load application.

A similar stress reversal has been observed in elastic-plastic torsion of some hollow bars [18, 19]. Shaw [20] has numerically solved the torsion of a bar whose cross section is a hollow rectangle with fillets. He has shown that plastic behavior starts on the inner boundary at the fillet, and that the stress vector there is in the direction of the torque. However, it was first pointed out in [18] and later numerically verified in [19] that at the limiting plastic torque the stress vector there must be in the opposite direction from the torque. An unfortunate result of this plastic stress reversal is that the Nadai sandhill—soapfilm analogy [21] is no longer applicable.

TABLE 6

Solution for truss of Fig. 13

Stage	1L	2L	3L	4L	5L
Status	all E	$1T$	$4T$	$4T, 1C$	$4T, 2T$
F_1 (kN)	1.00	1.09	−0.91	−1.17	0.83
F_2 (kN)	1.96	2.46	10.59	12.00	16.52
F_3 (kN)	0.92	1.23	7.73	8.82	11.07
F_4 (kN)	4.31	5.00	8.19	8.96	18.15
F_5 (kN)	−1.00	−1.09	0.91	1.17	−0.83
F_6 (kN)	−4.31	−5.00	−8.19	−8.96	−18.15
F_7 (kN)	1.41	1.54	−1.29	−1.66	1.17
F_8 (kN)	−1.41	−1.54	1.29	1.66	−1.17
F_9 (kN)	4.30	5.23	17.03	19.16	28.76
F_{10} (kN)	6.09	7.07	11.59	12.67	25.67
F_{11} (kN)	2.00	2.17	−1.83	−2.34	1.66
P (kN)	0.73	0.87	2.02	2.25	3.85
u (mm)	0.06	0.12	−1.62	−2.18	2.93
v (mm)	0.96	1.16	7.06	8.34	25.89

The phenomenon of non-uniqueness was encountered, apparently for the first time, in the development of a finite-element model with discontinuous displacements for use in plane-strain plasticity [22, 23]. Since this was a primarily numerical development, it was not clear whether the non-uniqueness was inherent in the *E/PP* model, or whether it was a peculiarity of the particular finite-element model. The analysis presented in Sec. 6 [24, 25] shows that it is, indeed, a possibility which must be considered in any problem using the *E/PP* model.

Figures 14 and 15 [24] show some other simple examples. The elastic solution for the truss in Fig. 14 is symmetric. It ends when the two side bars reach yield. Further increase of load can be associated with any non-negative change in the lengths of the vertical bars. Figure 14 shows the two extreme positions where one of the plastic bars does not change its length. Any intermediate solution can be obtained by a rigid-body rotation of the upper triangle about the center-point of the truss.

As the load on the frame in Fig. 15 is increased, hinges will form first at C, then at B, and then simultaneously at points A and D, but the frame will not collapse until the final hinges form at E. The solution is unique during the elastic and first two partly-plastic stages, but once the hinges form at A and D, the rotations at these new hinges are controlled only by inequalities, both being required to exhibit tensile strain at the top of the hinge. Figure 15 shows the two extreme positions. Any intermediate solution obtained by a rigid-body vertical motion of the central part of the frame is also possible.

The possibility of part of the solution being non-unique has important implications in finite-element programs. Even if the engineer is concerned

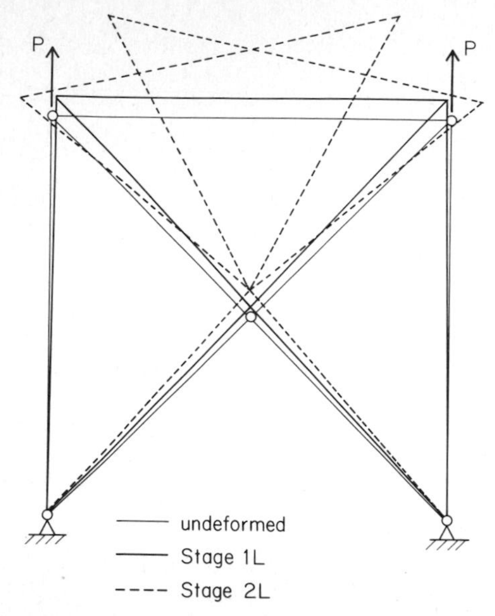

Fig. 14. Seven-bar truss.

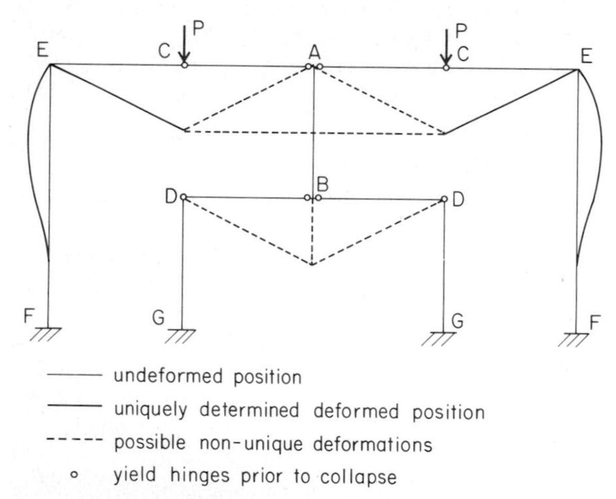

Fig. 15. Plane frame.

only with forces which are unique, a non-unique solution corresponds to a singular set of equations, and hence to a singular stiffness matrix. Thus it may be impossible to continue the solution past the point where the non-uniqueness occurs, even though the load is still below the yield-point load. Numerical round-off error may mask the singularity of the stiffness matrix, but the resulting problem would be highly ill-conditioned and might lead to misleading results. This problem is certainly worthy of further investigation.

References

1 W. Prager, Problem types in the theory of perfectly plastic materials. J. Aero. Sci., 15 (1948) 337—341.
2 A.M. Freudenthal, Effect of rheological behavior on thermal stresses. J. Appl. Phys., 25 (1954) 1110—1117.
3 P.G. Hodge, Jr., The practical significance of limit analysis. J. Aero-Sp. Sci., 25 (1958) 724—725.
4 W. Prager, Optimal plastic design of trusses for Bauschinger adaptation, in: Omaggio A Carlo Ferrari, Levrotto & Bella, Torino, 1974, pp. 625—632.
5 D.C. Drucker, Plasticity of metals — mathematical theory and structural applications. Trans. ASCE 116 (1959) 1059—1072.
6 W. Prager and P.G. Hodge, Jr., Theory of Perfectly Plastic Solids. J. Wiley & Sons, Inc., New York, 1951; Dover, New York, 1968.
7 P.G. Hodge, Jr., Plastic Analysis of Structures. McGraw-Hill Book Company, New York, 1959; Robert E. Krieger Publishing Co., Malabar, Fla., 1981.
8 J.B. Martin, Plasticity: Fundamentals and General Results. MIT Press, Cambridge, Mass., 1975.
9 L.M. Kachanov, Foundations of the Theory of Plasticity. North-Holland Publishing Co., 1971; Introduction to the Theory of Plasticity. Nauka, Moscow, 1969.
10 B.G. Neal, The Plastic Methods of Structural Analysis. Chapman & Hall, London, 1956.
11 P.G. Hodge, Jr., Boundary value problems in plasticity, in: E.H. Lee and P.S. Symonds (Eds.), Plasticity. Pergamon Press, New York, 1960, pp. 297—337.
12 E. Melan, Theorie statisch unbestimmter Systeme aus ideal-plastischen Baustoff. Sitz. Ber. As. Will., IIa (1936) 145—195.
13 P.S. Symonds, Shakedown in continuous media. J. Appl. Mech., 18 (1951) 85—89.
14 P.S. Symonds and B.G. Neal, Recent progress in the plastic methods of structural analysis. J. Franklin Inst., 252 (1951) 383—407, 469—492.
15 W.T. Koiter, A new general theorem on shakedown of elastic-plastic structures, Proc. Kon. Ned. Ak. Wet., 859 (1956) 24—42.
16 W.T. Koiter, General theorems for elastic-plastic solids, in: I.N. Sneddon and R. Hill (Eds.), Progress in Solid Mechanics. North-Holland Publishing Co., Amsterdam, 1960, Ch. 4.
17 D.A. Gokhfeld and O.F. Cherniavsky, Limit Analysis of Structures at Thermal Cycling. Sijthoff & Noordhoff, Alphen aan den Rijn, The Netherlands, 1980.
18 P.G. Hodge, Jr., On the soap-film—sand-hill analogy for elastic-plastic torsion, in: Progress in Applied Mechanics: The Prager Anniversary Volume. Macmillan Co., Boston, 1963.
19 C.T. Herakovich and P.G. Hodge, Jr., Elastic-plastic torsion of multiply-connected cylinders by quadratic programming. Int. J. Mech. Solids, 11 (1969) 53—63.
20 F.S. Shaw, The torsion of solid and hollow prisms in the elastic and plastic range by relaxation methods. Rep. ACA-11, Australian Council for Aeronautics, 1944.
21 A. Naidai, Der Beginn des Fliessvorgangs in einem tordierten Stab. Z. angew. Math. Mech., 3 (1923) 442—454.
22 H.M. Van Rij and P.G. Hodge, Jr., A slip model for finite-element plasticity. J. Appl. Mech., 45 (1978) 527—532.
23 P.G. Hodge., Jr., and H.M. Van Rij. A finite-element model for plane-strain plasticity. J. Appl. Mech., 46 (1979) 536—542.
24 P.G. Hodge, Jr. and D.L. White, Nonuniqueness in contained plastic deformation. J. Appl. Mech., 47 (1980) 273—277.
25 D.L. White and P.G. Hodge, Jr., Computation of non-unique solutions of elastic-plastic trusses. J. Comp. Struct., 12 (1980) 769—774.

Aspects of Damage to Buildings from Uncased Explosives

W. JOHNSON

University Engineering Department, Trumpington Street, Cambridge CB2 1PZ (England)

Abstract

The kinds and degrees of structural damage caused mainly to a few sorts of large office-type buildings by "point" explosive charges such as those typified by terrorists' car bombs are discussed.

To help structural designers appreciate how their usual type of building designs will stand up to placed, uncased bombs of about 100 kg T.N.T., the author reviews some of the useful "open" literature he has been able to find. There is now available specific non-classified "open" and helpful information of military origin but it is not widely known. Short references to fire following explosions in buildings and of repairs to damaged structures are made.

1. Introduction and scope

For over thirty years the scientific activities of Professor D.C. Drucker have been conspicuous to many overseas researchers, especially those engaged in the field of metal plasticity. Perhaps most prominently has he been seen to be associated with Limit Theorems and Load Bounding Methods, yield criteria and plastic stress—strain (increment) laws. However, those who are sufficiently old have also encountered papers of his in journals and conferences on photo-elasticity, stress analysis, the study of material properties, dynamic plasticity and the relationships between micro- and macro-plasticity — among many others. In later life he has also been seen to be prominently engaged in educational and professional roles.

The subject matter of this paper embraces many of the scientific and professional dimensions of the subjects described above with which Prof. Drucker has been concerned. Below, specific concern is with an impulsive load delivered to a building for which it was not initially designed. The topic we discuss is that of the uncased bomb exploded in or near to ordinary civil premises with the intention of deliberately inflicting damage upon them. In certain quarters, terrorist bomb attack is constant and frequent and even where it is not, the purchasers of buildings sometimes express anxiety about the behaviour of their structure should it become an object of attack by unperceived extremist groups. To judge the consequences of a deliberately placed car bomb explosion is not easy or straightforward. Not only do buildings vary greatly in their design details and construction — though of course Codes of Practice and the like do have to be subscribed to and therefore tend to enforce a kind of uniformity so that they all do have certain common "protective" characteristics — but the varieties of structural material employed or involved are large and their response and characteristics differ greatly. It is easily appreciated too that any anticipation of total probable structural performance is not easily to be had because every bomb-placing circumstance is different. Principally, we shall be interested in response to blast loading (and fragments), for example directly against window-perforated walls, steel skeletons and through foundations etc., which elicits elastic and plastic quasi-static, dynamic and wave response in metals, soils, concrete or glass and the like, either separately, through admixtures or composites of them. Some appreciation of shock wave propagation through air is also required. These remarks are made to emphasise the true engineering or applied nature of the multi-faceted situation addressed. The design engineer does not here encounter the simple one element — one phenomenon problem of his student days. He must grasp a multitude of quite different scientific phenomena using his imagination in the first instance to judge which of them will have value for him; the range of expected behaviour and the number of unknown parameters renders circumstances so complex that rigorous scientific calculation is of limited value. A multitude of physical interactions, geometrical and material discontinuities and frequently the sheer lack of knowledge of prevailing details sets a great premium on judgement and previous experience. As is often the case in medicine, confrontation with many previous similar situations gives an enormous degree of help and confidence in predicting an outcome.

One subsidiary reason for outlining the difficulties encountered in analysing the title problem, is to emphasise a difficulty or challenge which perenially confronts the educator — researcher, namely how to develop students concurrently in scientific depth and engineering range; the scientific part is usually easy but to promote an attitude of mind which, through imagination, will facilitate the ability to analyse a many-parameter problem (set around by incomplete scientific understanding, and fuzzy, imperfectly known empirical knowledge) is very difficult. At best students are introduced to such design problems through projects and through case studies. (And doctoral students, mostly, have very little exposure to such matters).

2. Purpose

The principal aim is to review for structural engineers sources of information and references to enable them to assess, broadly, the degree and kind of damage likely to be caused to premises, equipment, utilities and people by a placed high explosive uncased bomb whose weight is of the order of 100 kg. If this can be done then some degree of economic protection can be designed or planned against them. Also, if an event does occur, probable repair costs can be at least dimly foreseen and estimated.

Until recently there has been little easily accessible and openly published information for the guidance of the designer. Such knowledge as exists, is mostly held by defence, home security and police departments; its circulation tends to be restricted and confidential. Much of the information available is highly empirical and has been assembled from military and civil reports written about incidents which have occurred, and simple testing. Information derived from war situations too has often been compiled and then extended or added to by testing, eg. see refs. (i) and (iii) in Section 3 below. (Information and the discussion of programmes about reinforced concrete slab and containment vessel design when there has been impact by aircraft, deformable missiles and blast is to be found in the Transactions of the 6th Int. Conf. on Structural Mechanics in Reactor Technology, Aug. 1981, Vols. J (a) and J (b), Loading Conditions and Structural Analysis and Reactor Containment, North Holland Publishing Co. Unfortunately, in the author's opinion, some of these papers are not reported in a manner which gives great confidence).

Without a relevant body of knowledge designers are unable to make rational decisions. Besides the appreciation of a wide body of scientific information, empirical knowledge about several kinds of material behaviour and geometrically complex structural performance is mandatory.

It is worth noting that to a significant extent some authorities deliberately prevent an awareness of potential hazards from being known and circulated in order to evade (supposedly) excessive public apprehension and attention. Opinions differ about the desirability of adopting this viewpoint. It suffices to say that from the experience of the author, structural designers are often unable to do their job adequately without access to proper information and discussions of aspects of their task which need to be considered. They require to have access to useful sources of information and discussion*. In the author's view informed and critical public groups also need to, and should be able to have ready access to knowledge about environmental hazards the better to allay any qualms they may have about officially "reliable" designs.

*This standpoint was accepted, to a small degree, in the U.K. during World War II as reference to the Proceedings of the Inst. Mech. Engrs. (London) for the early 1940s will show.

3. Some general references

Since we have in mind high explosive attack against various kinds of premises, some guidance about the damaging effects (of the magnitude) of blast overpressure and fragmentation on a wide variety of structures can be had in the very useful book, Effects of Nuclear Weapons (Edited by S. Glasstone and P.J. Dolan, U.S. Dept. of Defence and Energy Research and Development Administration, 1977).

Four other major references are:

(i) Structural Defence, 1945*, by D.G. Christopherson, (U.K.) Ministry of Home Security, Research and Experiments Department.

(ii) The Nuclear Explosive Yields at Hiroshima and Nagasaki**, by Penny, Samuels and Scorgie, Trans. Roy. Soc. (London), Vol. 266, 1970, 357—424.

(iii) Fundamentals of Protective Designs (Non-nuclear), Dept. of the (U.S.) Army Technical Manual, TM5-855-1, 1965.

(iv) Structural Design for Dynamic Loads, C.H. Norris et al., McGraw-Hill Book Co., New York, 1959, pp. 453.

An older but simple and easy to read book is C.S. Robinson's Explosions: Their Anatomy and Destructiveness (McGraw-Hill Book Co.), 1944. Other helpful papers, often containing many references are [1—4]. Blast Injury studied up to 1956, (with 166 references) has been extensively discussed by Clemedson [5].

Observe that aside from differences in pressure—time shock profiles as between H.E.*** and nuclear bombs (and in other ways too) nuclear blast is predominantly uniformly plane when it strikes the kind of structure we have in mind. The placed bomb however creates a blast which does not everywhere strike structures as a plane wave; roughly it is spherical and therefore its local obliquity varies enormously over a given face. Its diffraction effects are generally unlikely to be serious.

Two relevant reports are those of Hobbs and Cubison [6, 7] in which the effects of removing one or more columns from a typical reinforced concrete office building are studied in some detail. Figure 1 is taken from [7] and shows the yield line pattern which follows the removal of two columns closest to a corner of a 1/20 scale model building uniformly loaded over the floors.

4. Blast pressures and ground shock

Bombs manifest themselves in the form of blast, fragments and by earth shock. Except when intended to injure — and then they can carry nails and the like — car bombs are essentially lightly cased or uncased.

---

*This was rated Confidential until very recently.

**Deductions of bomb yield were made from studying damage to poles, drums, safe doors and cabinets etc.

***High Explosive.

Fig. 1. Yield hinges in a model structure after the removal of two columns near to a corner. (a) Model prior to commencement of loading. (b) Completion of test after removal of loads in order to show the yield line pattern.

Typically, thin milk churns are loaded with H.E. and placed in the boot of cars; the objective is to ensure a high degree of blast and they perform as did non-penetrating aerial or parachute mines of W.W. II (save perhaps from the effects of the particular chemical charge and its shape). Such bombs are untamped to the structure or the earth; they are in fact stood-off or uncoupled from them, and especially the earth transmitted shock is generally relatively weak. Of course buried bombs may be likened to aerial ones which are fused to explode after penetrating the earth.

The blast wave from an essentially point charge of detonated H.E. is of the form shown in Fig. 2 *when fully developed*; its characteristics are discussed in detail by Kinney [8]. However note that detonation results in near-zone and far-zone processes and that they need to be well distinguished; the distinction takes place at 15—20 charge radii. In the former, detonation products and shocked air interact to give a composite

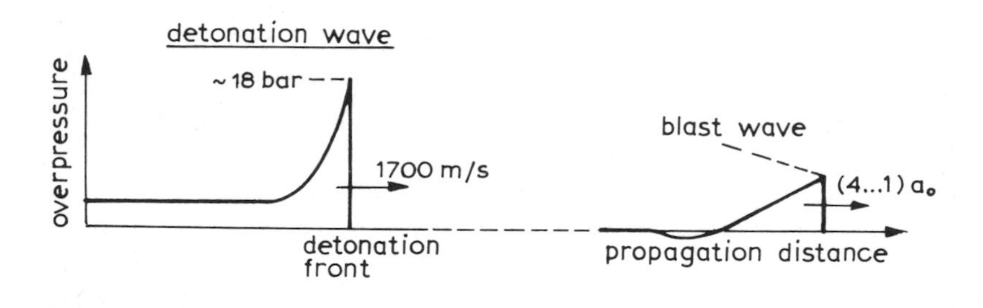

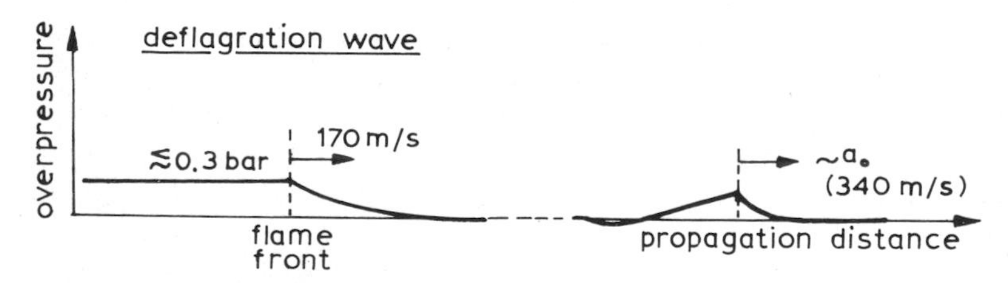

Fig. 2. Form of developed detonation and deflagration waves.

blast effect whilst in the latter only the atmosphere is involved; characteristics pertaining to one zone do not well apply to the other. Maximum compressive pressure p_{max} (the shock front or peak pressure) mostly determines damage level; the pressure falls off, at a given location according to

$$p/p_{max} = (1 - t/t_d) \exp(-ct/t_d). \tag{1}$$

t denotes time after the passing of the shock, t_d the duration of the positive pulse and c is a constant. It is only after a certain distance that a suction phase of small intensity (never less than about 0.77 bar) is developed; the magnitude of both the latter and p_{max} decrease with distance (or time), though their durations increase. Roughly, pressures fall off inversely as distance squared. Near to a building the former ($p_{max} > p > 0$) will push it inwards but further away, the suction effect may cause a pushing-out of the walls or windows. (The latter is probably due to the pressure internal to a building exceeding the suction pressure by an amount great enough to topple walls or fracture glass panes.)

Reference (iii) in Section 3 above gives,

$$p_{max,s} = \frac{4120}{z^3} - \frac{105}{z^2} + \frac{39.5}{z}, \tag{2}$$

where p_{max} is in lb/in^2 and $z = r/W^{1/3}$ where r is the distance from the charge W (in lbs) in ft; this gives for a 100 kg bomb, at 24 ft, $p_{max} \simeq 70\,lb/in^2$.

Expression (2) applies to side-on pressure. Normal or side-on reflection of the blast pressure from a rigid surface increases it to between 2 and 8 according to

$$\frac{p_r}{p_s} = 2 + \frac{6p_s}{p_s + 7p_0}; \tag{3}$$

p_r is the reflected pressure and p_0 ambient pressure.

The distance and angle of obliquity at which the shock wave from a ground placed bomb encounters points on the face of a large building varies enormously as previously remarked. Recourse to Tables, ref. [8], is needed to determine reflected pressure magnitudes. Note that data derived from tests using nuclear devices for relating blast overpressure to damage to specific building types generally refers to effectively plane waves, i.e. the whole of a building face is considered identically loaded. Also note that obstacles alter the impact effect of the two phases of blast and can have complicated consequences according to their shape though with buildings positive peak pressure tends to be reduced. Abrupt turns in corridors affect positive peak pressure significantly and suction little.

For values of $p_s > 70\,lb/in^2$, shock pressure is accompanied by a significant "wind" giving a drag overpressure; but these values of p_s are too large to be of interest to us directly, i.e. are irrecoverably destructive.

Peak overpressure magnitudes (i.e. perpendicular to the blast wave) have many times been related to structural damage; see Table 1 made up from references such as in Section 3 (iii), [3] and [8].

From careful tests on six large brick houses on two different soils, it has been found using charges of up to about 600 lb of H.E. in holes from 15 to 30 ft in depth, that a soil particle speed of 4—5 in per second was sufficient to cause building damage. A formula for damage threshold is $W^{2/3}/d = 0.3$; (d is the distance from the charge, W lb, to the building). (As this expression is said to agree with a Swedish recommendation for small charges, the equation should hold over a wide range of charges and

TABLE 1
Structural damage and peak over-pressure

Structural type	Damage	Overpressure (lb/in^2)
Wooden frame residences	Moderate	2–3
	Severe	3–4
Wall-bearing, masonry apartment houses	Moderate	3–4
	Severe	5–6
Multi-storey, wall bearing, monumental	Moderate	6–7
	Severe	8–11
Reinforced concrete: small window area	Moderate	8–10
	Severe	11–15
Steel frame office type	Moderate	4
	Severe	19

soils.) A charge of 250 lbs at 120 ft distance was found to effect no damage at all [10].

Bomb crater dimensions for a given charge depend on the depth of burial of the latter and the type of soil. Shallow burial gives shallow craters. The fall-back of earth into a crater from which it has been ejected may significantly alter the crater dimensions. For greatest amount of displaced material a common rule is to bury the charge at depth $D \approx 2W^{1/3}$ (W is in lb and D in ft.). Too great a burial results in camouflet formation.

Ground shock results in a violent upward jerk followed by a slow horizontal outward movement with partial recovery. Both horizontal and vertical motion fall off rapidly with distance. Roofs are often uplifted and thrown clear of walls. (With light roofs this behaviour following on internal explosion, reduces damage).

It may be recognised that whilst attack from car bombs is principally in mind, somewhat similar attack occurs when vapour cloud explosions and aircraft impact take place. Much too that is similar results from internal explosion i.e. dust and town gas explosions [11—14].

Useful experimental results where high explosives were detonated in contact, tamped or stood-off in water, from reinforced concrete dams have been given by Collins [15].

Earthquake studies have yielded an enormous amount of empirical damage information and many analytical design approaches have long been available, see for instances [16, 17] and [18].

5. Types of building damage

It is usual to divide damage into (i) primary: which occurs locally due to the direct effect of the blast and the pushing-over of part of the structure and (ii) secondary: due to subsequent collapse of other parts. Damage (i) is unavoidable and reduces with proper strengthening but in connection with (ii) instability should be anticipated and be preventable.

A framed structure is much less liable to collapse than one having solid load bearing walls. A principal factor making for anti-bomb worthiness is the establishment of a proper *continuity of reinforcement* or soundly *jointed structures* with beams and columns acting as one structure. This can only be found in steel or reinforced concrete framed buildings. "Knock-on" or "domino" effects are then avoided and increased resistance to collapse follows. Fully framed construction performs extremely well, but a composite building having internal steel beams and columns with external walls of brick to support the external ends of beams has a resistance of lower order.

The necessity of designing-in "continuity" and "ductility" for successfully combating blast is abundantly well made in the monograph by Baker [19]. "Ductility" allows large plastic deflection and the dissipation of strain energy due to blast without excessive cracking, any disintegration or loss of ability to carry (or re-distribute) the design loads.

Redundancy in any attacked structure is enormously helpful; the load carried by members which are removed by blast is then easily re-distributed. Permeability — the allowing of a blast wave to travel through the space between the main members of the structure — is equally desirable: the frame elements should have little area on which the blast pressure may act and be very rapidly surrounded or engulfed. With these features or principles in mind, high degrees of structural integrity can be obtained.

Experiments [20] have shown that 500 lb and 1000 lb (cased) bombs exploded in contact with initially undamaged surfaces of r. c. slabs (with a Brunswick system of reinforcement) require to be of 5 and 6½ ft thickness respectively in order to resist severe spalling. (The concrete would have a compressive strength after 28 days of 4000 lb/in^2. If spall plates are used the concrete thickness may need to be reduced a little.) Side-on pressure from 500 lb bombs is estimated to be 20 lb/in^2 at 40 ft distance and the impulse 0.05 lb s/in^2.

A French formula for mass explosions is

$$R = CW^{1/3};$$

R denotes the radius in metres through which heavy damage is sustained, C is a constant (depending on the explosive but ≈ 10), and W is the explosive weight in kgs. Generally the constant in the formula is believed to be too large. For $W = 100$ kg, the latter formula gives $R \approx 70$ ft whilst that on p. 3-47 in ref. 3 (iii) above, gives $R \approx 50$ ft.

Broadly, charge shape is of no influence when its distance from a wall, r_c is greater than $1.5\,W^{1/3}$ ft; if $W = 100$ kg, $r_c \approx 9$ ft [21]. In the same reference results from a W.W. II investigation by Bernal are given for charge weights which cause slight cracks to brick walls of cement mortar, of one, two and three thicknesses, due to blast; a 110 lb charge at 17 ft is said to cause such a damage level. Also an "average" brick wall-bearing construction suffers demolition out to a distance of about $W^{2/3}$ or of damage to $2\,W^{2/3}$; for $W = 100$ lb, the distances are ~ 20 ft and 40 ft respectively.

Fragment penetration is treated in the first four general references mentioned in Section 3 above but refs. [22, 23] contain data pertaining to relatively large weight missiles impinging with velocities of up to 600 ft/s against concrete slabs.

GENERAL DAMAGE FEATURES

If the blast is sufficiently intense, progressive collapse may arise following an explosion or after the structural damage-causing-agent has initiated it in the case of high rise masonry or block-work brick load-bearing structures. Some differences in structural behaviour follow according as the explosion is internal or not and whether venting is possible or not.

Spread of collapse downwards may occur due to impact loads from falling débris, e.g. when side load-bearing walls are blown out, such as happened in the 22-storey Ronan Point disaster. It may also occur sideways and/or upwards due to the removal of supports* or columns; see Fig. 1 and [6, 7].

The ability to bridge-over severe structural damage is highly desirable** and in essence is due to redundancy in the structure; rooms should be enclosed by substantial walls and there is value in allowing for (rapid) venting to limit major structural damage to just a few rooms.

The venting of blast via doors, windows, chimneys and other apertures is helpful in allowing confined air to escape. For an internal explosion, lines of low resistance for the escape or relief of expanding gases are useful, see [2].

Particles of glass penetrate many items e.g. insulation, and can puncture to admit moisture. Generally, elasticity in anchorages is desirable (the tying-down of equipment), otherwise they evoke resistances which lead to excessively high stresses. Remember that cast iron is especially vulnerable to shock especially for instance the feet of motors, brackets and bearings. Vertical piping runs are particularly blast prone if very securely held [24].

The cellular treatment of floors to localise explosion and fire effects was adopted in W.W. II. Dividing walls should be "elastic" to "absorb" blast pressure without communicating vibration to the structure and sufficiently thick to arrest fragments.

False ceilings, copings, parapets and heavy hung items should be avoided.

---

*Specifically, sufficient 'vertical' structure.

**Ends and corners are often very vulnerable points.

WINDOWS: GLASS

Many glasses break into needles and hence even at small speeds of projection may be very injurous to people and may penetrate equipment etc. Various proprietory, tough, adhesive films and fabrics are available for application to glass windows to help reduce any shattering or indeed hold it together without seriously reducing its light transmitting qualities. Other "anti-scatter" treatments have included liquids, (e.g. rubber latex and resin laquer), adhesive paper and netting.

To offset these effects, additional "supports" can be devised to increase the pressure at which rupture occurs (though this may so increase the natural frequency of a glass plate that the prospect of fracture is actually increased), or even yield easily and soon, i.e. vent, without a large impulse and hence speed being given to it. Also, of course, more ductile translucent materials can be substituted for glass.

FOUNDATION DAMAGE

The earth shock encountered is usually considered in terms of (i) crater damage, (ii) the soil permanently displaced, and (iii) the elastic zone which simply transmits waves. Foundation damage if observed requires a soil survey. Loss of bearing capacity through (ii) is thought to be improbable unless water pipes, drains or ground drainage have been disturbed to lead to some degree of flooding.

Piled foundations only suffer damage to the heads of piles and by the cracking of their shanks at a lower level. Shoring-up, the cutting away of heads and replacement by mass concrete is possible.

Vertical and horizontal movement due to explosion results in the shift of equipment and services, e.g. boilers and steam pipes.

6. Repair of damaged buildings [25]

Once the kind of damage undergone by buildings has been identified the method of repair is often fairly obvious. Reference [25] considers and then discusses repairs by reference to the damage inflicted on particular materials of construction and the form in which they are used. Some interesting comments which emerge are as follows.

WALLS

Walls cracked vertically or diagonally may well retain adequate load-bearing capacity without repair though weather resistance and appearance may demand attention. However, loss of stability in walls which are tilted, bulged or have moved laterally* usually proves to be more important than any reduction in strength which is undergone; and complications due to foundation settlement add to this. Bulging or tilting can be arrested by adding buttresses or piers and in some cases internal tie rods may be allowable.

ROOFS AND FLOORS

Timber roofs though properly triangulated and tied together, which may have been lifted off their supports without being damaged, can be jacked back into position after repairing broken elements.

STRUCTURAL STEEL FRAMEWORK

Structural steel frameworks can suffer some small degree of plastic deformation to stanchions and beams without creating great concern. Stanchions in multi-storey buildings can be fractured in a damaged upper portion and yet remain suspended, i.e. without there being general collapse. Concrete casing damage may need to be carefully examined to disclose the degree of hidden damage to steel work.

REINFORCED CONCRETE STRUCTURES

Columns at intermediate distances from the explosion may crack by sideways bending causing a shattering of concrete and requiring attention to longitudinal rods and a re-casting of concrete. It may also cause stretching between floors — usually at the head or foot of a column.

Loosened concrete needs to be cut away and replaced. Floors may need shoring-up with concrete beams being cut-out and renewed. For concrete cracks as caused by reversed loading, grouting may be sufficient.

Reference [26] refers to two 60 m-high framed structures (with external insulating brick walls), one in r. c. and the other a steel skeleton; both survived heavy artillery bombing in Warsaw in 1944 and restoration was economically achieved.

7. Fire damage

Regrettably, the academic fraternity at large pays little or no attention to the study of fire, i.e. the combustion of materials and the fluid flow of hot gases in burning buildings, at undergraduate or post-graduate level and this is a matter which should be improved**. A conspicuous exception is

*With internal blast, upper storeys are more vulnerable than those lower down because of lower compressive restraints to wall displacement and bowing.

**These words were written before the Falkland Islands War revealed the apparently poor fire performance of British naval ships when hit by Exocet-like missiles.

the review "Fire", treated by H.W. Emmons in terms of applied fluid mechanics [27]. The latter, however, treated a single room or cell rather than an office-block type complex — of which there appears to be little or no theoretical study. There are none-the-less many diverse articles on aspects of industrial fire and those typified by Murgai [28, 29] and Sax [30] on structural and industrial fire protection are very useful in many respects. Helpful information and references are to be had from national research stations directly, (e.g. the British Fire Research Station) and for instance from a perusal of occasional "References to Scientific Literature on Fire" [31].

Some books at levels from the elementary (and meant for fire fighters), see [32] and [33], to the scientific/research monograph, see [28], may be consulted for up-to-date data, results, discussions and analyses pertaining to fire in buildings.

FIRE DAMAGE TO STEELWORK

This has two aspects: (i) the overheating of metal and (ii) the distortion of members and connections by unequal expansion*. Creep in stanchions especially can lead to rapid distortion and collapse. Also protective coverings can be destroyed and steel exposed to undergo softening, whilst burnt steel (prolonged high temperature heating) will lead to scale formation.

Where the in-situ straightening of bulged stanchions is inappropriate additional welding plates or angles can be added, or for severer cases, splicing-in of new pieces can be carried out or a reinforced concrete column cast around them; inserting secondary stanchions is of course optional.

FIRE DAMAGE TO REINFORCED CONCRETE

Reinforced concrete buildings are best suited to resist fire. Heated concrete surfaces expand and crack and portions then detach or spall, fall away and ultimately may expose reinforcement.

Concrete strength typically falls off with temperature, negligibly at 200°C, but at 300°C having deteriorated by 20%; its colour then changes from grey to pink or red. At 600°C the concrete again becomes grey and its strength is 1/3 its original value. At 900°C all strength is lost and the colour is buff, whilst at 1200°C, the concrete sinters. (Colour changes across a section and reveals the temperatures reached.) Thus concrete discolouration is expected often with some buckling of vertical rods between hoops.

It is often emphasised that fire passes upwards through openings in

*Again, recall the criticism launched on naval architects for allowing Falkland War battle frigates to have aluminium super-structures.

concrete floors and that had they been sealed (as are vertical pipe runs) conflagration would have been avoided.

A point often made in the literature is that damage caused by water for use in the extinguishment of fire can exceed by many times that caused by the fire. Drying out and deterioration (especially of electrical gear) due to exposed conditions are important features.

Equipment, instruments and tools (jigs and gauges) suffer more damage by fire than shock.

FIRE ATTACK ON SERVICES

The occurrence of many large building fires has recently drawn international attention and it will suffice to recall that if fire succeeds blast damage, then further large scale damage and difficult to manage rescues may ensue. Considerable here are,

(a) Highly flammable polymer-covered* electric insulation and furnishings*; if these have been used then black, choking, toxic smoke which is very rapidly created may be expected.
(b) Fire can be spread enormously fast* in the vicinity of machines or equipment which take in or circulate large volumes of air.
(c) Roof louvres to mitigate fire spread are very worthwhile. An increased number of artificial fire breaks to prevent the communication of fire near to points of potential explosive attack may be considered. Lift shaft-heads probably need special attention.
(d) Structural materials such as aluminium can quickly distort for relatively small rises in temperature. (Recall the footnote on the previous page.)

8. Design to resist external blast

DESIGN PROCEDURES

Newmark [16] has earlier remarked on and considered the effects of external blast on buildings in terms of the effective duration of the loading and the fundamental period of the structure, rather than actual damage. When the duration of blast is short — less than 1/3 the natural period — design is to be based on impulse and momentum but when it is long, say four times the period, then a quasi-static analysis — a step load of infinite duration — can be considered applicable. Wind and earthquake design criteria constitute first approximation approaches. Newmark's paper appears however to have in mind bomb sizes much larger than those envisaged here. The earlier remarks in Section 4 need however always to be kept in mind. But recall the refs. in Section 3 and see ref. [18].

---

*Deaths from such furnishings in buildings and vehicles (aircraft, ships, underground, coaches and buses etc.) are frequently reported in the press: no more than two minutes may be available to effect escape from such situations.

GENERAL SUGGESTIONS

It has been said that "Steel-framed buildings encased with concrete, well-designed, well put together, well anchored and mainly with flat roofs . . . withstand the effects of bombs best of all. In steel-frame buildings internal partitions should be anchored, reinforced and tied-in to maintain a cohesive structure; the same applies to external panel walls. Piping should run in conduits and generally be not secured if underground or rested on beams; then services underground also should be kept from fixing to the earth. Cables should be put in ducts and then filled with sand and covered with cement."

Small buildings built with lime mortar are highly damage proof but cement mortar much less so.

Codes of Practice, regulations

Much (quasi) legislation exists in the U.K. (and in U.S.A.?) to facilitate designing against internal gas explosions or for facilitating the safe demolition of prestressed, high-rise structures*, see British Standard Code of Practice 94, 1971.

Conclusion

Within the space available, a discussion document and useful references have been provided about an extensive multi-disciplinary subject which the late 20th century structural design engineer may, unfortunately, have to examine. It is hoped that the remarks, implicit suggestions, recommendations and information will prove useful to him in this task.

Acknowledgement

I am indebted to Mrs. Sarah Purlan for typing this paper.

References

1 W. Johnson and S.K. Ghosh, Demolition and Dismantling. I.J.M.E.E., 8 (1980) 111—126.
2 R.J. Mainstone, Accidental Explosions and Impacts. Brit. Building Res. Est., CP 58/78, 1978.

---

*For some remarks and references on this topic, see pp. 377—340 in Mechanical Properties at High Rates of Strain, (1979), Inst. of Physics, London, Conf. Series No. 47.

3 M. Kornhauser, Structural Effects of Impacts. Spartan Books Inc., 1964, pp. 205.
4 J. Henrych, The Dynamics of Explosion. Elsevier Scientific Pub. Co., 1979, pp. 558.
5 C.J. Clemedson, Blast Injury. Brit. Med. J., 36 (1956) 335—354.
6 B. Hobbs and S.J. Cubison, Analysis of R.C. Buildings Subjected to Localised Damage. ACI Int. Sym. on Rehabilitation of Structures, Dec. 1981.
7 B. Hobbs and S.J. Cubison, The Effect of Damage on the Strength and Stability of Structures. Internal Report, Univ. of Sheffield, Dept. of Civil and Structural Eng., May 1979.
8 G.F. Kinney, Explosive Shocks in Air. MacMillan Co., 1962, pp. 198.
9 S.B. Hamilton, Repair of bomb-damaged buildings. The Structural Engineer, (i) XXIII(Feb. 1945) 77—92, and (ii) XXIII(August 1945) 376—390.
10 A.T. Edwards and T.D. Northwood, Experimental studies of the effects of blasting on structures. The Engineer, 210 (1960) 538—546.
11 K. Gugan, Unconfined Vapour Cloud Explosions. Inst. Chem. Engrs., Geo. Godwin Ltd., 1980, pp. 168.
12 T.R. Anton and J.H. Pickles, Deflagration of Heavy Flammable Vapours. Inst. Maths. and its Applications, 16 (1980) 126—133.
13 R.J. Mainstone, Internal Blast. Tech. Comm., No. 8, Fire and Blast, 1973.
14 W.C. Griffith, Dust Explosions. Ann. Rev. Fluid Mech., 10 (1978) 93—105.
15 A.R. Collins, The origins and design of the attack on the German dams. Proc. Instn. Civ. Engrs., Pt. II, 73 (1982) 282—405.
16 N.M. Newmark, External Blast. Tech. Comm. No. 8, Fire and Blast, 1973.
17 R.L. Wiegel (Ed.), Earthquake Engineering, Prentice Hall Inc., 1970, pp. 518.
18 D.G. Fertis, Blast and Earthquakes in Dynamic and Vibration of Structures. John Wiley, 1973 Chap. 11.
19 (Lord) J.F. Baker, Enterprise Versus Bureaucracy. Pergamon Press, 1978, pp. 123.
20 (Lord) J.F. Baker, Memorandum on the Design of Bomb-Resisting Structures against H.E. Attack. Structural Engineering Branch, Ministry of Works, U.K., 1954.
21 T.D. Northwood and R. Crawford, Blasting and Building Damage. Can. Bldg. Digest No. 63, NRC Div. Building Res., Ottawa, 1965.
22 R.P. Kennedy, A review of procedures for the analysis and design of concrete structures. Nuclear Engineering and Design, 37 (1976) 183—203.
23 P.P. Degen, Perforation of reinforced concrete slabs by rigid missiles. J. Struct. Div. A.S.C.E., (July 1980) 1623—1642.
24 H. Gutteridge, Proneness to damage of plant through enemy action. Proc. I. Mech. E., 147 (1942) 99 and 120.
25 H. Gutteridge, The repair of bomb-damaged buildings. The Structural Engineer, XXIII (Feb. and Aug. 1945) 77 and 376.
26 S. Kajfasz, External blast. Tech. Comm., No. 8, Fire and Blast, Discussion No. 6, 1973.
27 H.W. Emmons, Fire. Proc. of 8th U.S. Cong. of App. Mech., 1978.
28 M.P. Murgai, Natural Convection from Combustion Sources. Mohan Primlani, Oxford and IBH Publishing Co., New Delhi, 1976, pp. 132.
29 M.P. Murgai, Similarity Analysis in Fire Research. Mohan Primlani, Oxford and IBH Publishing Co., New Delhi, 1976, pp. 377.
30 N.I. Sax, Industrial Fire Protection, Dangerous Properties of Engineering Materials. Van Nostrand Reinhold, London, 1975.
31 N.I. Sax, References to Scientific Literature on Fire (July—December, 1980), ISSN 0306 5766. Building Research Establishment, U.K.
32 E.W. Marchant, Fire and Buildings. Medical and Technical Publishing Co. Ltd., 1973, pp. 268.
33 M.F. Dennett, Fire Investigation. Pergamon Press, 1980, pp. 104.

On the Brittle Fracture of a Thin Plastic Interlayer in Creep Conditions

L.M. KACHANOV

Boston University, Boston, MA (U.S.A.)

Summary

The creep under tension of a thin plastic layer joined to two parallel "rigid plates" in the cases of plane strain and axial symmetry is analyzed. In the central part of the layer the state of three-axial tension is developed; its level is significantly higher than the average tensile stress. The time to brittle fracture due to damage accumulation in the layer is determined.

1. Introduction

Drucker [1] considered the mechanism of brittle fracture under tension of a thin plastic layer joined to more rigid parts of the specimen.

This paper deals with the problem of brittle fracture of a thin plastic interlayer under tension in creep conditions. Such "soft" (in the sense of creep resistance) interlayers can appear in metals with inhomogeneities subjected to hot working and also in welds where they appear as a result of metallurgical changes caused by high temperatures. The deformation of a soft interlayer is hindered by adjacent more rigid layers, which results in high tensile stresses appearing in the interlayer.

The peculiarities of stress state of a thin soft interlayer were first investigated by Prandtl in his well-known work on the flow of a thin ideally-plastic layer compressed between parallel rigid plates. His solution to this problem can be found in text books on the theory of plasticity (see, for example, [2]).

2. Steady creep of a thin interlayer under tension (Plane strain)

Consider a plane strain problem of steady creep in a thin interlayer $|x| \leqslant l$, $|y| \leqslant h$ (Fig. 1) under tension; it is assumed that

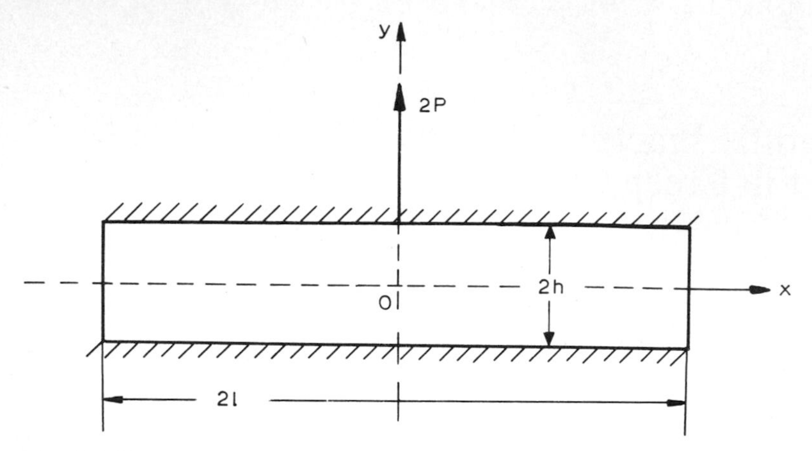

Fig. 1. Thin interlayer under tension.

$$\delta \equiv h/l \ll 1.$$

We assume that no creep deformation occurs in rigid plates.

Introduce dimensionless coordinates

$$\xi = x/l, \qquad \eta = y/l.$$

The shear stress τ_{xy} is an odd function of η. We assume that

$$\tau_{xy} = R(\xi)\eta/\delta \tag{1}$$

where $R(\xi)$ is an unknown function. In the case of plane strain the stress components τ_{xz} and τ_{yz} equal zero. The velocity w (in the z-direction) also equals zero. The sections $\eta = 0$, $\eta = \pm\,\delta$ remain plane. The layer is thin, therefore it is natural to assume that all sections $\eta =$ const. remain plane, i.e.

$$v = v(\eta) \tag{2}$$

where v is the velocity of the particles of the layer in the y-direction. It is obvious that $v(\eta)$ is an odd function since the middle section $\eta = 0$ is a symmetry plane. The velocity component u (in the x-direction), according to the adhesion condition on contact planes, is

$$u = 0 \quad \text{for} \quad \eta = \pm\,\delta. \tag{3}$$

In the case of plane strain the relations of steady creep are (see, for example, [3])

$$\dot{\epsilon}_z = 0; \qquad \dot{\epsilon}_x + \dot{\epsilon}_y = 0; \qquad \sigma_z = \tfrac{1}{2}(\sigma_x + \sigma_y); \tag{4}$$

$$\sigma_y - \sigma_x = 4g(H)\dot{\epsilon}_y; \qquad \tau_{xy} = g(H)\dot{\gamma}_{xy}; \tag{5}$$

where $\sigma_x, \sigma_y, \ldots, \tau_{xy}$ are stress components; $\dot{\epsilon}_x, \dot{\epsilon}_y, \ldots, \dot{\gamma}_{xy}$ are strain rate components; $g(H)$ is a characteristic function (for the given temperature) of the intensity of shear strain rates

$$H = (4\dot{\epsilon}_x^2 + \dot{\gamma}_{xy}^2)^{1/2}. \tag{6}$$

It is possible to assume that in creep conditions the metal is incompressible, i.e.

$$\partial u/\partial\xi + \partial v/\partial\eta = 0. \tag{7}$$

We shall use Norton's power law, then

$$g(H) = BH^{\mu-1}, \qquad 0 \leqslant \mu \leqslant 1, \tag{8}$$

where $B > 0$ and μ are constants. Substituting $v(\eta)$ in the equation of incompressibility and integrating, we obtain

$$u = -v'(\eta)\xi. \tag{9}$$

The arbitrary function of η can be ignored due to the symmetry condition ($u = 0$ for $\xi = 0$), so

$$l\dot{\epsilon}_x = -v'(\eta); \qquad l\dot{\gamma}_{xy} = -v''(\eta)\xi. \tag{10}$$

From the adhesion condition (3) it follows that

$$v' = 0 \quad \text{as} \quad \eta = \pm\,\delta. \tag{11}$$

We seek the solution in the form

$$v''(\eta) = A|\eta|^{s-1}\eta \tag{12}$$

where A, s are some constants. Determining $v'(\eta)$ and substituting $\dot{\epsilon}_x, \dot{\gamma}_{xy}$ and τ_{xy} in the equation (5) we obtain ($\eta \geqslant 0$)

$$R(\xi)\frac{\eta}{\delta} = \frac{AB}{l}\xi\eta^s\left[\frac{2|A|\delta^s}{(s+1)l}\right]^{\mu-1}$$

$$\left\{\left(\frac{s+1}{2}\right)^2\left(\frac{\eta}{\delta}\right)^{2s}\xi^2 + \delta^2\left[\left(\frac{\eta}{\delta}\right)^{s+1} - 1\right]^2\right\}^{\frac{\mu-1}{2}}. \tag{13}$$

The second term in braces, containing the factor δ^2, is small. Neglecting this term and comparing both sides of the latter equation, we obtain that $s = m$ and

$$R(\xi) = -\frac{AB}{l}\left|\frac{A}{l}\right|^{\mu-1}\delta\,|\xi|^{\mu-1}\xi, \qquad (\mu = 1/m). \tag{14}$$

Because of the adhesion condition (11) we obtain that along the contact plane $\dot{\epsilon}_y = 0$, so

$$\sigma_y = \sigma_x \quad \text{at} \quad \eta = \pm\,\delta. \tag{15}$$

The layer is thin, therefore it is possible to assume that the normal stress σ_y is constant in the direction normal to the layer. Substituting the shear stress τ_{xy} into the differential equation of equilibrium

$$\frac{\partial \sigma_x}{\partial \xi} + \frac{\partial \tau_{xy}}{\partial \eta} = 0$$

and integrating along the contact line we obtain

$$\sigma_x = \frac{B}{1+\mu} \frac{A}{l} \left| \frac{A}{l} \right|^{\mu - 1} \xi^{1+\mu} + C_1 \tag{16}$$

where C_1 is an arbitrary constant. The stress σ_y, as mentioned before, can be determined by (16) in the whole layer. Then the stress σ_x, inside the layer, can be found from the first equation (5). But it is easy to see that the right side of the equation equals zero at $\eta = \pm\, \delta$ and contains the small factor δ^2. Therefore it is possible to assume with sufficient accuracy that the stress σ_x is also constant in the direction normal to the layer.

The boundary conditions at the ends $\xi = \pm\, 1$ of the layer will be satisfied according to Saint-Venant's principle, i.e.

$$\int_{-\delta}^{\delta} \tau_{xy}\, \mathrm{d}\eta = 0; \qquad \int_{-\delta}^{\delta} \sigma_x\, \mathrm{d}\eta = 0. \tag{17}$$

The first condition is satisfied. The second condition determines the constant C_1. Finally, the stresses σ_y are equivalent to the tensile force $2P$. This condition determines the constant A. As a result of calculations we obtain

$$\begin{aligned} \sigma_y &= \frac{2+\mu}{1+\mu}(1 - |\xi|^{1+\mu})p; \qquad \sigma_x = \sigma_z = \sigma_y; \\ \tau_{xy} &= (2+\mu)|\xi|^{\mu-1}\, \xi \eta p \end{aligned} \tag{18}$$

where $p = P/l$ is the mean stress; in the case of compression $p < 0$.

From this solution it follows that a three-axial tension takes place in the central part of the layer. The biggest tensile stress (at $\xi = 0$) is

$$\sigma_{\max}/p = 1 + 1/(1+\mu). \tag{19}$$

The distribution of the normal stress σ_y is shown in Fig. 2 for $\mu = 0; 0.5; 1.0$. The shear stress on the contact line $\eta = \delta$ is: if $\mu = 0$, then $\tau_{xy} = +2p\delta$ for $\xi > 0$ and $-2p\delta$ for $\xi < 0$. If $\mu = 1$ (linear viscosity), τ_{xy} is proportional to ξ.

For $\mu = 0$ and $p < 0$ the stress distribution is analogous to the stress distribution in the above-mentioned Prandtl's problem. But in the case of creep it occurs at any load p.

In creep problems there is no yield condition and, consequently, no limit load.

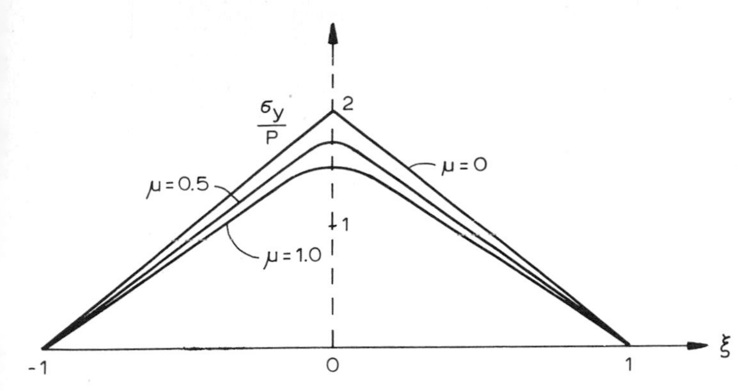

Fig. 2. Distribution of normal stress σ_y.

3. Unsteady creep

At the initial moment $t = 0$ the layer is elastic. It is possible to assume that the layer and the "rigid plates" have the same elastic constants. Then the initial elastic stress state of the layer (we denote this state by a prime $'$) is

$$\sigma'_x = 0; \qquad \sigma'_y = p; \qquad \tau_{xy} = 0. \tag{20}$$

The stresses in the steady creep state, considered above, will be denoted $(\sigma''_x; \sigma''_y; \tau''_{xy})$. In unsteady creep under constant load the stress state changes from the initial elastic state to the steady creep state. According to the general method of solution (see [3]) the stresses can be represented in the form

$$\begin{aligned} \sigma_x &= \sigma'_x + \tau(t)(\sigma''_x - \sigma'_x), \\ \sigma_y &= \sigma'_y + \tau(t)(\sigma''_y - \sigma'_y), \\ \tau_{xy} &= \tau'_{xy} + \tau(t)(\tau''_{xy} - \tau'_{xy}). \end{aligned} \tag{21}$$

The function of time $\tau(t)$ is determined by the variational equation of unsteady creep; $\tau(t)$ is a monotonically increasing function (Fig. 3); it equals zero at the initial moment $t = 0$ and tends to 1 as $t \to \infty$.

The function $\tau(t)$ has the approximation [3]

$$\tau(t) = 1 - e^{-t_0} \tag{22}$$

where the dimensionless time

$$t_0 = \frac{Q(0)}{2\widetilde{\Pi}_-} \Omega(t) \tag{23}$$

is introduced. The factor $Q(0) > 0$ is calculated from the stress fields in the initial state and in the final steady state; $\widetilde{\Pi}_-$ is the elastic energy of the

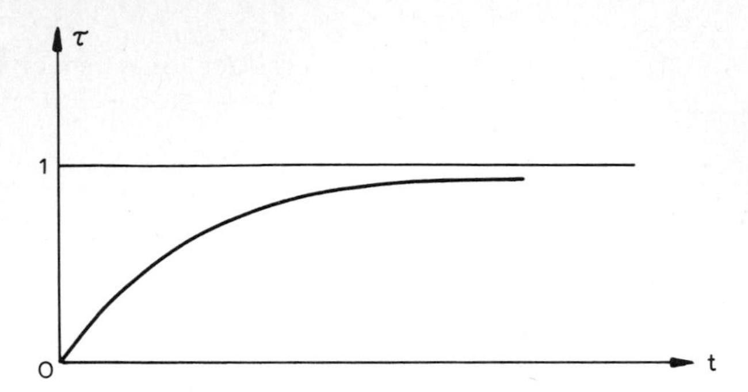

Fig. 3. τ as a function of time.

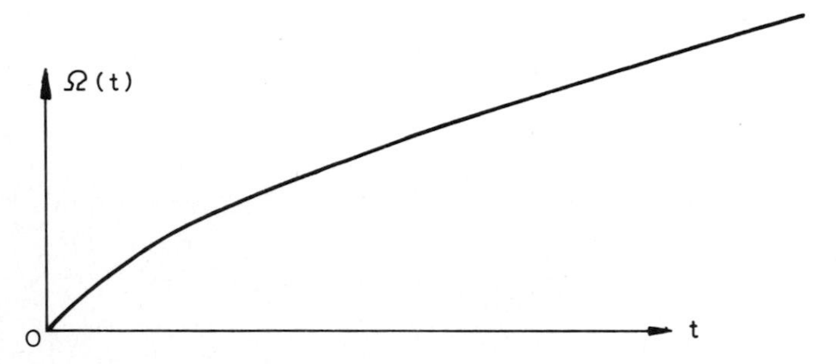

Fig. 4. Ω as a function of time.

layer for the stress differences enclosed in parentheses (21). The function $\Omega(t)$ is proportional to the creep curve of the uniaxial tension under a constant load, Fig. 4. The function $\Omega(t)$ is approximately linear when dealing with long periods of time.

According to the solution (21) the maximum tensile stress (at $\xi = 0$) is

$$\sigma_{\max} = \left[1 + \frac{\tau(t)}{1+\mu}\right] p. \tag{24}$$

The ratio $\sigma_{\max}/p$ monotonically increases (Fig. 5) from the initial value 1 to the steady (for $t \to \infty$) value $\alpha = (2+\mu)/(1+\mu)$. Note that $1.5 \leqslant \alpha \leqslant 2.0$.

4. Time to brittle fracture

The time to brittle fracture of the layer is determined by the damage accumulation described by the kinetic equation [3]

$$\mathrm{d}\psi/\mathrm{d}t = -A(\sigma_{\max}/\psi)^n \tag{25}$$

where $A > 0$, $n > 1$ are constants. The function ψ ("continuity") characterizes the level of damage. In the initial undamaged state $\psi = 1$; the

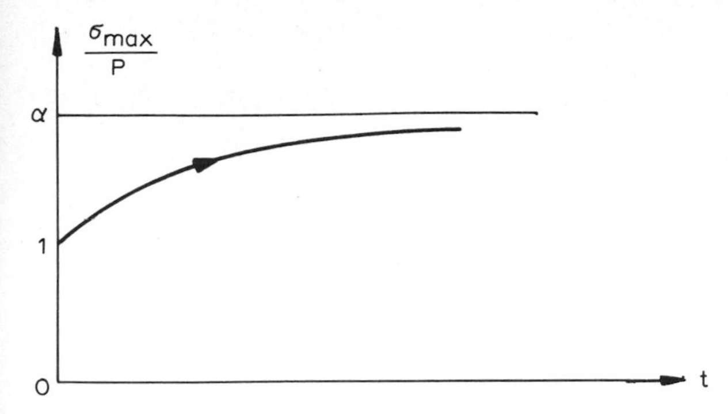

Fig. 5. σ_{max}/p as a function of time.

material is fractured at $\psi = 0$. From (25) it follows that the time to fracture t_* can be found from the equation

$$1/A(n+1) = \int_0^{t_*} \sigma^n_{\max}(t)\mathrm{d}t \qquad (26)$$

It is easy to see that there exists a unique root t_* of the equation (26). Since $p \leqslant \sigma_{\max} \leqslant p\alpha$, we have

$$t'' \leqslant t_* \leqslant t' \qquad (27)$$

where $t' = [(n+1)Ap^n]^{-1}$ is the time to fracture for the elastic stress field, and $t'' = t'\alpha^{-n}$ is the time to fracture for steady creep stress field.

5. Tension of a thin axisymmetric interlayer

Consider the problem of tension of a thin "soft" axisymmetric layer in the case of steady creep.

Let r, ϕ, z be cylindrical coordinates, u, v, w — radial, circular and axial velocity components, respectively. As before, the layer ($0 \leqslant r \leqslant a$; $|z| \leqslant h$) is joined to parallel rigid plates. We again assume that $\delta \equiv h/a \ll 1$ and introduce the dimensionless coordinates

$$\rho = r/a; \qquad \zeta = z/a.$$

The case of torsion is excluded from consideration here; therefore $v = 0$. Assuming that the sections $\zeta = \text{const.}$ remain plane we have $w = w(\zeta)$. The strain rate components are as follows

$$\dot{\epsilon}_r = \frac{1}{a}\frac{\partial u}{\partial \rho}; \dot{\epsilon}_\phi = \frac{1}{a}\frac{u}{\rho}; \dot{\epsilon}_z = \frac{1}{a}\frac{\partial w}{\partial \zeta}; \dot{\gamma}_{rz} = \frac{1}{a}\frac{\partial u}{\partial \zeta}.$$

Substituting the strain rates in the condition of incompressibility and integrating, we obtain

$$u = -\tfrac{1}{2}\rho w'(\zeta) + C(\zeta)/\rho \tag{28}$$

where $C(\zeta)$ is an arbitrary function. The velocity u is bounded, hence $C(\zeta) = 0$. Then

$$\dot{\epsilon}_r = \dot{\epsilon}_\phi = -\frac{1}{2}\dot{\epsilon}_z = -\frac{1}{2a}w'(\zeta);\ \dot{\gamma}_{rz} = -\frac{1}{2a}\rho w''(\zeta)$$

and from the relations of steady creep it follows that $\sigma_r = \sigma_\phi$ and

$$\sigma_z - \sigma_r = \frac{3}{a}g(H)w'(\zeta), \tag{29}$$

$$\tau_{rz} = \frac{1}{2a}g(H)\rho w''(\zeta) \tag{30}$$

As before, we assume

$$\tau_{rz} = R(\rho)\zeta/\delta. \tag{31}$$

Substituting the strain rate components into the relation (30), using the power law (8), assuming that

$$w''(\zeta) = A|\zeta|^{s-1}\zeta \tag{32}$$

and comparing the formula (31) to the formula (30) and neglecting a small summand, we find that $s = m$ and

$$\tau_{rz} = A_2\rho^\mu\zeta \tag{33}$$

where A_2 is a constant. Substituting the obtained values into (29) we can see that its right hand part is small (of the order of δ^2). Therefore, according to the order of approximation used, we have

$$\sigma_r = \sigma_\phi = \sigma_z. \tag{34}$$

Consider now the differential equation of equilibrium

$$\frac{\partial\sigma_r}{\partial\rho} + \frac{\sigma_r - \sigma_\phi}{\rho} + \frac{\partial\tau_{rz}}{\partial\zeta} = 0. \tag{35}$$

Using (33) and (34) and integrating we find the stress σ_r. The constant of integration is determined from the boundary condition:

$$\sigma_r = 0 \quad \text{for} \quad \rho = 1. \tag{36}$$

The stress σ_z is assumed to be constant in the direction normal to the layer; then the stress components σ_r, σ_ϕ are also constant in this direction.

Determining the constant A_2 from the condition that the stresses σ_z are statically equivalent to the tensile force P, we obtain

$$\sigma_z = \frac{3+\mu}{1+\mu}p(1-\rho^{1+\mu});\quad \sigma_r = \sigma_\phi = \sigma_z\,;$$

$$\tau_{rz} = (3+\mu)p\,\rho^{\mu}\zeta$$

where $p = P/\pi a^2$.

These formulae are analogous to the formulae (18) except for the factor $(3+\mu)$ here instead of $(2+\mu)$ there. Hence the stresses in the axisymmetric case are $(3+\mu)/(2+\mu)$ times higher than in the case of plane strain.

Using the same method we can also consider the unsteady creep. The initial elastic stress state of the layer is

$$\sigma_z' = p; \ \sigma_r' = \sigma_\phi' = 0; \ \tau_{rz}' = 0.$$

The time to brittle fracture of the axisymmetric interlayer is shorter than the time t_* in (27).

6. Conclusion

In the case of plane strain considered above, the triaxial tension is realized in the central part of the layer. The maximum stress is $(2+\mu)/(1+\mu)$ times higher than the average stress p. According to Prandtl's solution the level of the triaxial pressure in a thin ideally plastic layer compressed between rigid plates can be significantly higher than the average stress. It can be accounted for by the value of shear stress on the contact planes. For an ideally plastic layer it is usually assumed that shear stress is equal to the shear yield stress τ_y. Here it must be noted that the actual value of the contact shear stress is unknown, but on the average it is smaller than τ_y.

In the case of creep considered above the contact shear stress is equal to $2p\delta$ (for $\mu = 0$). This value follows from the more realistic condition that adhesion is susbstantially smaller than τ_y. Using the same assumptions as in the case of creep under constant load P considered above, it is possible to analyze the layer in relaxation in terms of creep and fracture. It is also possible to consider the plain strain problem of bending a thin layer [4].

References

1 D.C. Drucker, in: H. Liebowitz (Ed.), Fracture, Vol. 1, Acad. Press, 1968.
2 W. Prager and P. Hodge, Theory of Perfectly Plastic Solids, Wiley and Sons, 1961.
3 L.M. Kachanov, The Theory of Creep, Moscow, 1960. (English translation by Kennedy (Ed.) National Lending Library, Boston Spa, England, 1967.)
4 L.M. Kachanov, Izv. Akad. Nauk S.S.S.R. Otd. Techn Nauk, 4 (1963) 86—91.

Dynamic Loading of Fiber-reinforced Beams

H. KOLSKY and J.M. MOSQUERA

Division of Applied Mathematics, Brown University, Providence, Rhode Island 02912 (U.S.A.)

Abstract

This paper describes the results of experiments on the dynamic mechanical response of beams of fiber-reinforced materials when these are subjected to transverse loading. The response both when elastic deformations and plastic deformations are observed is described, and it is shown that the elastic response corresponds very closely to the theoretical predictions of Sayir whose analysis is developed by the methods of limit analysis. The response when plastic deformations take place was found to conform reasonably well with the idealized treatment of Spencer and his group. Spencer models a fiber-reinforced metallic composite as a perfectly plastic metal matrix which is unidirectionally reinforced by completely inextensible fibers. The specimens used in these experiments were generally prepared in the laboratory. They consisted of steel piano wires embedded in metal matrices of lead and lead tin alloys. Some beam specimens were also prepared by embedding steel wires in a rubber matrix and some were made from fiber-reinforced, glass epoxy resin (Scotchply 1002) which is available commericially.

1. Introduction

As a result of the increasing use of fiber-reinforced solids a large and growing literature on the mechanical response of such composite materials has been published. The elastic response of fiber-reinforced beams subjected to transverse loading has received special attention.

Flexural wave propagation in an isotropic elastic beam is inherently dispersive, thus so long as the wavelength of the flexural disturbance is large compared with the lateral dimensions of the beam the velocity of propagation of flexural sinusoidal waves is inversely proportional to the wavelength, so that a flexural pulse disperses rapidly as it travels along an isotropic elastic beam.

For fiber-reinforced solids the situation can be quite different, thus in the extreme case where the fibers are assumed to be completely inextensible (such a material is called an *ideal* fiber-reinforced solid) the only type of strain that can occur is the shear of the matrix material. Thus transverse disturbances produce only shear in the matrix and travel at the constant shear velocity, so that no dispersion takes place.

The purpose of the present investigation was to carry out an experimental program on the mechanical response of fiber-reinforced beams to transverse dynamic loading. In order to study this mechanical response, a resonance method was employed. Fiber-reinforced beams with fibers running longitudinally were freely supported, sinusoidal oscillations were generated at one end and the motion of the middle of the beams was monitored. For simplicity in the calculations circular beams were used (0.5″ diameter, lengths between 10″ and 15″). A series of the resonant peaks was obtained for each rod, from the values of the resonant frequencies the velocities were computed.

The mechanical response under transverse dynamic loading was also studied indirectly, by noting the changes of the resonant frequency with temperature. Experiments were carried out with composite beams where the reinforcing fibers had elastic moduli which were very insensitive to change of temperature (metal wires, glass fibers) and a matrix whose elastic moduli were very temperature sensitive. Thus if the flexural modulus depended primarily on the longitudinal elastic modulus of the beam, as it does for isotropic beams, there should be little change in flexural modulus with temperature. If however the transverse oscillations were primarily produced by shearing the beam, the transverse modulus would be highly sensitive to temperature change.

A convenient measure of anisotropy is the ratio E/G, where E is the extensional modulus and G is the transverse shear modulus. Specimens of different degrees of anisotropy were studied (very weak, i.e. $E/G < 10$ and very strong, i.e. $E/G > 5 \times 10^4$). Since specimens having the desired properties are not available commercially, it was necessary to prepare our own specimens. The longitudinal and the shear modulus were measured dynamically for all these specimens.

Spencer [1], Jones [2] and others have developed theoretical treatments to examine the response to transverse dynamic loading of rigid-plastic beams reinforced by ideal fibers (i.e. fibers which may be regarded as inextensible). The purpose of the second part of this experimental program was to test the validity of these predictions to enable us to gauge the magnitude and nature of the corrections which need to be made to the theoretical treatment to allow for the finite extensions which take place in the fibers.

2. Theoretical considerations

TRANSVERSE ELASTIC VIBRATIONS OF FIBER-REINFORCED BEAMS

The dispersive nature of flexural waves in isotropic elastic beams is well known [3, 12]. However, for highly anisotropic fiber-reinforced beams with the fiber direction parallel to the axis of the beam transverse deformation of the beam results in the strains being largely in the form of shear of the matrix material, and one might therefore expect the velocity of flexural waves to be close to the velocity of shear waves in the matrix.

Mahir Sayir [5] has developed explicit analytical expressions for the distribution of stress and displacement, as well as for the transverse wave velocity in beams of all degrees of anisotropy. He did this by the asymptotic expansion of the basic three dimensional equations of dynamic elastic response.

The results are summarized here. For an isotropic beam the phase velocity C_{II} for a train of sinusoidal flexural waves of wavelength λ travelling along an elastic beam of circular cross-section, radius R, is given approximately by

$$C_{II} = C_I \left\{1 + \frac{\pi^2}{3} \cdot \frac{R^2}{\lambda^2} \left(10 + 6\nu - \frac{\nu(1 + 2\nu)}{(1 + \nu)}\right)\right\}^{-1/2} \tag{1}$$

where $C_I = 2\pi(K/\lambda)C_0$; $C_0 = \sqrt{E/\rho}$; ν = Poisson's ratio; E = Young's modulus; K = radius of gyration of the cross-section of the beam about the neutral axis; ρ = density (assumed to be constant).

This expression contains corrections for the shearing (Timoshenko) effect, the rotary inertia (Rayleigh) effect, the lateral contraction effect and the effect of lateral normal stresses. Among these effects only the first two seem to be numerically important. In order to consider the anisotropic cases one has to define first a characteristic parameter of anisotropy. Sayir chose the parameter

$$p = \pi(2R/\lambda) \sqrt{E/G}, \tag{2}$$

where G is the transverse shear modulus.

If p is small with respect to 1, the composite can be considered as weakly or moderately anisotropic. The beam behavior is then similar to the isotropic case, except that corrections for the shearing effect become more important. The velocity can be computed as

$$C_{II} = C_I(1 + (7/24)p^2)^{-1/2} \tag{3}$$

If p is large compared to 1, the composite may be considered as strongly anisotropic. The wave velocity is almost equal to the shear velocity, even for long wavelengths. Thus

$$C_{II} = \sqrt{G/\rho}\,(1 - 1/2p).$$

BOUNDARY CONDITIONS

The relation between the phase velocity of flexural waves C_{II} and the resonant frequencies for flexural oscillations of elastic beams is given in most books on acoustics. These relations can readily be modified to allow for the effect of the mass T of a driving coil mounted at the end of the beam.

For the normal modes of oscillation $u = f(x) \sin(\omega t + \alpha)$; where u is the lateral displacement, α is a constant. The equation of motion reduces to

$$\frac{d^4 f(x)}{dx^4} = m^4 f(x), \text{ where } m^4 = \frac{\omega^2}{C_0^2 k^2}$$

and the solution for $f(x)$ is of the form

$$f(x) = A \cos mx + B \sin mx + C \cosh mx + D \sinh mx.$$

Thus for an unloaded free-free beam of length l we get the condition

$$\cos ml \cosh ml - 1 = 0. \tag{5}$$

For a cantilever beam (i.e. a fixed free bar) the condition becomes

$$\cos ml \cosh ml + 1 = 0, \tag{6}$$

The boundary conditions used to obtain these results are that at a fixed end the displacement u and its gradient $\partial u/\partial x$ vanish identically, while at a free end the shearing force F and the bending moment M both vanish, i.e. $\partial^2 u/\partial x^2$ and $\partial^3 u/\partial x^3$ are both zero.

To allow for the effect of the mass of the coil T at the free end we put $T\, \partial^2 u/\partial t^2$ equal to the shearing force. This leads to the relation

$$ml\, T\, \{\sinh ml \cos ml - \sin ml \cosh ml\} = H\{1 - \cos ml \cosh ml\}$$

for a free-free bar where H is the total mass of the bar. For a fixed-free beam the expression is

$$ml\, T\, \{\cosh ml \sin ml - \sinh ml \cos ml\} = H\{1 + \cos ml \cosh ml\} \tag{8}$$

IDEAL FIBER-REINFORCED RIGID-PLASTIC BEAMS

The theory of fiber reinforced rigid plastic beams has been formulated by Spencer [6, 7], Jones [2], Laudiero and Jones [9], Rogers and Pipkin [4].

The composite is called an ideal fiber-reinforced beam when the material is assumed to be inextensible in the fiber direction and also to be incompressible. The response of structures composed of such materials is discussed fully by Spencer [6] and simple theoretical solutions have been found in many cases. Spencer [1, 7], Shaw and Spencer [8], Jones [2], and Laudiero and Jones [9] have given theoretical treatments of the dynamic response of ideal reinforced beams. Spencer and Jones employed the same general assumptions which are customarily made to obtain the response of rigid-plastic beams of isotropic materials.

It has been found from many studies that the material elasticity hardly influences the overall response of an isotropic structure subjected to an amount of external dynamic energy considerably larger than the maximum amount of strain energy which can be stored in a wholly elastic manner by the material [10, 11]. If in addition, the duration of loading is short compared to the natural period of elastic vibrations, the influence of elasticity may be disregarded altogether and the material idealized as rigid-plastic. It is evident that the requirement of a large external dynamic energy to elastic strain energy ratio may conflict with the limitations of infinitesimal displacements, which are assumed in the theoretical methods used in linear theory. In some cases, this conflict is not too important as, for example, in cantilever beams. However in other cases the influence of finite displacements or geometry changes play a vital role in the structural response. Thus, the theoretical analyses in references [1, 2, 7–9] for ideal fiber reinforced beams were simplified by neglecting material elasticity and assuming that transverse displacements remained infinitesimal. In addition transverse wave propagation effects were neglected.

The theoretical behavior of these ideal fiber-reinforced beams was also compared to the corresponding dynamic response of beams which were made from a rigid perfectly plastic isotropic material. Generally speaking, it appears that the permanent transverse deflections and response durations are less than the corresponding values for isotropic beams. Furthermore, the characteristics of energy dissipation in the isotropic beams were quite different from those in the ideal fiber-reinforced ones. The initial kinetic energy in the isotropic beams was dissipated through bending at plastic hinges, whereas the initial energy in the ideal fiber-reinforced beams was dissipated as a result of shear deformation in the metal matrix.

The theoretical predictions of Spencer [7] and Jones [2] were developed for a rigid linear strain-hardening ideal fiber-reinforced material. So that

$$N_p = (N_0 + N_i\,|\gamma|)$$

where N_0 and N_i are positive constants, $\gamma(X, t) = \partial U/\partial X\ (X, t)$ is the shear strain, and N_p is the yield shear force, that depends on the history of γ.

The transverse equation of motion,

$$\frac{\partial \sigma_{xy}}{\partial x} + \frac{\partial \sigma_{yy}}{\partial y} = \rho\, \partial V/\partial t \text{ (where } V \text{ is the particle velocity)}$$

may be integrated across the cross section $X =$ constant of the beam to give

$$\frac{\partial N}{\partial X} + p(X, T) = m(\partial V/\partial t).$$

Now, if we consider a discontinuity in slope traveling in the beam we have by the conservation of momentum

$$[N(X, t)] = -m\dot{a}\,[\sigma_2(X, t)]$$

where $N = \int_s \sigma_{xy}\, ds$ and $\dot{a}$ is the speed of the discontinuity in slope traveling along the X axis (square brackets indicate the value of the "jump" in the function across the discontinuity).

For the particular case of a perfectly plastic material, Spencer [7] has shown that the discontinuity cannot propogate i.e. $\dot{a} = 0$. By the continuity of the displacement we have that

$$\dot{a}\left[\frac{\partial U_2}{\partial X_1}\right] = -\left[\frac{\partial U_2}{\partial t}\right] = -[V_2]$$

$$\dot{a}[\gamma] = -[V].$$

CANTILEVER BEAM

Shaw and Spencer [8c] have applied the theory of ideal fiber-reinforced materials to the problem of a cantilever beam struck transversally at any point by a mass which subsequently adheres to the beam. In the subsequent motion, slope and velocity discontinuities propagate outwards from the point of the impact. They point out that the discontinuity that propagates towards the free end of the beam always comes to rest before it reaches this end, but for sufficiently high values of impact mass and velocity, and strain-hardening parameter, one or more reflections of the discontinuity may occur at the fixed end of the beam and at the point of impact. Some particular solutions are given for different final forms of the deformation. Simplification of the solutions is given when the deformation is complete before the discontinuity reaches the fixed end of the beam.

3. Experimental procedure

(a) SPECIMEN PREPARATION

Since specimens having the desired properties are not readily available commercially, it was necessary to prepare our own specimens. The first set of specimens were prepared by embedding phosphor bronze wires, 0.01 and 0.02 in. in diameter in a matrix of lead alloy (50% lead--50% tin). The volume of "fibers" was 40% of the total volume. Special attention was given to keeping the wires straight during the melting process of the lead, so that the cross section remained symmetrical, since coupling of the flexural vibrations with both longitudinal and torsional modes might otherwise occur. The main reason for using phosphor bronze wires to reinforce the specimens was to ensure a reliable bond between the wires and the matrix. A liquid flux was later found which enabled piano wires to be used instead, and the later experiments were carried out with specimens which used such wires for reinforcement. Thus steel piano wires, 0.02 in. in diameter were embedded in the same matrix of lead alloy with densities of 40% and 50% of fibers to the total volume. In order to achieve good bonding between the wires and the matrix material a soldering flux (MG 860) was used. A strongly-anisotropic material was also made by embedding steel piano wires into a rubber matrix.

(b) MEASUREMENT OF ELASTIC CONSTANTS

There are 5 basic material elastic constants for a specimen reinforced in the X-direction, with a symmetrical cross section (see Fig. 1): E_1, ν_{21} for the anisotropic direction X; E_2, ν_{22}, G_{12} for the plane of isotropy YZ.

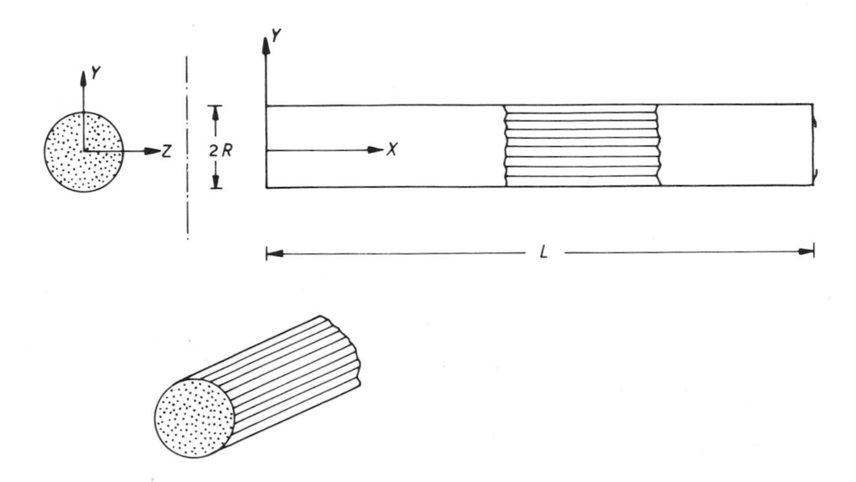

Fig. 1. Beam reinforced in the X-direction.

Among these constants, we were basically concerned with E_1 and G_{12}. The longitudinal modulus in the fiber direction E_1 and the transverse shear modulus G_{12} characterize the dynamic flexural behavior of the beam. E_1 was determined both by a pulse propagation experiment and from resonance tests.

The first set of experiments was done by propagating a longitudinal pulse along circular and square bars of 0.5 in. cross section, 10 to 15 in. long (Fig. 2). The pulse was generated by hitting the specimen with a small steel ball 0.25 to 0.5 in. in diameter. The ball was suspended by a fine wire 15 in. long and was held by an A.C. electromagnet. When the current was reduced the ball hit the specimen. The specimens were suspended at both ends by fine 38 gauge wires. The strain was generally recorded at the mid-point of the bar. Semiconductor strain gauges were used to measure the transit times of the longitudinal pulses, the outputs from these were amplified and recorded on cathode ray oscillographs which were triggered by electrical signals generated at the instant of impact. The pulse widths were sufficiently long for no appreciable dispersion to be observed and the velocities were deduced by dividing the length of the bar by the time of transit of the longitudinal pulse. The pulse velocity C_0 gives the value of E_1 since $C_0 = (E_1/\rho)^{1/2}$. These values were checked by setting bars into longitudinal resonance where the phase velocity is given by the expression

$$C_p = n(2l/k)$$

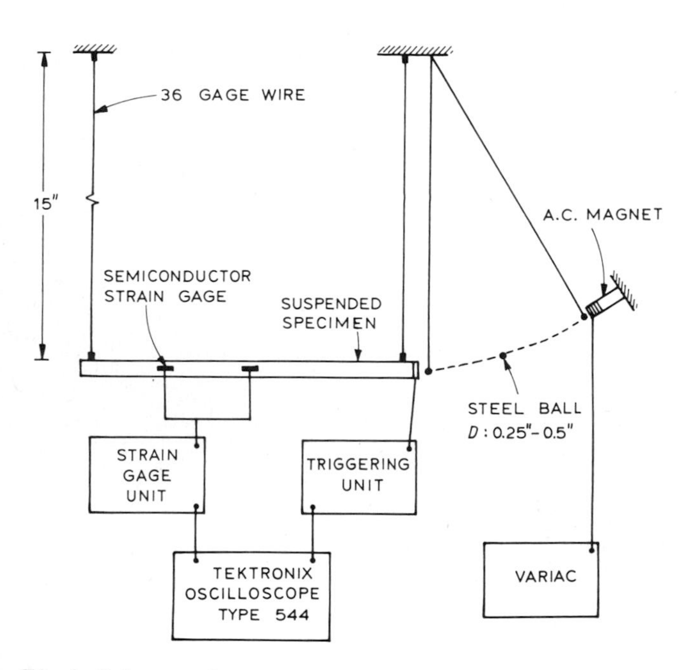

Fig. 2. Set-up to determine E.

where n is the resonant frequency, l is the length of the rod and k is the number of the harmonic mode (thus $k = 1$ for the fundamental mode of oscillation). Since for the fundamental mode and for the first two or three harmonics the wavelength is large compared with the radius of the bar, the above expression is equal to C_0 which enables E_1 to be calculated. The values of E_1 determined by this method and by pulse propagation were found to be in close agreement.

G_{12} was determined by torsional oscillations. The specimens studied were 3—10 in. long and of circular cross section (0.5 in radius).

A torque was applied to the bottom of the specimen while the upper end was clamped rigid (see Fig. 3). The torque was applied to the specimen by a metal rod, which had a coil wound around one end. A permanent magnet was placed close to the coil and an oscillating electric current was passed through the coil. To detect the displacement, a capacitor detector (condensor microphone) was used. The other end of the rod was made flat and a micrometer head was placed close to it. An insulated

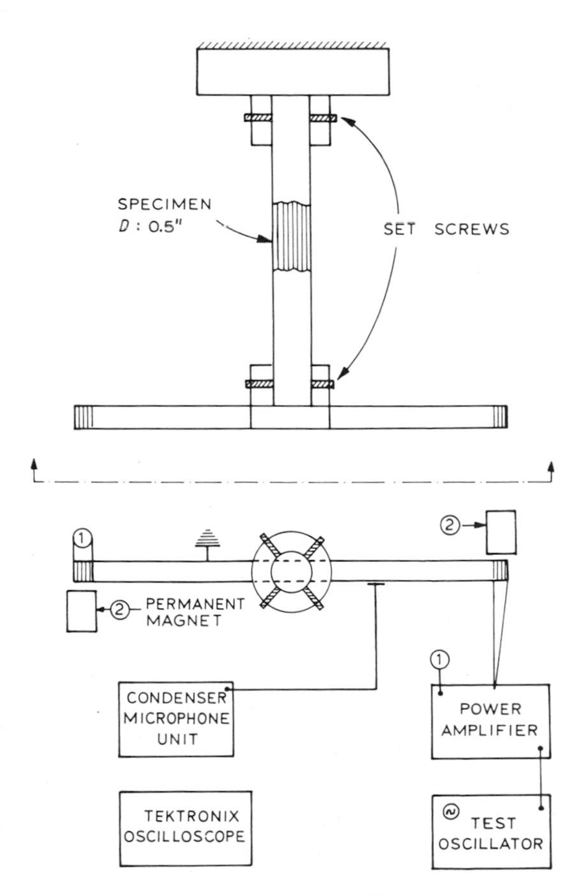

Fig. 3. Apparatus to determine G.

probe was mounted on the head. This probe was charged to a high voltage through a high resistance. The metal specimen was grounded so that the probe and the flat portion of the specimen formed a capacitor. When the rod moved, the capacity changed and this change in capacity resulted in a change in voltage which was amplified and fed onto a cathode ray oscillograph. A ballast fixed capacity was connected across the detector. By varying the frequency in the driving coil, and maintaining constant amplitude, the resonant frequency of the sample was found by observing the amplitude of the detected signal.

The value of G_{12} could be obtained from the equation

$$G_{12} = 2l\omega^2 J/\pi R$$

where l is the length of the specimen, ω is the resonant frequency, R the radius of the specimen and J the polar moment of inertia.

The same procedure was used for the composite materials (matrix and fibers) and it was found that the longitudinal modulus can be expressed as $E = qE_f + (1-q)E_m$ in which E_f is the longitudinal modulus of the fibers, E_m is the longitudinal modulus of the matrix and q is the fraction of the total volume occupied by fibers. The shear modulus $G\,[= G_{12}]$ can be expressed as $1/G \approx q/G_f + (1-q)/G_m$ in which G_f is the shear modulus of the fibers and G_m is the shear modulus of the matrix.

The following values were obtained for the elastic constants.

Isotropic materials

(1) 50% lead 50% tin (used as a matrix):
$E = 3.58 \times 10^{11}$ dyn/cm^2; $G = 1.5 \times 10^{11}$ dyn/cm^2; $\nu = 0.39$.
(2) Steel (piano wires) used as fibers:
$E = 20.75 \times 10^{11}$ dyn/cm^2; $G = 8.05 \times 10^{11}$ dyn/cm^2; $\nu = 0.29$.
(3) Phosphor bronze (wires) used as fibers:
$E = 10.72 \times 10^{11}$ dyn/cm^2; $G = 3.98 \times 10^{11}$ dyn/cm^2; $\nu = 0.34$.

Composites

(1) Lead—tin; phosphor bronze wires, 40% fibers—60% matrix:
$E = 6.40 \times 10^{11}$ dyn/cm^2; $G = 1.95 \times 10^{11}$ dyn/cm^2; $E/G = 3.2$.
(2) Lead—tin; steel piano wires, 40% fibers—60%matrix:
$E = 10.55 \times 10^{11}$ dyn/cm^2; $G = 2.52 \times 10^{11}$ dyn/cm^2; $E/G = 4.75$.
(3) Lead—tin; steel piano wires, 50% fibers—50% matrix:
$E = 12.29 \times 10^{11}$ dyn/cm^2; $G = 2.52 \times 10^{11}$ dyn/cm^2; $E/G = 4.88$.
(4) Rubber; steel piano wires, 50% fibers—50% matrix:
$E = 10.53 \times 10^{11}$ dyn/cm^2; $G = 1.37 \times 10^{7}$ dyn/cm^2; $E/G = 7.69 \times 10^4$.
(5) Epoxy; glass fibers:
$E = 3.69 \times 10^{11}$ dyn/cm^2; $G = 0.38 \times 10^{11}$ dyn/cm^2; $E/G = 9.73$.

(c) RESONANCE TESTS

Since the velocity dispersion of flexural pulses in the beam was so high it was decided to measure their dynamic response by setting the beams into flexural resonance under sinusoidal loading.

The specimens used were 0.5 in. in diameter and between 10 and 15 in. long. The rods were driven electromagnetically, a small coil was wound around one end and a permanent bar magnet was placed close to the coil (usually about 1/10 in. away). A oscillating voltage was applied across the coil, so that it acted as an A.C. magnet, with the amplitude proportional to the voltage input. For bars made with steel wires the process was reversed. The specimen was magnetized and an A.C. electromagnet was placed close to it, so that it acted as an oscillating magnet.

The dynamic strains in the vibrating specimens were detected by semiconductor strain gauges. These were generally mounted at the middle of the beams. The experimental set-up is shown in Fig. 4. The beams were suspended by flexible 36 gauge wire and the driving coil was energized by the amplified output from a Hewlett Packard oscillator. Up to 10 W in electrical power could be fed to the driving coil. It was found that at these small inputs the coupling between the coil and the magnet did not change the value of the resonant frequency. Small corrections had to be made to allow for the mass of the driving coil which was about

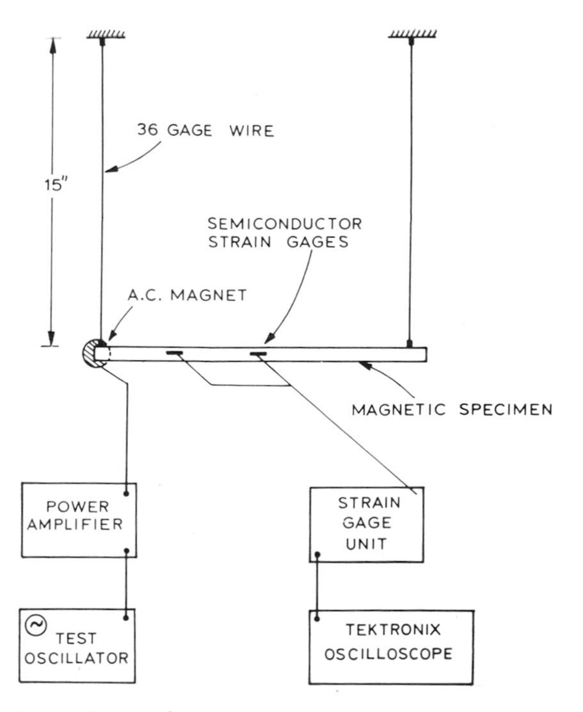

Fig. 4. Set-up for resonance test.

1% of the mass of the beam. The semi-conductor strain gauges used were type DDP-350-500 Kalite.

The wavelength was computed with help of the values ml determinated from eqns. (5) and (6), and for those specimens with a coil attached to them, for eqns. (7) and (8). The wavelength then can be expressed as $\lambda = 2\pi l/ml$. The values of the velocity are shown in Figs. 5 and 6.

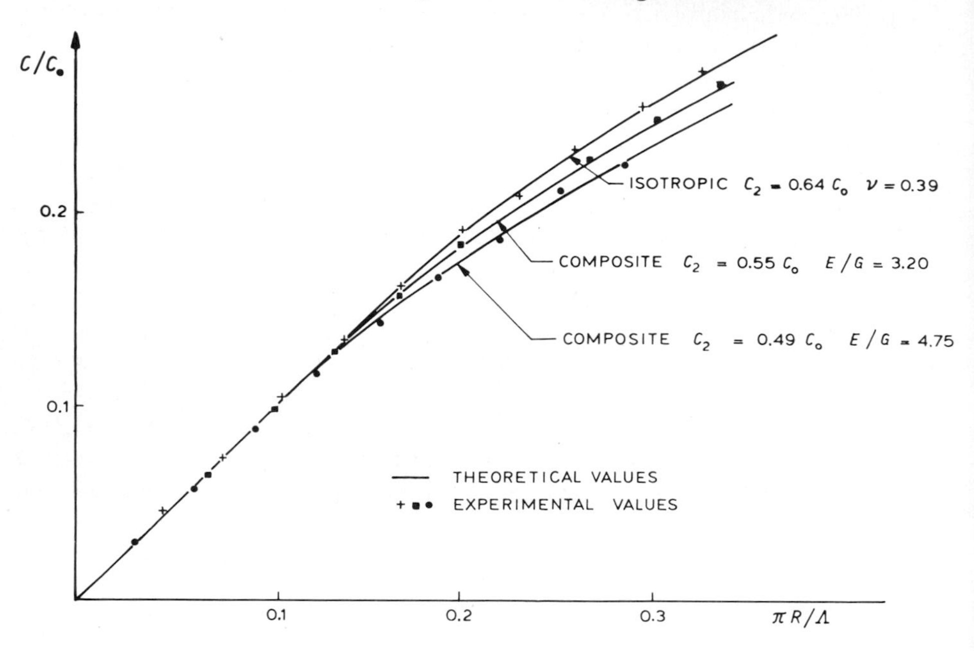

Fig. 5. Flexural phase velocities.

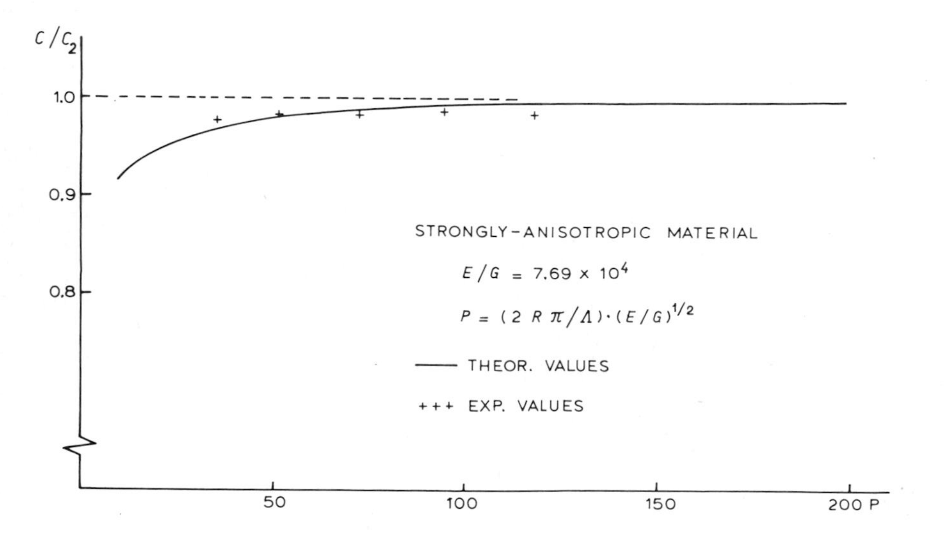

Fig. 6. Flexural wave velocity as a function of p.

(d) TEMPERATURE TEST

In these experiments the wavelength of flexural waves was not determined directly, instead a series of experiments were carried out to observe the change in resonant frequency of bars in flexural vibration over a range of temperatures.

Specimens 12 in. long and 0.5 in. in diameter were set in resonance as described above. The temperature was then changed by steps of 10°C and for each step the frequency was recorded. Enough time (at least one hour) was allowed to elapse between readings, so that the specimen would come to thermal equilibrium.

Elementary beam theory states that for wavelengths many times larger than the thickness of the beam, the phase velocity or its related resonant frequency may be expressed as ΩC_0 where Ω is a function of the wavelength and the dimensions of the cross section, and C_0 is $\sqrt{E/\rho}$. However, for a beam which is strongly-anisotropic, one would expect flexural waves of reasonably short wavelength to travel at a velocity very close to the shear wave velocity. Indeed, as Sayir has pointed out, the velocity should be given by θC_2, where θ is a function of the wavelength, degree of anisotropy and dimensions of the cross section, and $C_2 = \sqrt{G/\rho}$. For high degrees of anisotropy $\theta \approx 1$.

In our special case, as a result of the composition of the material, C_0 is not very dependent on temperature, since the modulus of the fibers is not temperature sensitive. However, C_2 is very sensitive to the change

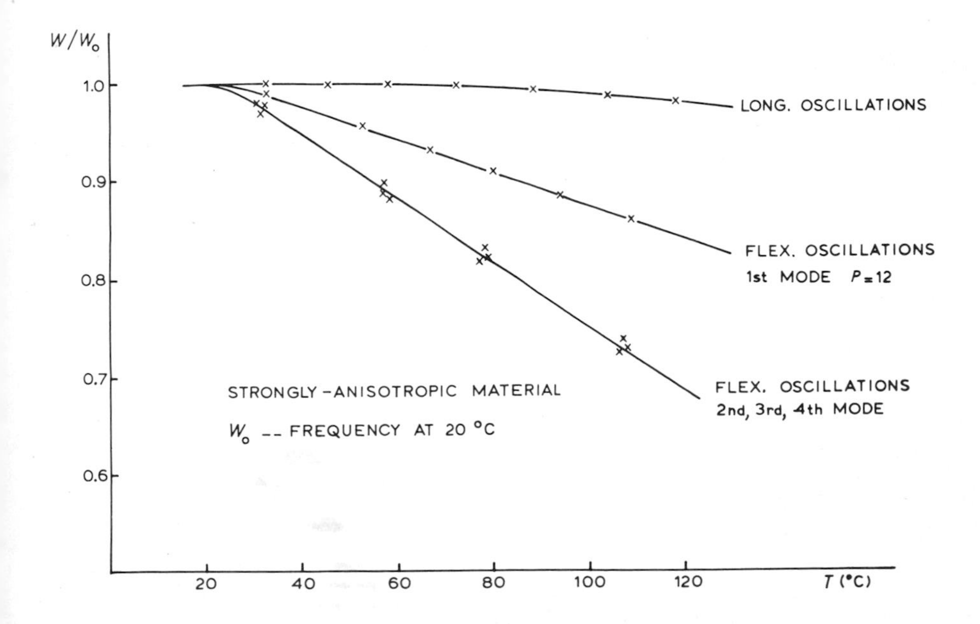

in temperature, because it basically depends on the shear properties of the rubber matrix. The results showed that as the temperature increased there was very little change in the value of the longitudinal velocity since this depended primarily on the extensional modulus of the steel piano wires. The velocity of transverse waves however decreased rapidly since this was close to the value of the shear velocity of rubber which decreases rapidly with increasing temperatures (see Fig. 7).

(e) QUASI-STATIC LOAD-DEFLECTION TESTS

In the theoretical treatment of the plastic response of ideal fiber-reinforced beams to transverse loads, Spencer has assumed that the matrix material behaves in a rigid plastic manner with respect to shear deformations. In order to see how closely the specimens we had prepared conformed to this assumption we applied transverse forces to reinforced beams of various lengths, composed of steel piano wires embedded in a lead matrix. The deformations were carried out in an "Instron" testing machine and the specimens were mounted as cantilevers. They were deformed monotonically and the strain gages were mounted at the tip of the cantilever and at points halfway along it. The strain gages used were of the post-yield resistance type. It was found that for low applied forces response was elastic, and the strains were proportional to the force. The values of the applied transverse force for which this linear relation broke down were determined and as can be seen from Fig. 8

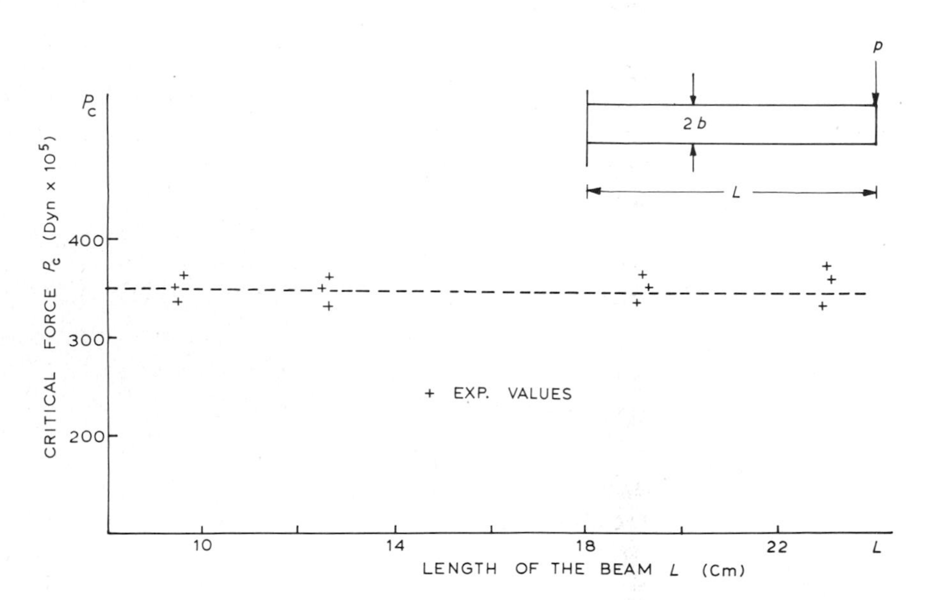

Fig. 8. Static tests of plastic yield in transverse loading of beams of various lengths L.

the value of this critical force was almost independent of the length of the beam. When yield occured it did so simultaneously along the whole beam and not just at the supported end which is what occurs when beams of isotropic metals are loaded.

(f) DYNAMIC LOADING TESTS

In order to investigate the response of the fiber-reinforced beams to large transverse dynamic loads beam specimens were mounted as cantilevers in a "Hyge" shock testing machine. They were impacted by "hammers" travelling at velocities up to 100 ft./s. The deformation of the beam was observed by cinephotography. For this purpose a Fastax high speed cine camera was employed, this camera enabling records to be taken at speeds up to 10,000 frames/s. The speed of the hammer just before impact was recorded by arranging for the hammer to make two electrical contacts a known distance apart. The electrical impulse from the first electrical contact started a Hewlett Packard microsecond counter while the second stopped it. The experimental set-up is shown schematically in Fig. 9. A composite beam specimen 10 in. in length and of rectangular cross section ($1'' \times \frac{1}{2}''$) was placed in the machine, one end being clamped. Post yield resistance strain gages were mounted at points 1/4 and 7/8 of the beam length from the fixed end. Figure 10 shows the records for low velocity and high velocity impacts. At low velocity the deformation is purely elastic and here the extensional strain is considerably larger than the shear strain (Fig. 10a), while at high velocities plastic deformation takes place and here the shear strain is considerably larger than the extensional strain (Fig. 10b).

Figure 11 shows a photograph of the deformed specimen after the impact. It may be seen that the plastic strain is mainly in the form of shear and that as predicted by Spencer the plastic strain that propagates out from the point of impact does not reach the free end of the cantilever but stops about halfway along the beam. Quantitative agreement is not very good but this is to be expected since the theory is based on two assumptions, namely that the fibers are completely inextensible and that the matrix material behaves in a rigid plastic manner, neither of which is realized in real specimens.

4. Results and conclusions

The velocity of propagation of transverse waves in anisotropic materials is very dependent of the grade of anisotropy. For materials in which E/G is small (<10), the dynamic behavior of the beam is very similar to that in the isotropic case, where for large wavelengths shear effects may be neglected. This is shown in Fig. 5, with the results for lead as a

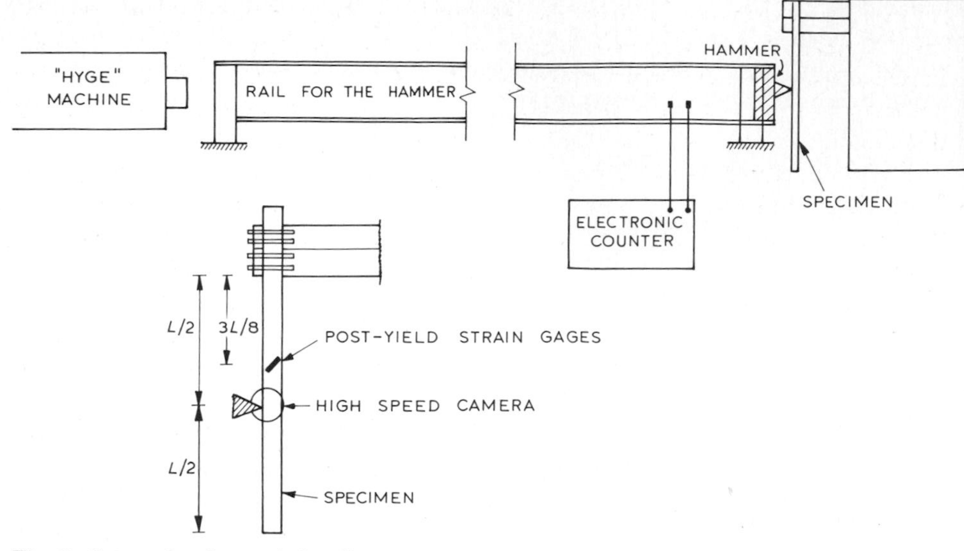

Fig. 9. Set-up for dynamic loading.

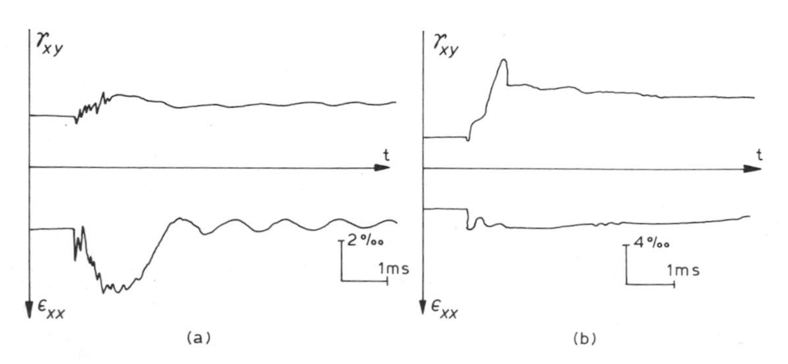

Fig. 10. Strain gage records at 3/8 of beam length from clamped end. (a) Low velocity impact (elastic); (b) high velocity impact (plastic).

Fig. 11. Deformed specimen.

reference. Experiments have shown that for these materials the elementary theory for isotropic beams represents the velocity very closely. However if the specimen is strongly anisotropic ($E/G > 10^5$) the wave velocity is equal to the shear velocity even for long wavelengths, and no dispersion is observed. The results are shown in Figs. 6 and 7.

For a specimen with this degree of anisotropy the strain energy results from shear deformation, and it may be regarded as an ideal fiber-reinforced material, in which no extension takes place in the fibers. However, in practice this condition is very restrictive and only specimens with very high degree of anisotropy can be considered as ideal, elastic fiber reinforced beams. For our particular case the condition for the effects of shear to dominate over those due to flexure is that $2L/H < 40$ with a degree of anisotropy of 7.5×10^4 or $2L/H < \alpha(E/G)^{1/2}$ where $\alpha \approx 1.5 \times 10^{-1}$. However, when the beam is deformed well above the elastic limit, the condition of inextensibility is less restrictive. The behavior of the beam for large transverse deformation results from the plastic shear in the matrix. The deformation of fibers, however, remains elastic. Dynamic tests show that the shear plastic strain is larger than the flexural axial strain (Fig. 10). This is in qualitative agreement with the theoretical prediction of Spencer and Jones.

Acknowledgements

The authors wish to express their gratitude to Mrs. E. Fonseca for preparing the manuscript of this paper and to Messrs. W. Carey and P. Russo for help with the experiments. They also wish to thank the National Science Foundation and the Materials Research Laboratory for support under grants 5-26548 and 5-26906.

References

1 A.J.M. Spencer, A note on ideal fibre-reinforced rigid-plastic beams brought to rest by transverse impact. Mech. Res. Comm., 3 (1976) 55.
2 N. Jones, Dynamic behavior of ideal fibre-reinforced rigid-plastic beams. J. Appl. Mech., *43*, (1976) 319.
3 H. Kolsky, Stress Waves in Solids. Clarendon Press Oxford, 1953 (Dover reprint 1963).
4 T.G. Rogers and A.C. Pipkin, Small deflections of fibre-reinforced beams or slabs. J. Appl. Mech., 38 (1971).
5 M. Sayir, Flexural vibrations of strong anisotropic beams. Ing. Archiv., 49 (1980) 309—30.
6 A.J.M. Spencer, Deformation of Fibre-Reinforced Materials. Clarendon Press, Oxford, 1972.
7 A.J.M. Spencer, Dynamics of ideal fibre-reinforced rigid-plastic beams. J. Mech. Phys. Sol., 22 (1974) 147.

8 L. Shaw and A.J.M. Spencer, Impulsive loading of ideal fibre-reinforced rigid-plastic beams.
(a) Free beam under central impact. Int. J. Solids and Structures, 13 (1977) 823.
(b) Beam with supports. ibid., 13 (1977) 833.
(c) Cantilever beam. ibid., 13 (1977) 845.
9 F. Laudiero and N. Jones, Impulsive loading of an ideal fibre-reinforced rigid-plastic beam. J. Struct. Mech., 5 (1977) 369—382.
10 P.S. Symonds, Survey of Methods of Analysis for Plastic Deformation of Structures under Dynamic Loadings. Brown University Rept. BU/NSRDC/1-67, 1967.
11 N. Jones A literature review of dynamic plastic response of structures. The Shock and Vibration Digest, 7(8) (1975) 89—105.
12 P. Arseneaux, An Experimental Investigation of Stress Waves in Rods of Fiber-Reinforced Composite. Tech. Rept. 21, Div. Appl. Math., Brown University, 1973.

Limit Concepts and High Temperature Design

F.A. LECKIE

Departments of Mechanical and Industrial Engineering and of Theoretical and Applied Mechanics, University of Illinois at Urbana-Champaign, Urbana, IL 61801 (U.S.A.)

1. Introduction

The need to base the design of high temperature plants on time-dependent deformation and rupture material properties has long been recognized. One of the first successful attempts to provide design criteria, taking into account creep deformation of the material, dates from the publications by Bailey [1] in 1935. The early studies were stimulated by the desire to increase the efficiency of turbo-generator plants which implied higher operating temperatures. Similar considerations in chemical and processing plants resulted in the expansion of these early studies into the boiler and pressure vessel fields. Since then, there has been a steady development of methods to analyze time-dependent stresses and strains in components. Initially these were restricted to solutions for simple shapes, using analytical techniques. With the advent of digital computers, solutions became available for a wide range of axisymmetric thin shelled structures. Calculation techniques have now been developed to the stage that stress and strain states can be determined for the most complex components, subjected to temperatures and load cycles. This progress is deceptive however, being less substantial than first might be thought for the following reasons:

(1) The constitutive equations which describe material behavior at high temperatures are generally deficient. While the state variable approach suggested by Onat [2] provides a means of developing systematic programs of testing to describe the state space, investigations to date are overwhelmingly concerned with the uniaxial stress state. The response to multiaxial non-proportional loading is much less well understood and the normal procedure currently used is to postulate a plausible multiaxial response from uniaxial data.

(2) Even for uniaxial loading states, extensive long-term testing is required before a rather complete description of material behavior is possible. Some metals such as the stainless steels and some nickel based alloys have been subject to rather complete testing but this is not true for many metals in current use. The most common situation faced by designers is a shortage of the material properties required to complete a

finite element analysis. To remedy this situation requires both time and money. While the time element can be minimized to some extent by judicious selection of test conditions, it will remain a troublesome aspect, especially in conditions involving failure.

(3) A full time-dependent analysis is often very expensive and time consuming. While the results can be useful at the final stages of design, they are unlikely to provide the insight and guidance which are so necessary at the conceptual and early stages of design.

It may be that the increasing power and availability of computers has been responsible for an apparent preference in high temperature applications for computational methods over the traditional methods used in design. In the traditional methods, a small number of dominant material parameters, together with limited but critical calculations, form the basis of rational design procedures. In addition to providing the basis for the development of rational design procedures, the approach also provides a quantitative means of communicating desirable material properties and their balance to metallurgists. To argue the need for the development of theory based on a few material properties is not to deny the important role of computer calculation. An interesting example of the imaginative combined use of theory and computers is to be found in the modern application of the theory of plasticity to pressure vessel design. The effective application of the theorems of plasticity to complex components requires a skill normally outside the repertoire of the designer. The determination of limit and shakedown loads using finite element procedures is certainly possible but the calculations are so demanding and expensive they have not been adopted in design procedures. Instead, the theorems of plasticity are used as a means of interpreting the results of elastic calculations in terms of limit and shakedown loads. Elastic finite element calculations even for very complex components, are routine and cheap and it is the result of these calculations interpreted in a plasticity context which now form the basis of modern design. Such procedures could not have been developed without a clear understanding of plasticity theory but it is now evident with the passage of time that the origins of the method have become somewhat obscured by an imagined exactness of elastic finite element calculations.

In this paper, procedures are developed for high temperature design when the effects of time-dependent deformation and material rupture must be considered. It will be shown that plasticity concepts can be conveniently used provided the yield stress is replaced by new material properties. Consequently, the extension of existing codes to high temperature design is a simple matter of introducing into existing procedures two new material properties. It is indeed interesting to note that the pioneering work of Drucker [3] in the application of plasticity theory to pressure vessels had the potential for such extensive development.

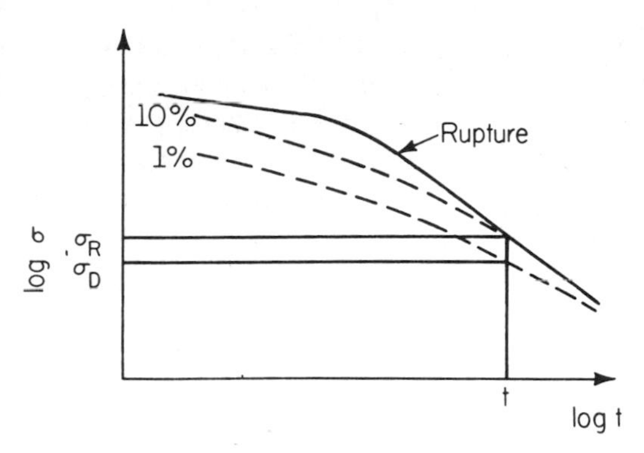

Fig. 1. Creep rupture and strains.

2. Design criteria

Two of the important design criteria associated with high temperature components are associated with deformation and rupture. Fatigue, creep-fatigue, and fracture criteria are also included in the total design consideration but these are concerned with the behavior of details rather than with global concepts which are the topic of this paper.

The criteria used for deformation normally include a condition on average strain and on the maximum strain. For example, in the ASME N-47 Code Case, the average accumulation of strain in the lifetime of the component is limited to 1% and the maximum strain to 5%. This maximum strain requirement also provides a criterion on material ductility. The rupture criterion requires that material damage is limited to all points so that local leakage does not occur.

The material data usually available are presented in the form shown in Fig. 1 in which log stress is plotted against log strain and from which both the rupture condition and the accumulation of strain may be reconstructed. However, this form of presentation is particularly useful for the determination of the material properties used in the procedures which are to be described. These are (1) the deformation stress σ_D which is the stress which causes a 1% strain in the required lifetime and (2) the rupture stress σ_R which is the stress which results in rupture at the required lifetime.

3. Deformation studies

The simplest constitutive equations assume that the total strain ϵ is the sum of elastic ϵ_{el}, creep ϵ_c, and time-independent plastic strain ϵ_p so that

$$\epsilon = \epsilon_{el} + \epsilon_c + \epsilon_p . \qquad (3.1)$$

The separation of plastic and creep strains is clearly an artificial device since they are both related to the movement of dislocations. However, it has been shown by Onat [2] that this is a reasonable and convenient assumption when ϵ_p is associated with high rates of loading. More sophisticated equations which do not make this assumption and which are physically more satisfactory can be formulated but since they yield essentially the same results as the simple constitutive equations, it proves convenient to retain the simplification. The constitutive equations then have the form

$$\dot{\epsilon}_{el} = \dot{\sigma}/E; \qquad \dot{\epsilon}_c/\dot{\epsilon}_0 = (\sigma/\sigma_0)^n \tag{3.2a, b}$$

with suitable modification for multiaxial stress states. $\dot{\epsilon}_0$ is the creep rate associated with an applied stress σ_0, and n is a constant. The expression for creep energy dissipation density rate is

$$\dot{D}(\sigma_{ij})/\sigma_0\dot{\epsilon}_0 = (\bar{\sigma}/\sigma_0)^{n+1} \tag{3.3}$$

where $\bar{\sigma}$ is the effective stress.

The yield condition is $f(\sigma_{ij}) = 0$ and reduces to $\sigma = \sigma_y$ in the uniaxial state where σ_y is the uniaxial yield stress. For convenience, it is assumed that the shape of the yield surface and constant energy dissipation rate surfaces are identical and governed by the Mises condition. Using these constitutive equations, it is possible to determine results which are of particular use in design and these are now briefly described.

For a component of volume V subjected to constant generalized loading P, the stress state within the component changes with time from its initial distribution to a so-called steady-state distribution σ^s which remains constant with time. The total displacement $\Delta(t)$ of the local P is given by

$$\Delta(t) = \Delta_i + \Delta_s(t) + \delta \tag{3.4}$$

where Δ_i is the initial displacement, $\Delta_s(t)$ is the steady-state deformation associated with a constant stress distribution σ^s and δ is an additional term due to stress redistribution which can be shown to be a small proportion of Δ. Hence, it is the steady-state condition which dominates the description of the deformation state under constant loading. Concentrating on the steady-state solution, it is possible to obtain the result (Ponter and Leckie [4])

$$P\dot{\Delta}_s/\sigma_0\dot{\epsilon}_0 = \int_V \dot{D}(\sigma^s)\,\mathrm{d}V \leqslant \int_V \dot{D}(\sigma^*)\,\mathrm{d}V \tag{3.5}$$

where σ^s the steady-state stress distribution, and σ^* is any stress distribution in equilibrium with the applied load. The stress distribution σ^* must also satisfy the inequality

$$f[(n+1)/n\sigma^*] \leqslant 0. \tag{3.6a}$$

The implication of this result is that for the theorem to be applicable, the applied loads P must be no greater than $n/(n+1)P^{L}$ where P^{L} is the limit load of the component with yield stress σ_{y}. Note that plastic terms do not appear in the dissipation of energy rate and the physical interpretation is that for loads less than $n/(n+1)P^{L}$, the time-dependent plastic deformation may be neglected and that for loading levels in excess of this amount, the influence of plastic strain will become evident. This effect has been observed in experiment (Leckie and Ponter [5]) but clearly the result will be conservative for materials which exhibit considerable hardening.

For cyclic loading, it can be shown that the stress reaches a steady cyclic distribution σ^{sc} and that an equation similar to eqn. (3.4) can be produced from which it can be concluded that the additional deflections resulting from stress redistribution may again be neglected. Hence, the problem is reduced to that of studying the steady cyclic state. The corresponding energy theorem involves energy quantities integrated over the cycle time Δt so that

$$\int_{\Delta\tau} P\dot{\Delta}\mathrm{d}\tau = \int_{\Delta\tau}\int_{V} \dot{D}(\sigma^{sc})\mathrm{d}V\mathrm{d}t \leqslant \int_{V} \dot{D}(\sigma^{**})\mathrm{d}V\mathrm{d}t. \tag{3.6b}$$

In this expression, $\sigma^{**} = \sigma^{E} + \rho$ where $\sigma^{E}(t)$ is the instantaneous elastic response in equilibrium corresponding to the current load $P(t)$ and ρ is a constant self-equilibrating stress distribution. The stress distribution σ^{**} is subjected to the constraint

$$f[(n+1)/n\sigma^{**}] \leqslant 0. \tag{3.7}$$

This constraint implies that the theorem may only be applied for loads less than $n/(n+1)$ of the shakedown load P^{s} and again the deduction is that plasticity has little effect at loads less than this value and may be neglected. For loads in excess of this value, plasticity effects will become evident and this has been confirmed by experiment (Leckie and Ponter [5]).

The theorems provide the means of performing approximate calculations on components but a greater significance is the identification of the strong role played by plasticity concepts already familiar to engineers. They also provide an indication of how to extend to high temperatures procedures already developed for plastic design.

4. Rupture studies

When metals are subjected to temperature and stress, deterioration of the metal takes place. This may be in the form of phase changes or of voids which nucleate and grow on the grain boundaries of the metallic crystals. In order to estimate the rupture time, it was formerly considered satisfactory to determine the steady-state stress distribution and to use the value of the maximum principal stress together with uniaxial rupture data. It is now known that this procedure has proved to be deficient on two accounts. First of all, the form of the multiaxial stress state can have a profound effect on the rupture time. A good example illustrating this point (Henderson and Sneddon [6]; Henderson, [7]) demonstrates that the experimental and computed rupture times for solid bars in torsion can be in error by orders of magnitude if the incorrect rupture condition is selected. Furthermore, many metals exhibit a softening effect in the tertiary creep regime when for constant stress the strain rate continuously increases. This means that at points of high stress concentration in a component, damage rates will be high which produce material softening and which in turn can be expected to reduce the stress at such point thereby increasing the stress elsewhere. A measure of the importance of stress redistribution can be deduced from the results of experiments on plates in plane stress subjected to uniaxial tension (Leckie and Hayhurst [8]). Stress raisers were introduced by manufacturing plates with circular holes and narrow slits. The maximum steady-state stress for the circular hole was 1.2 σ and for the slit 1.4 σ where σ is the average stress at the cross-section. Test results on copper and aluminum plates show that the rupture life is dictated not by the maximum steady-state stress but by the average stress σ, so that the form of the stress raisers appears to be unimportant in these cases.

In order to overcome the deficiencies of the previous procedures, extensive use has been made of the equations of Kachanov and Rabotnov [9] which introduce the concept of damage into the constitutive relationships. These can be generalized so that the strain rate is given by

$$\dot{\epsilon}_{ij}/\dot{\epsilon}_0 = (3/2)(\bar{\sigma}/\sigma_0)^{n-1}(s_{ij}/\sigma_0)/(1-\omega)^n \tag{4.1}$$

and the damage rate is

$$\dot{\omega} = [(\lambda-1)/\lambda n t_0]\,\Delta^n(\sigma_{ij}/\sigma_0)/(1-\omega)^{\eta},$$

where $\eta = [\lambda n/(n-1)-1]$, $\bar{\sigma}$ denotes the effective stress, s_{ij} the stress deviator, while the other terms are readily obtained experimental constants. The function $\Delta(\sigma_{ij}/\sigma_0)$ and the constant λ deserve special attention. Constant values of the function $\Delta(\sigma_{ij}/\sigma_0)$ give rupture times of the same value. Because of this property, the functions are referred to as isochronous surfaces. It has been found that the isochronous surface for rupture t_0 can be expressed in the form

$$\Delta(\sigma_{ij}/\sigma_0) = [\alpha(\sigma_1/\sigma_0) + (1-\alpha)(\bar{\sigma}/\sigma_0)] = 1 \tag{4.2}$$

where α is an experimentally determined constant and σ_1 is the maximum principal stress. The value of α lies between the limits 0 and 1.

In the constitutive equations, it will be noted that damage appears as a scalar quantity. This is clearly not the case as visual inspection of damaged materials readily indicates. However, the equations do describe in a satisfactory manner the behavior of materials which have been tested under constant stress conditions. In order to determine the tensorial nature of the damage, experiments involving non-proportional loading tests have been performed by Trampczynski, Hayhurst, and Leckie [10]. It can be deduced from these experimental results that the scalar description of damage provides conservative results on life estimates which is, of course, satisfactory for design.

The constitutive equations which have been discussed may be used in computer calculations but the calculations are apparently quite demanding as demonstrated by Hayhurst, Dimmer and Chernuka [11]. The other approach has been to attempt to determine approximate procedures of the type developed for deformation studies. The success to date has been somewhat limited but such results that have been obtained have fortunately proved to be quite useful in design.

A particularly useful result (Goodall and Cockcroft [12]) for constant loading may be obtained if it is assumed that the isochronous surface Δ is convex (an assumption which is supported by all existing data known to the author). The limit load P_0^L corresponding to a yield surface $\Delta(\sigma_{ij}/\sigma_0) = 1$ is determined. Then a time to failure t_u is calculated by reading from the stress—time rupture curve the time corresponding to a stress σ_u where

$$\sigma_u/\sigma_0 = (P/P_0^L) \tag{4.3}$$

and P is the current load. This result gives an upper load on the time to fracture which makes the result less useful for design but it is in a particularly convenient form for later discussion.

Attempts have also been made to attain bounds for the rupture lives of components subjected to cyclic loading which indicate similar results (Leckie and Wojewodski [13], Ponter [14]) except that the shakedown load replaces the role of the limit load.

It was noted that the result (eqn. (4.3)) gives an upper bound on rupture time and is consequently non-conservative. This fact was at first rather troublesome but experiments on a variety of components both in the laboratory and in practice have confirmed the predictions of eqn. (4.3) to be very good (Goodall and Ainsworth [15], Leckie and Hayhurst [8]).

When dealing with problems of deformation, it is immediately evident that the material must demonstrate sufficient ductility to accommodate the strain level defined in the specifications. The question which is less obvious is how to define ductility in terms suitable for creep rupture. The material must be able to accommodate tertiary deformation at points of concentration so that sufficient stress redistribution can occur for local failure to be avoided. The quantity λ illustrated in Fig. 2 is the ratio of the failure rupture strain to the product of the steady state creep rate and the failure time. It has been suggested by Goodall and Cockcroft [12]

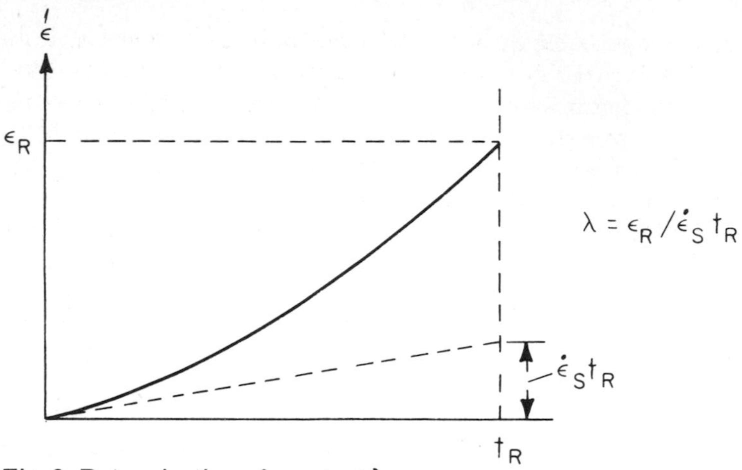

Fig. 2. Determination of constant λ.

that it is this value of λ which is the correct measure of the creep ductility since it represents a means of describing the ability of the material to accommodate stress redistribution without local failure taking place. To illustrate this point, it is generally observed that the highly stressed regions in components tend to be subjected to the kinematic constraint of constant strain rate. If the steady-state stress is σ_0 then the corresponding steady strain rate is $\dot{\epsilon}_0$. Because of kinematic constraint a constant steady strain rate $\dot{\epsilon}_0$ is maintained, and it can be shown (with modest approximation) that the rupture time for local failures is λt_0, compared to t_0 if the stress σ_0 remained constant. Hence, λ should be made as large as possible and it is suggested by Goodall and Cockcroft [12] that components with stress concentration factors of about 3, values of λ of at least 10 should be achieved in order to attain the full beneficial effect of stress redistribution.

5. Implications for design

The implications of the results for design are investigated in terms of the criteria discussed in Section 2. It is recalled that the criteria are given in terms of limitation in values of average strain (of 1%), a maximum strain (of 5%) and of rupture. It is also convenient to recall the conditions of the deformation stress σ_D and the rupture stress σ_R (Fig. 1). In a uniaxial constant stress creep test, the deformation stress is that stress which results in a 1% strain in the required lifetime. The rupture stress σ_R is the stress which causes rupture in the required lifetime.

Consider the component to be subjected to a constant load P. The limit load P^L remains unaffected by the presence of creep and it is convenient to consider states of loading $P = \lambda P^L$ where $\lambda < 1$.

As the upper bound of eqn. (3.5) involves only the creep energy dissipation rate, it implies for $\lambda < n/(n+1)$ that the plastic energy dissipation rate remains small compared with the creep energy dissipation rate. Thus the effects of stress concentration have little effect on the total deformation rates. If it is assumed that plastic deformation occurs at this load and that the position of maximum stresses remains constant and on the yield surface, then the stress concentration given by a creep solution is proportional to $(1 + 1/n)$ for varying n. This correlates with the interpolation formula given by Calladine [16].

For a structure where all points of the body are in the state of plastic yield at collapse, the stress state $\lambda\sigma_{ij}^{L}$ when used in inequality (3.5) gives the result

$$\int P_{\mathrm{i}}\dot{\Delta}_{\mathrm{s}}\mathrm{d}s < \int_{V}\dot{D}_{\mathrm{c}}(\lambda\sigma_{ij}^{\mathrm{L}})\mathrm{d}v = V\dot{D}_{\mathrm{c}}(\lambda\sigma_{\mathrm{y}}) \qquad (5.1)$$

On reflection, it is evident that (5.1) is valid in all cases as σ_{ij}^{L} either lies on the yield surface or within it. An upper bound on the average strain is obtained by computing the stress

$$\sigma = P\sigma_{\mathrm{y}}/P^{\mathrm{L}} \qquad (5.2)$$

If σ is set equal to σ_{D} (the stress for 1% creep), then the equation gives

$$P/\sigma_{\mathrm{D}} = P^{\mathrm{L}}/\sigma_{\mathrm{y}} \qquad (5.3)$$

Hence, a safe bound is obtained by using σ_{D} in the creep calculation in the same way as the yield stress σ_{y} is used in limit load calculations.

Limitations are placed on the maximum strain levels in the low temperature codes by applying a limiting stress concentration factor of approximately 2.5 or by applying the shakedown rule which has the same effect of avoiding excessive stress concentrations. At elevated temperatures, the simple interpolation of Calladine [16] referred to earlier in this section can be used to determine the relationship between the maximum creep strain ϵ_{M} and the average strain ϵ_{D}. Using this result, it is possible to determine the formula $\epsilon_{\mathrm{M}} = \epsilon_{\mathrm{D}} \exp{(k_1 - 1)}$ where k_1 is the elastic stress concentration.

If $k_1 = 2.5$, $\epsilon_{\mathrm{M}}/\epsilon_{\mathrm{D}} = 4.5$, so that if ϵ_{D} is restricted to 1%, a component designed elastically with elastic SCF of 2.5 should satisfy the maximum strain condition of 5%.

When designing against rupture use is made of eqn. (4.3) together with the rupture stress σ_{R} to give the result

$$P/\sigma_{\mathrm{R}} = P_{\Delta}^{\mathrm{L}}/\sigma_{\mathrm{y}} \qquad (5.4)$$

where P_{Δ}^{L} is the limit load corresponding to a yield surface which coincides with the isochronous rupture surface.

In practice, many components operate under conditions of plane stress. It is then conservative to assume the isochronous rupture surface has a

Mises form so that in eqn. (4.2) with $\alpha = 0$

$$\Delta(\sigma_{ij}/\sigma_0) = \bar{\sigma}/\sigma_0 \tag{5.5}$$

In these circumstances, eqn. (5.4) becomes

$$P/\sigma_R = P^L/\sigma_y \tag{5.6}$$

which is identical to eqn. (5.3) except that σ_R replaces σ_D.

When the effects of cyclic loading are considered arguments similar to those outlined above can be followed except that limit load stress states are replaced by shake down stress states with the yield stress σ_y being replaced as appropriate by the deformation stress σ_D and the rupture stress σ_R.

6. Conclusions

(1) The concepts of plasticity form the basis of many modern design procedures. The methods which have evolved can be readily applied to practical design.

(2) The concepts of limit and shakedown loads are found to be appropriate concepts in high temperature design when time-dependent deformation and rupture are important design considerations.

(3) In high temperature deformation studies, plasticity procedures can be applied directly if the yield stress σ_y is replaced by the deformation stress σ_D.

(4) In high temperature rupture studies, plasticity procedures can be applied directly if the yield stress σ_y is replaced by the rupture stress σ_R and if the yield condition is replaced by the isochronous surface Δ. In plane stress conditions, this additional condition is unnecessary.

Acknowledgement

The author wishes to acknowledge the support of the National Science Foundation under Grant NSF ENG 78-23549.

References

1 R.W. Bailey, The utilization of creep test data in engineering design. Prog. Inst. Mech. Engr., 131 (1935) 131—149.

2 E.T. Onat, Representation of Inelastic Behavior. Yale Univ. Report ORNL-SUB-3863/2, 1976.

3 D.C. Drucker, Limit analysis of cylindrical shells under axially-symmetric loading. Proc., 1st Mid-Western Conf. on Solid Mechanics, Urbana, Illinois, 1953, pp. 158—163.

4 A.R.S. Ponter and F.A. Leckie, The application of energy theorems to bodies which creep in the plastic range. J. Appl. Mech., 37 (1953) 753—758.
5 F.A. Leckie and A.R.S. Ponter, Theoretical and experimental investigation between plastic and creep deformation of structures. Archives of Mechanics, 24 (1972) 419—437.
6 J. Henderson and J.D. Sneddon, Creep fracture of torsional components. Creep and fatigue in elevated temperature applications. J. Inst. Mech. Engr., 190 (1975) 172. 1—172. 16.
7 J. Henderson, An investigation of multi-axial creep characteristics of metals. J. Engr. Materials and Technology, 101 (1979) 356—364.
8 F.A. Leckie and D.R. Hayhurst, Creep rupture of structures. Proc. Royal Soc. London, A340 (1974) 323—347.
9 Y.N. Rabotnov, Creep Problems in Structural Members. North Holland Press, Amsterdam, 1969.
10 W.A. Trampczynski, D.R. Hayhurst and F.A. Leckie, Creep rupture of copper and aluminium under non-proportional loading. J. Mech. Phys. Solids, 29 (1981) 353—379.
11 D.R. Hayhurst, P. Dimmer and M. Chernuka, Estimates of the creep rupture lifetime of structures using the finite element method. J. Mech. Phys. Solids, 23 (1975) 335—355.
12 I.W. Goodall and R.D.H. Cockcroft, On bounding the life of structures subjected to steady load and operating within the creep range. Int. J. Mech. Sci., 15 (1973) 251—260.
13 F.A. Leckie and V. Wojewodski, Estimates of the rupture life of structural components subjected to proportional cyclic loading. J. Mech. Phys. Solids, 24 (1976) 239—250.
14 A.R.S. Ponter, Upper bounds on the creep rupture life of structures subjected to variable load and temperature. Int. J. Mech. Sci., 19 (1977) 79—92.
15 I.W. Goodall and R.A. Ainsworth, Failure of structures by creep. Proc. Third. Int. Conf. on Pressure Vessel Tech., Vol. II, Tokyo, ASME, New York, 1977, pp. 871—885.
16 C.R. Calladine, A rapid method of estimating the greatest stress in a structure subject to creep. Proc. Inst. Mech. Engr., 78 (1963) 198—206.

Finite Deformation Effects in Elastic-Plastic Analysis

E.H. LEE

Department of Mechanical Engineering, Aeronautical Engineering and Mechanics, Rensselaer Polytechnic Institute, Troy, NY 12181 (U.S.A.)

Abstract

The non-linear kinematics of the combination of elastic and plastic deformations at finite strain to produce the resultant total deformation is discussed. It provides the mathematical structure to examine aspects of elastic-plastic analysis more succinctly than is possible with the common currently used approach based on infinitesimal elastic strain.

1. Introduction

In order to anticipate and so prevent failure of engineering structures or the generation in manufacturing processes of forming defects such as regions of high residual stress or internal or surface cracking, rational design requires stress and deformation analysis in the presence of finite strain. Structural metals are commonly ductile and so they can often be deformed to large strains while remaining intact. The history of stress throughout a structure or workpiece must be predicted since the location of the initiation of a defect and its motion while growing is not known in advance. Therefore, even though the plastic strains may be much larger than the elastic strain components, it is necessary to adopt elastic-plastic theory, for if the elastic strain components are neglected compared to plastic strains, resulting in rigid-plastic theory, stresses can only be predicted in the regions currently flowing plastically which may comprise only a small part of the body under consideration.

The early development of elastic-plastic theory was built on the assumption of infinitesimal deformation on the basis of which the total strain could be considered equal to the sum of elastic and plastic components and hence with a similar summation law applying also for strain rates. At finite strains there is a coupling between elastic and plastic deformation since plastic flow occurs in a material already stressed to yield and hence strained elastically, and these two components interact in the nonlinear kinematics of finite-deformation theory. In the current commonly used approach to finite-deformation analysis, the summation of elastic and plastic strain rates to give the total strain rate is adopted. The significance of this assumption is examined in the light of the non-linear kinematic theory. The latter gives a precision to the kinematics which permits aspects of the theory to be investigated more succinctly.

In the undeformed reference configuration material elements or particles are labelled using rectangular Cartesian coordinates expressed by the column vector $\mathbf{X} = (X_1, X_2, X_3)^T$ where the superscript T denotes the transpose of the row matrix. After deformation, which involves both elastic and plastics strain, the particles of the body occupy the configuration $\mathbf{x}$ using the same Cartesian axes, the body having been deformed at time t according to the motion

$$\mathbf{x} = \mathbf{x}(\mathbf{X}, t). \tag{1.1}$$

In order to uncouple the elastic and plastic deformations, the body is considered to be destressed from the configuration $\mathbf{x}$ to the configuration $\mathbf{p}$. It is assumed that the destressing to zero stress involves change of the elastic strain only, although the theory can be modified to include materials which exhibit a strong Bauschinger effect involving plastic flow during the destressing to zero stress [1]. Since the stress is zero in the configuration $\mathbf{p}$, the elastic strain is zero and hence the configuration $\mathbf{p}$ displays purely plastic strain. This is the same as the plastic strain present in the configuration $\mathbf{x}$ since only the elastic strain changed during the destressing.

The elastic-plastic deformation which took place during the motion (1.1) is expressed by the deformation gradient $\mathbf{F}$

$$\mathbf{F} = \partial \mathbf{x}/\partial \mathbf{X} \qquad (F_{ij} = \partial x_i/\partial X_j). \tag{1.2}$$

The mapping $\mathbf{X} \to \mathbf{p}$ with deformation gradient $\mathbf{F}^p$ expresses the plastic deformation in both the configurations $\mathbf{p}$ and $\mathbf{x}$. The mapping $\mathbf{p} \to \mathbf{x}$ with the deformation gradient $\mathbf{F}^e$ constitutes the elastic deformation in the configuration $\mathbf{x}$. Because removing the surface tractions from a body which has been subjected to non-homogeneous plastic strain leaves it in a state of residual stress, destressing involves considering the body to be sectioned into vanishingly small elements which means that mappings to configuration $\mathbf{p}$ from $\mathbf{X}$ and from $\mathbf{x}$ are not differentiable. $\mathbf{F}^e$ and $\mathbf{F}^p$ are then not deformation gradients but are matrix point functions of position defined by the partition procedure. This causes no difficulty since the objective of introducing the configuration $\mathbf{p}$ is to generate variables through which to express the constitutive relation for the stress in the configuration $\mathbf{x}$ following general elastic-plastic deformation. The mapping $\mathbf{X} \to \mathbf{x}$ is differentiable so that the deformation gradient $\mathbf{F}$ does exist. Considering the body sectioned into small elements to achieve destressing to zero stress is analogous to machining away parts of a specimen to measure residual stresses. Such partition does not affect the basic elastic law so that using this in combination with measurements of change in strain, the stresses in the original whole component can be deduced. In practice, test specimens subjected to homogeneous strain are commonly used and then unloading the specimen does indeed destress it.

The configurations **X**, **x** and **p** can be obtained and measured experimentally, and the mapping $\mathbf{X} \rightarrow \mathbf{x}$ is identical geometrically with the resultant of the sequence $\mathbf{X} \rightarrow \mathbf{p} \rightarrow \mathbf{x}$. The chain rule then generates the relation

$$\mathbf{F} = \mathbf{F}^{e}\mathbf{F}^{p} \tag{1.3}$$

The generally non-commutative structure of this matrix product relation indicates that it is not compatible with the summation law for elastic and plastic strain or strain rate. It should perhaps be pointed out that the sequence of maps $\mathbf{X} \rightarrow \mathbf{p}(\mathbf{X}, t) \rightarrow \mathbf{x}(\mathbf{X}, t)$ corresponding to a fixed time t cannot be carried out physically, because, for example, plastic flow cannot be generated at zero stress, but since, by passing through **x**, the configuration **p** can be achieved and then measured, the factors in (1.3) can in principle be evaluated.

The unstressed state **p** is not unique since arbitrary rotation leaves it unstressed. Thus, without loss of generality, unstressing and hence the inverse, elastic deformation $\mathbf{F}^{e}$, can be chosen to be pure deformation without rotation and thus can be expressed by the symmetric deformation gradient matrix $\mathbf{V}^{e}$. This choice simplifies some parts of the analysis.

Substituting (1.3) into the expression for total finite Lagrange strain

$$\mathbf{E} = (\mathbf{F}^{T}\mathbf{F} - \mathbf{I})/2 \tag{1.4}$$

(where **I** is the unit matrix) expresses the total strain in terms of the elastic and plastic strains calculated from $\mathbf{F}^{e}$ and $\mathbf{F}^{p}$ by relations analogous to (1.4) [1]

$$\mathbf{E} = \mathbf{F}^{pT}\mathbf{E}^{e}\mathbf{F}^{p} + \mathbf{E}^{p} \tag{1.5}$$

In the case of large plastic strains for which $\mathbf{F}^{p}$ is very different from **I**, the right-hand side of (1.5) is clearly far removed from the summation of elastic and plastic strains.

In plasticity theory strain rate is appropriately defined in terms of the velocity field ($\mathbf{v}(\mathbf{x}, t)$ in the current elastically-plastically deformed configuration **x**) as the symmetric part of the velocity gradient **L**:

$$\mathbf{L} = \partial\mathbf{v}/\partial\mathbf{x} = (\partial\mathbf{v}/\partial\mathbf{X})(\partial\mathbf{X}/\partial\mathbf{x}) = \dot{\mathbf{F}}\mathbf{F}^{-1} \tag{1.6}$$

where $\dot{\mathbf{F}}$ expresses a time derivative at fixed **X**, that is at a fixed material particle. Expressing **L** as the sum of its symmetric and anti-symmetric parts

$$\mathbf{L} = \mathbf{D} + \mathbf{W} \tag{1.7}$$

determines the deformation rate or stretching tensor **D** and the spin **W**. The former expresses the rate of strain about the current configuration **x** which is appropriate for plasticity analysis. Plasticity theory is commonly termed incremental or flow type, more akin to fluid behavior than to solid, the initial configuration $\mathbf{X}$ playing a minor role unlike the case of elasticity theory.

Substituting (1.3) with $\mathbf{V}^e$ into (1.6), incorporating (1.7), yields

$$\mathbf{D} = \mathbf{D}^e + \mathbf{V}^e \mathbf{D}^p \mathbf{V}^{e-1}|_S + \mathbf{V}^e \mathbf{W}^p \mathbf{V}^{e-1}|_S \tag{1.8}$$

$\mathbf{D}^e$, $\mathbf{W}^e$, $\mathbf{D}^p$ and $\mathbf{W}^p$ are the symmetric and anti-symmetric parts of $\dot{\mathbf{V}}^e \mathbf{V}^{e-1}$ and $\dot{\mathbf{F}}^p \mathbf{F}^{p-1}$ respectively and subscript S denotes the symmetric part.

Now $\mathbf{V}^e$ is the symmetric part of $(\mathbf{I} + \partial \mathbf{u}^e/\partial \mathbf{p})$ where $\mathbf{u}^e = \mathbf{x} - \mathbf{p}$ is the elastic displacement. Since elastic strains are usually of the order (yield stress divided by elastic modulus) $\sim 10^{-3}$, $\mathbf{V}^e$ is commonly close to the unit matrix $\mathbf{I}$. There are exceptions to this circumstance, for example in explosively generated shock waves in metals when volumetric elastic strain can be of the order unity. When $\mathbf{V}^e \approx \mathbf{I}$, (1.8) approximates the strain rate summation relation

$$\mathbf{D} = \mathbf{D}^e + \mathbf{D}^p \tag{1.9}$$

since $\mathbf{D}^p$ is symmetric and $\mathbf{W}^p$ anti-symmetric. Thus the common assumption in finite-deformation elastic-plastic theory usually provides a good approximation to the theory based on nonlinear kinematics. However, in considering certain aspects of the structure of the theory, the precision of the nonlinear kinematical theory permits a more incisive investigation.

2. Elastic-plastic constitutive relations

Plasticity exhibits a "flow" or "incremental" type of response to stress such that in the case of strain hardening the plastic strain rate is determined by the stress variation, so that the plastic strain is determined by integrating the rate relations through the stress history. The final deformation thus depends on the stress history and similarly the stress depends on the deformation history. For a strain hardening material the strain-rate is given in terms of the current state (stress and hardening parameters) and the stress rate. For deformation of most structural metals at ambient temperatures, the plasticity law is of first order in time rates on each side of the constitutive equation so that any monotonically increasing function of time can replace the time variable without modifying the resulting stress and strain histories. For example a monotonically increasing displacement or strain variable would be appropriate. Materials which exhibit such a response are termed time-rate independent though they are of flow or incremental type. This paper is concerned with such materials, although the finite-deformation kinematics discussed applies equally well to rate dependent laws which usually apply at higher temperatures.

Since the plasticity law involves plastic strain rate, the formulation of elastic-plastic theory requires that the plasticity law and elasticity law be substituted into the total strain rate relation (1.8), or the approximation to it (1.9), to provide an expression for the resultant strain rate D in terms of stress, stress rate and variables which express the influence of the previous history of the stress or deformation. Since elasticity is usually expressed as a function relation for the stress in terms of the deformation, a rate form of this relation must be derived. This procedure was carried out in [2] using the nonlinear kinematics equation (1.8), the isotropic elasticity law valid for finite deformation and not modified by plastic flow, and isotropic strain hardening based on a Mises type yield condition

$$J_2(\boldsymbol{\tau}') = c \tag{2.1}$$

where $\boldsymbol{\tau}$ is the Kirchhoff stress, the prime indicates the deviator and c is a scalar function of the history of plastic deformation. The theory could be readily generalized for a yield function of the two stress-deviator invariants J_2 and J_3: $f(J_2, J_3) = c$.

As discussed in [3], the last term in (1.8) comprises a change in elastic strain associated with the rotation of the body taking place. In the case of applying an increment of stress and then removing it (as in the usual experimental program for measuring elastic and plastic increments of strain) the same term generates a residual elastic strain increment because unloading, associated with the pure deformation $\mathbf{V}^e$, is considered to occur without rotation. Of course most testing configurations seek to avoid rotation. It is necessary to have rotation included in order to permit the analysis of general elastic-plastic deformation and it is convenient to avoid it in unloading to define most simply the elastic and plastic deformations. To have to prescribe some particular rotation other than zero with unloading would unnecessarily complicate the analysis.

In [2], since the spin term in (1.8) expresses an elastic rate of strain it is combined with the $\mathbf{D}^e$ to form the total elastic component which takes the form

$$\mathbf{D}^e + (\mathbf{V}^e \mathbf{W}^p \mathbf{V}^{e\,-1})_S = \mathbf{V}^{e\,-1} \overset{\triangledown}{\mathbf{C}}{}^e \mathbf{V}^{e\,-1}/2 \tag{2.2}$$

where $\overset{\triangledown}{\mathbf{C}}{}^e$ is a Jaumann type derivative of the right Cauchy—Green tensor, $\mathbf{C}^e = \mathbf{F}^{eT}\mathbf{F}^e$, associated with the spin $\mathbf{W}^p$.

For deformation without rotation, which defines the elasticity law because of the choice of $\mathbf{V}^e$ for destressing, the finite-deformation-valid elasticity constitutive relation takes the form [2]

$$\boldsymbol{\tau} = 2\mathbf{C}^e(\partial\psi/\partial\mathbf{C}^e) \tag{2.3}$$

where ψ is the strain-energy function per unit undeformed volume. This is objective, so that operating with the same derivative on both sides will produce an objective law in rate form appropriate for substitution into (1.8). Applying the Jaumann derivative ($\overset{\triangledown}{\ }$) in conjunction with (2.2) produces such a rate form law which introduces the stress derivative $\overset{\triangledown}{\boldsymbol{\tau}}$ into the equation for **D**.

The plasticity constitutive relation is already in flow-type form:

$$D_{ij}^{\mathrm{p}} = \frac{1}{h}\frac{\partial J_2}{\partial \tau_{ij}}\frac{\partial J_2}{\partial \tau_{mn}}\overset{\vee}{\tau}_{mn} = \frac{1}{h}\tau'_{ij}(\tau'_{mn}\overset{\vee}{\tau}'_{mn}) \tag{2.4}$$

where h is a scalar function of the history of plastic deformation and $\overset{\vee}{\boldsymbol{\tau}}$ is a time derivative. It is appropriate to use the Jaumann time derivative ($^{\triangledown}$) adopted for the elasticity law in rate form so that the same stress derivative appears in all terms in (1.8) expressed in stress-rate form.

In the special case considered in [2], the Mises type flow law (2.4) prescribes that $\mathbf{D}^{\mathrm{p}}$ and $\boldsymbol{\tau}$ have the same principal directions. For an isotropic elastic medium so also does the elastic deformation gradient $\mathbf{V}^{\mathrm{e}}$. Thus the matrix products in the term containing $\mathbf{D}^{\mathrm{p}}$ in (1.8) are all commutative, hence the $\mathbf{V}^{\mathrm{e}}$'s cancel and that term reduces to $\mathbf{D}^{\mathrm{p}}$.

The usual approach to formulating the elastic-plastic constitutive relation for finite strain applications is based on the concept of small elastic strains and Hooke's law is used in the form

$$\epsilon_{ij}^{\mathrm{e}} = \frac{1+\nu}{E}\tau_{ij} - \frac{\nu}{E}\tau_{kk}\delta_{ij} \tag{2.5}$$

where $\boldsymbol{\varepsilon}$ is the infinitesimal strain and $\boldsymbol{\tau}$ the Kirchhoff stress [5, 6]. A rate form of this is commonly taken to be

$$D_{ij}^{\mathrm{e}} = \frac{1+\nu}{E}\overset{\circ}{\tau}_{ij} - \frac{\nu}{E}\overset{\circ}{\tau}_{kk}\delta_{ij} \tag{2.6}$$

where $\overset{\circ}{\boldsymbol{\tau}}$ is a time derivative of the stress tensor.

Because the final constitutive relation must be valid for large plastic deformation, it must be so for large total strain and rotation. It is thus important to check that the constitutive relation is objective, i.e. that if a time-dependent rigid body motion is superimposed on the deformed configuration $\mathbf{x}$, the effects on the stress at time t must be simply to rotate it according to the value of the superimposed rotation at that time. If (2.6) is substituted into (1.9) and combined with an objective plasticity law it is clear that the stress rate $\overset{\circ}{\boldsymbol{\tau}}$ must be objective. There is an infinity of such rates to choose from [4], but the infinitesimal approximation built into (2.6) prevents a deductive choice. In current finite-element elastic-plastic programs two such rates have been adopted for finite-deformation analysis, the Jaumann rate and the Truesdell rate.

In the transformation of (2.5) to rate form (2.6), (2.5) could not be formally differentiated because it is not a valid equality in the context of finite deformation theory. Rather a derivative of a type not uniquely selected was applied to the right-hand side, and the velocity field was introduced with which to express the "rate of strain" for the left-hand side. In contrast we have seen that the nonlinear kinematical relation (1.8) is combined with the finite-deformation-valid rubber elasticity law to permit formal differentiation of both sides of the elastic constitutive relation to provide a strain rate term which can be substituted into (1.8). It is necessary that the plasticity law will also mesh conveniently into (1.8). These components all meshed together in [2] to yield an elastic-plastic constitutive relation and a corresponding variational principle. A simplification of the final result valid for small elastic strain yielded the conventional structure.

3. Discussion

The terms in (1.8) which involve coupling between the elastic deformation gradient $\mathbf{V}^e$ and terms containing the plastic kinematic variables $\mathbf{D}^p$ and $\mathbf{W}^p$ have a bearing on the analysis of material stability as developed by Drucker in many publications and summarized in the recent report by Palgen [7]. In the Workshop : Plasticity of Metals at Finite Strain, held at Stanford University in 1981, Drucker pointed out [8] that what he terms plastic strain increments (residual increments of strain following application and removal of the stress increment) are the appropriate variables for discussing stability and normality. This definition differs from that of plastic strain increment in the literature on which the present paper is based which adopts the residual strain increment when the stress is reduced to zero for this quantity. However the latter two terms in (1.8) do express the rate of residual strain (corresponding to the strain increment associated with adding and removing a stress increment) factored into elastic effects and effects associated with plasticity in the formulation presented in this paper.

For isotropic elastic response $\mathbf{V}^e$ has the same principal directions as $\boldsymbol{\tau}$ and the scalar product in nine-dimensional space of $\boldsymbol{\tau}$ and $\mathbf{V}^e\mathbf{W}^p\mathbf{V}^{e-1}$ is

$$\mathrm{Tr}(\boldsymbol{\tau}\mathbf{V}^e\mathbf{W}^p\mathbf{V}^{e-1}) = \mathrm{Tr}(\mathbf{V}^{e-1}\boldsymbol{\tau}\mathbf{V}^e\mathbf{W}^p) = \mathrm{Tr}(\boldsymbol{\tau}\mathbf{W}^p) = 0 \qquad (3.1)$$

where Tr denotes the trace; the first equality arising since cyclic permutation does not affect the trace, the second since products of matrices with the same principal axes are commutative, and the third since $\boldsymbol{\tau}$ is symmetric and $\mathbf{W}^p$ anti-symmetric. Thus the corresponding strain rate is tangential to the yield surface, the normal to which is parallel to $\boldsymbol{\tau}'$. The $\mathbf{V}^e$ coupling does not influence this result. Note that (3.1) can be interpreted as the power expended on this component of strain rate. These results would not arise in the case of anisotropic elastic properties when $\mathbf{V}^e$ and $\boldsymbol{\tau}$ are not parallel matrices which is in conformity with analysis presented in [7].

As already pointed out in Section 2, for isotropic elasticity simplifications also arise for the plastic flow term in (1.8) for which the V^e terms cancel. It turns out that this is also true for a yield condition containing J_3.

In considering power relations which arise in the study of material stability and normality, those associated with plastic flow related to the theory presented in this paper were expressed in terms of products of the stress $\boldsymbol{\tau}$ which acts in the $\mathbf{x}$ configuration with the strain rate $\mathbf{D}^p$ defined in the purely plastically deformed $\mathbf{p}$ configuration. This approach introduces the question of physical appropriateness and a physically more natural formulation is to consider the plastic strain rate $\mathbf{D}^p$ to produce a strain rate in the current configuration $\mathbf{x}$ given by $\mathbf{V}^e\mathbf{D}^p\mathbf{V}^{e-1}$. Rate of work terms associated with this can be expected to be appropriate for work and related stability considerations. This is in conformity with Drucker's suggestion [8] and as mentioned in the continued discussion, was being considered for the anisotropic hardening case. Considering the separate factors in this term associated independently with plastic and elastic deformation could well help to clarify aspects of this question. We have seen that in the case of isotropic elasticity the elastic terms cancel but more involved coupling appears to arise in the case of anisotropic elasticity.

Acknowledgement

The results presented in this paper were obtained in the course of research sponsored by the U.S. Army Research Office under Contract No. DAAG 29-82-K-0016 with Rensselaer Polytechnic Institute. The author gratefully acknowledges this support.

References

1 E.H. Lee, Some comments on elastic-plastic analysis. Int. J. Solids Structures, 17 (1981) 859—872.
2 V.A. Lubarda and E.H. Lee, A correct definition of elastic and plastic deformation and its computational significance. J. Appl. Mech., 48 (1981) 35—40.
3 E.H. Lee and R.M. McMeeking, Concerning elastic and plastic components of deformation. Int. J. Solids Structures, 16 (1980) 715—721.
4 W. Prager, An elementary discussion of definitions of stress rate. Quart. Appl. Math., 18 (1961) 403—407.
5 R. Hill, Some basic principles in the mechanics of solids without a natural time. J. Mech. Phys. Solids, 7 (1959) 209—225.
6 R.M. McMeeking and J.R. Rice, Finite-element formulations for problems of large elastic-plastic deformation. Int. J. Solids Structures, 11 (1975) 601—616.
7 L. Palgen, The Structure of Stress-Strain Relations in Finite Elasto-Plasticity. Ph.D. Thesis, University of Illinois at Urbana-Champaign, October 1981; T. & A.M. Report 452, UILU-ENG 81-6006, October 1981.
8 D.C. Drucker, Discussion, In: E.H. Lee and R.L. Mallett (Eds.), Plasticity of Metals at Finite Strain, Div. Appl. Mech., Stanford Univ. and Dept. Mech. Engg., Rensselaer Polytechnic Inst., 1982, p. 121.
9 E.H. Lee, Elastic-plastic deformation at finite strains. J. Appl. Mech., 36 (1969) 1—6.

Experimental Evaluation of the Overall Anisotropic Material Response on Continuous Damage

A. LITEWKA

Technical University, Piotrowo 5, 60—950 Poznań (Poland)

A. SAWCZUK

Polish Academy of Sciences and Université Aix-Marseille III, Faculté des Sciences et Techniques de Saint-Jérôme, 13397 Marseille Cedex 04 (France)

Summary

Internal continuum damage of solids is modelled by an arrangement of oriented cracks. The observed overall mechanical response of directionally damaged materials is anisotropic. Differences in elastic, plastic and failure anisotropies are illustrated for a homogenized damaged material and uniaxial stress experiments were performed on continuously damaged alumium alloy sheets. The stress principal directions varied with respect to those of the introduced damage tensor. Experimentally obtained values of material constants for an orthotropic homogenized solid are given. The preliminary tests allow an approach to mathematical modelling of anisotropic continuous-damage mechanical response within the tensor functions theory for two independent tensor variables.

1. Introduction

Internal damage due to fissuration and void growth in strained solids exhibits directional characteristics even before a localized failure occurs. The continuum damage mechanics employing a scalar parameter is unable to describe fully the deterioration growth resulting in an anisotropic material response. The overall anisotropy of homogenized damaged solids influences elasticity and plasticity as well as the orientation of the final localized rupture.

Two points seem to be essential in continuous-damage mechanics. On one hand the question arises of an appropriate mathematical description of the damage evolution and the overall mechanical response of deteriorating solids. The second point concerns experimental evidence as to the presence and evolution of oriented damage. It also concerns specifications of the material constants or functions entering the constitutive equations involving both the stress—strain relations and the evolution of the damage.

References regarding the continuous-damage theory, originated by Kachanov [1] for a uniaxial stress and further developed employing a scalar damage parameter, can be found in [2] and [3]. Several proposals also exist for anisotropic continuous-damage mechanics. They introduce and often define internal fissuration tensors. Kachanov Jr. [4] and Murakami and Ohno [5] specified second-order damage tensors but an account of other attempts can be found in [2] and [6]. Once a damage tensor is introduced it is straightforward to account for the overall anisotropic response of deteriorating solids. Polynomial representations of tensor functions seem to be suitable for mathematically modelling the oriented continuum damage [6]. This specific point will not, however, be elaborated upon in this note, as principles and examples of an inelastic, non-linear material response of structured media are included, for example, in [7—9].

Our attention will be focussed on experimental simulations of the oriented damage and on an evaluation of the directional overall material response of an originally isotropic solid. For perforated solids, both experimental and mathematical modelling questions have been asked in [10] as well as employed for creep damage in [11].

This paper concerns oriented damage simulation. Oriented openings of prescribed geometry and spatial arrangement are introduced into an isotropic sheet. Tensile tests are made at various orientations of the principal overall stress trajectories with respect to the direction of cracks. The tests are performed in order to evaluate the overall elastic material properties as well as to estimate the directional plastic response. The failure modes are recorded and their dependence on the internal damage orientation is shown. Constitutive laws of anisotropic damage can thus be outlined. The preliminary tests presented in this note are restricted to a uniaxial tension and a defined size of cracks. A developed experimental study of anisotropic continuum-damage evolution and modes of failure under combined stress and concerning a simulation of oriented crack growth will be reported elsewhere [12].

The objective of this note is to illustrate the directional character of the internal damage, to indicate a method of simulation of an anisotropic continuum-damage growth and to outline possible modelling of such damage as regards the stress—strain relations as well as the evolution equation for the internal deterioration tensor.

2. Experimental setting

To simulate an overall anisotropic mechanical response of a damaged solid, tests were made on aluminium alloy sheets with regularly distributed

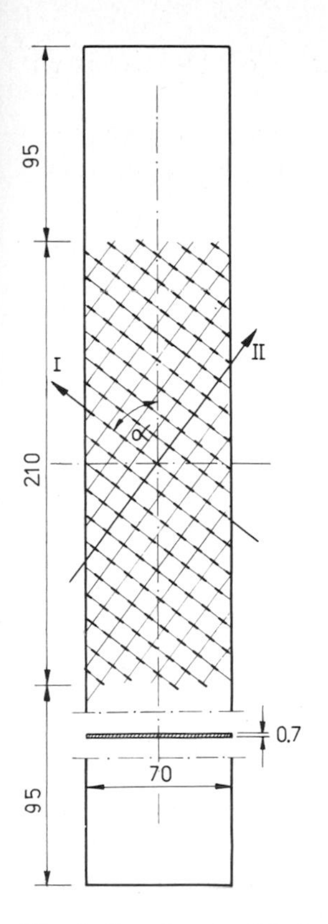

Fig. 1. Specimen with simulated damage.

rectilinear cracks. Specimens were machined from a homogeneous and presumably isotropic sheet, their longitudinal axes coinciding with the rolling direction so as to eliminate the effect, if any, of the production-induced anisotropy. The specimens' shape and dimensions are given in Fig. 1. The known mechanical characteristics of the test material are as follows: Young's modulus $E = 69\,550$ MPa; Poisson ratio $\nu = 0.337$; conventional yield stress at 0.1% axial elongation $\sigma_0 = 132$ MPa.

Rectangular openings 5×1 mm were punched into the specimens. The holes were arranged in two square patterns with 10 mm pitch, Fig. 2. The difference in the crack arrangement is that in the first case the longitudinal axis of a crack coincides with the pitch, whereas in the second situation the crack length makes an angle $\alpha = \pi/4$ with the pitch. The reason for testing two arrangements of damage orientation is to introduce a difference between the symmetries on the macro scale and the principal directions of the openings. Such information is of pertinence when devising the tensor function representations for materials with an internal orientation [7]. The practical purpose of such a comparison consists in attempting to get information as to the future mathematical modelling of

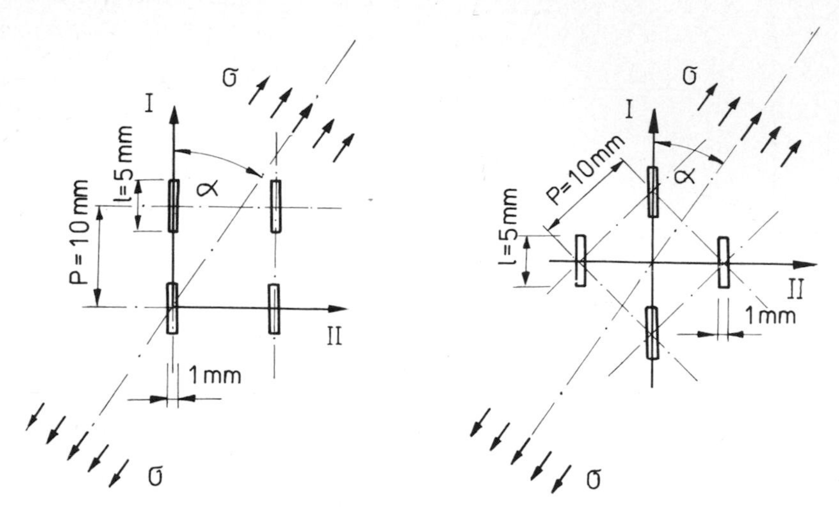

Fig. 2. Studied crack arrangements and sizes.

the mechanical response, either as a simple orthotropy or with a need for more involved accounts on the symmetries of a homogenized material.

In both cases the density ratio of homogenized materials ρ_H to the original one ρ_0, $\mu = \rho_H/\rho_0$, is the same. In further tests the crack length l will be varied for a fixed pitch P in order to simulate the damage growth and to furnish the data regarding appropriate polynomial representations for the respective evolution equations [12]. The dimensionless length of crack as referred to the pitch is denoted by $\lambda = l/P$. The tests described in this paper concern $\lambda = 0.5$ and $\mu = 0.95$.

The arrangement of rectangular openings in the specimens is such as to allow variation in the principal directions of stresses in the homogenized material with respect to the crack orientation, thus with the principal directions of a damage tensor we intend to introduce when modelling the overall anisotropic response. These directions of crack length and width are marked in the Figures as I and II, respectively. The specimens are subjected to an axial load at various inclinations with respect to the directions of cracks. The test program includes seven angles the overall tensile stress direction makes with the crack direction, namely $\alpha = 0, \pi/12, \pi/4, \pi/3, 5\pi/12$ and $\pi/2$.

As a homogenization of directionally damaged materials is attempted, the crack pitch is several times smaller than the specimen width.

3. Test results

The stress—strain curves obtained for the specimens loaded at various directions with respect to the crack orientation are given in Fig. 3 and 4 for the damage arrangement indicated. The recorded axial strain and the

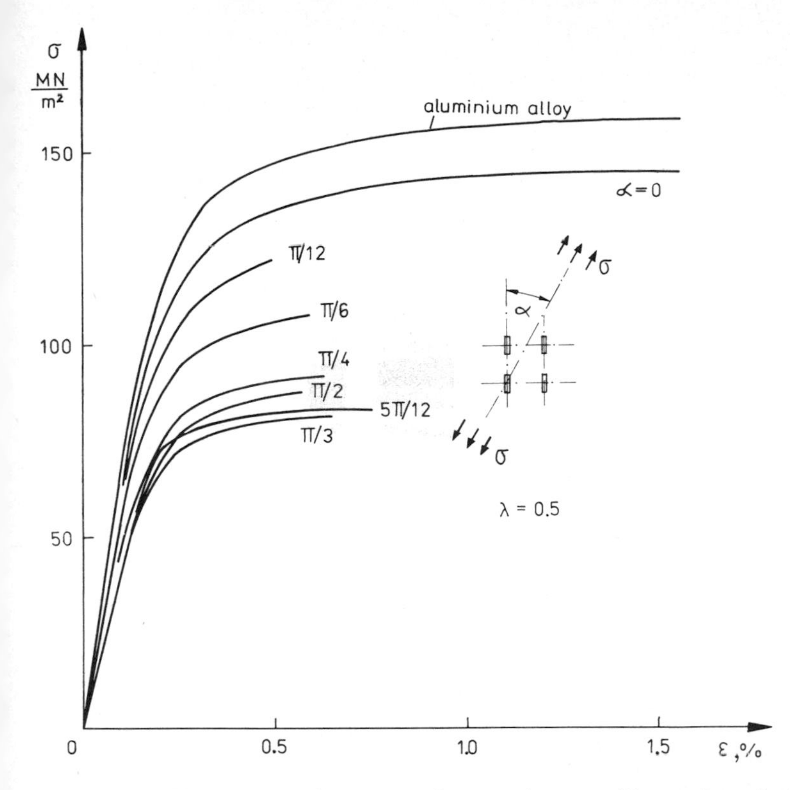

Fig. 3. Overall stress—strain curves for specimens with cracks oriented in the pitch direction.

computed stress refer to the undeformed and undamaged cross-sections. Hence they represent the response of "homogenized" materials.

Note that, both in the linearly elastic and in the ductile deformation range, the overall material response is direction dependent. The force-controlled tests terminate in an abrupt failure. Both the overall yield point and the overall ductility vary with the stress orientation as referred to the direction of damage.

The failure modes are presented in Figs. 5 and 6. Note that, in the load-controlled tests, brittle failure dominates at small overall extensions. The modes of failure differ significantly with damage orientation. Both rupture by extension and by slip is observed. The mutual orientation of the principal directions of stress and damage decides whether the deformation instability is associated with shear or with thinning. The material presented here qualitative only but the purpose of this paper is to visualize certain phenomena prior to their mathematical modelling. For example, it is seen by comparing the results for $\alpha = 0$ and for $\alpha = \pi/2$ in Figs. 5 and 6 that the scalar damage parameter ω, defined in the Kachanov theory [1], is not sufficient to describe the damage growth

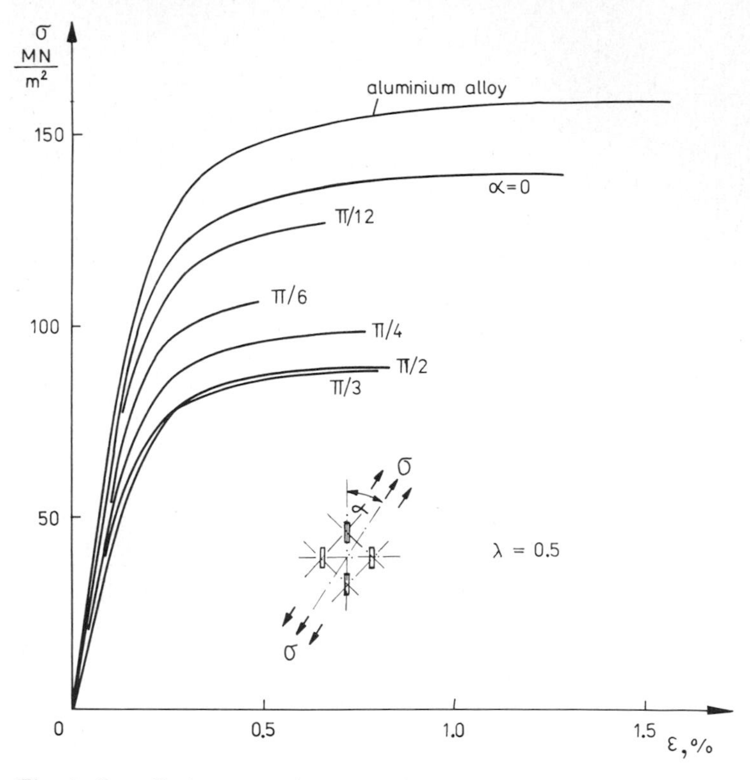

Fig. 4. Overall stress—strain curves for specimens with cracks oriented in the diagonal direction.

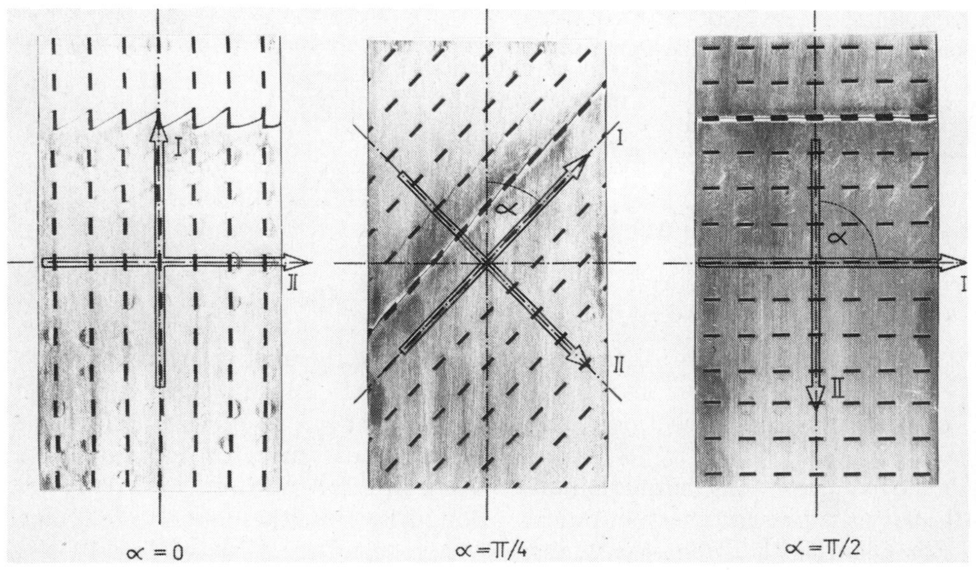

Fig. 5. Failure mode variation with the direction of overall stress at cracks aligned with the pitch.

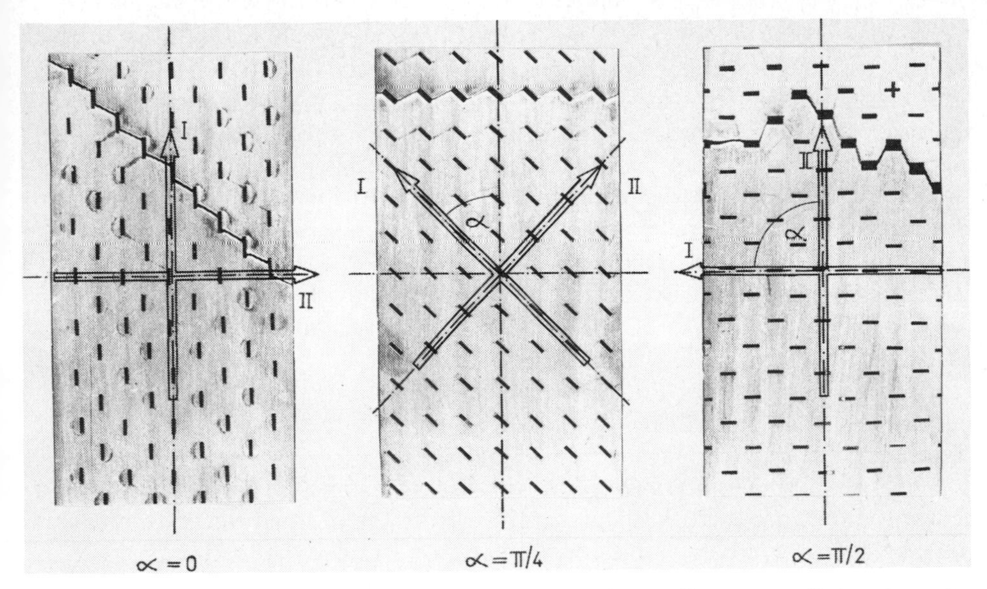

Fig. 6. Failure mode variation with the direction of overall stress at diagonal crack arrangement with respect to the pitch.

and the failure mode. This parameter is defined as follows

$$\omega = 1 - A/A_0 \tag{3.1}$$

where A_0 denotes the original area and A stands for the current one.

Let us compare the results for $\alpha = 0$ and $\alpha = \pi/2$. The failure modes for $\alpha = 0$, $\omega = 0.90$ and $\alpha = \pi/2$, $\omega = 0.50$, are different. For example, Fig. 5 shows that at oriented damage, $\omega = 0.90$ is associated with a shear failure, whereas $\omega = 0.50$ results in an elongation type of separation. However, according to the scalar damage theory the "homogenized" stress σ for an element in tension is that specified by the net area. Hence $\sigma(\alpha = 0)/\ \sigma(\alpha = \pi/2) = \omega(\alpha = 0)/\omega(\alpha = \pi/2)$. With reference to Fig. 5, the damage parameter ratio is $0.9/0.5 = 1.8$, whereas from Fig. 3 one can conclude that the rupture stress ratio is $148/89 = 1.65$. These remarks are intended to justify the attempt to simulate the damage growth and to specify the deformability and failure modes associated with an oriented continuous-damage process.

4. Elastic anisotropy

As the observed elastic deformability is linear we shall evaluate the material constants within the theory of infinitesimal linear anisotropic elasticity [13]. The overall mechanical response of a sheet with the regular crack systems considered is orthotropic. In the plane stress four constants have to be determined.

A theoretical analysis of the overall orthotropic behaviour of elastic sheets with regular crack systems is given by Gross [14, 15]. He calculates the elastic energy in the presence of cracks and deduces the equivalent elastic constants from the requirement that this energy is the same both for the cracked and for the homogenized material [15]. Specific computations have not yet been made; they involve the stress concentration factors at the crack tips.

Our analysis is restricted to presenting experimentally evaluated constants. With reference to the orthotropy directions indicated in Figs. 5 and 6, thus related to $\alpha = 0$ and $\alpha = \pi/2$, we have to determine the Young's moduli E_1, E_2, the Poisson ratios ν_{12}, ν_{21}, and the Kirchhoff modulus G_{12}. The constants are related, hence in plane stress four constants are to be established. This can be done by means of simple uniaxial tests.

The elastic overall deformability was measured in longitudinal and lateral directions with respect to the direction of loading at $\alpha = 0$ and $\alpha = \pi/2$. This enables the determination of E_1, E_2, ν_{12} and ν_{21}, whereas the shear modulus G_{12} is given by the appropriate equation relating the longitudinal effective Young's modulus $E(\alpha)$ for specimens loaded in the direction inclined at the angle α with respect to the crack orientation [13]. The obtained values of the equivalent material constants at $\mu = 0.95$ and $\lambda = 0.5$ are given in Table 1.

TABLE 1

Elastic moduli at simulated oriented damage specified by $\mu = 0.95$, $\lambda = 0.5$

Crack arrangement	E_1/E	E_2/E	ν_{21}/ν	ν_{12}/ν	G_{12}/G
Pitch	0.874	0.602	0.875	0.614	0.750
Diagonal	0.882	0.643	0.842	0.623	0.781

In Fig. 7, the effective Young's modulus variation is shown for specimens loaded in the directions defined by the angles α with respect to the crack orientation. The recorded longitudinal overall strains concerned $\alpha = \pi/12$, $\pi/6$, $\pi/4$, $\pi/3$ and $5\pi/12$, in addition to those corresponding to the principal directions of openings. Two points have to be made: the first concerns a continuous and regular decrease of the specimen axial rigidity when passing from the maximal principal direction of the opening to the minimal one. Such a situation will not be observed with regard to the plastic response. This apparently indicates a difference between the overall elastic and plastic anisotropies of an originally elastically and plastically isotropic material. The second remark concerns a rather small influence of the crack arrangement on a square grid as to the effective orthotropy. Both the data in Table 1 and in Fig. 7 differ a

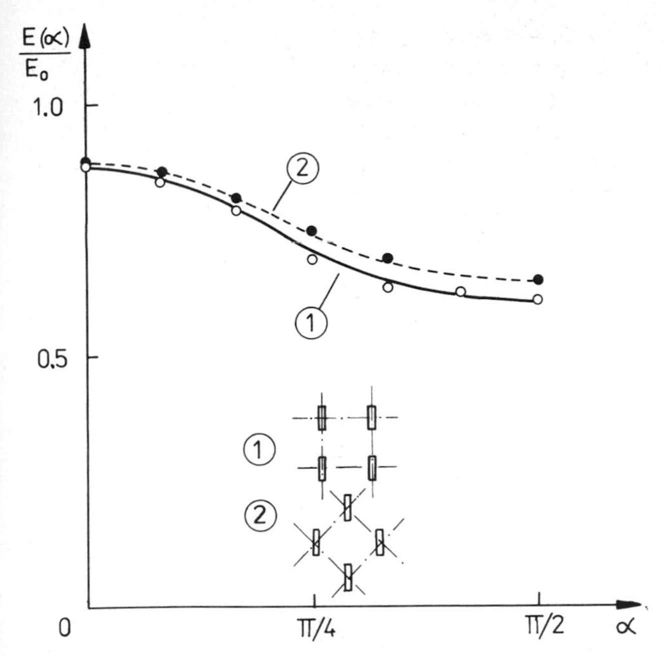

Fig. 7. Effective Young's modulus of directionally damaged sheet.

little for the considered hole patterns so that the symmetries imposed by crack arrangements can be disregarded when developing a theory of mechanical response involving a continuous-damage tensor.

5. Plastic anisotropy

For plastically orthotropic materials the simplest criterion of yielding generalizing the Huber—Mises yield condition employs the fourth-order tensor of material constants. It can be written in the form [16]

$$A_{ijkl}\sigma_{ij}\sigma_{kl} - 1 = 0. \tag{5.1}$$

For deteriorating solids the volume variation due to void growth has to be accounted for in the yielding criterion. Thus, the Hill yield condition [17] appears to be unsuitable in the considered case of an overall plastic flow for deteriorating solids.

In plane stress, six material constants enter the yield condition (5.1). Simple uniaxial stress tests are not sufficient for the purpose and tubular specimens have to considered. To develop a yielding criterion for fissurated materials the approach employing the stress and the internal structure tensors seems preferable. Such a procedure was used for perforated solids in [10, 18] and, in general, for materials with an oriented internal structure in [7, 19].

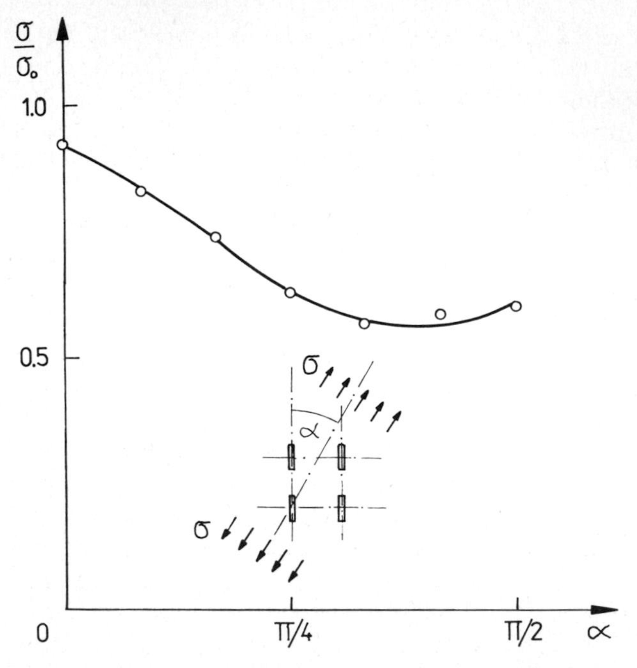

Fig. 8. Uniaxial overall yield stress variation depending on the mutual orientation of the stress and the fissuration directions; pitch alignment of cracks.

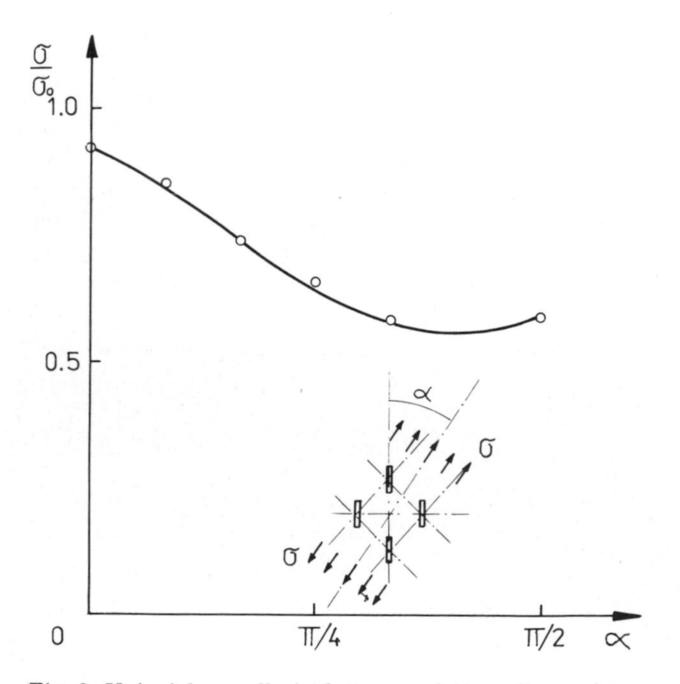

Fig. 9. Uniaxial overall yield stress variation; diagonal arrangement of cracks.

At this juncture we restrict our attention to experimental results concerning a conventional yield stress corresponding to 0.1% permanent overall strain and employing the data presented in Figs. 3 and 4. The uniaxial conventional dimensionless yield stress σ/σ_0 is plotted in Figs. 8 and 9 for the square penetration patterns considered, at the same values of $\mu = 0.95$ and $\lambda = 0.5$.

For the respective inclinations of the stress and crack directions in both cases of crack symmetries the uniaxial homgenized yield stresses almost coincide. This means that the internal macrostructures studied have but little influence on the plastic overall behaviour. This fact should be taken into account when devising a yield criterion employing the strain, damage and mixed stress-damage invariants, and not necessarily restricting the attention to a particular form (5.1).

Another feature, specific to the directionally dependent yield stress seen in Figs. 8 and 9, is to be remarked upon. The direction-dependent conventional yield stress does not uniformly when α varies continuously. The smallest yield stress is noticed for $\pi/3 < \alpha < \pi/2$ but not for $\alpha = \pi/2$ as was the case for the elastic stiffness variation. A similar effect was observed for perforated materials [10, 18] as well as for stratified solids under combined stress [20]. This indicates that a possibly full set of stress and mixed stress-damage invariants has to be studied instead of the criterion (5.1), even if A_{ijkl} is considered to be damage-dependent.

6. Anisotropic damage

The test results presented are not sufficient to advance any specific constitutive law governing the stress—strain relation in the presence of oriented damage nor can they give an evolution law for damage growth. The experimental results on combined stress states and simulated damage growth as attained by varying μ and λ and reported in [12] will probably permit such specification. At this point we indicate only the general approach to the continuum-damage mechanics within the tensor-function representations.

Let us introduce a damage tensor **D**, considered to be symmetric and of the second order. The basic assumption here is that continuum damage develops in three-dimensional continua as a result of void and crack growth. Thus a tensor quantity with three real principal directions and values can be conceived. For our purpose it is sufficient to state that the internal damage can be described by a second-order tensor. Various definitions of the damage tensor are possible e.g. [4, 5].

In a continuum-damage theory we have as independent variables the second-order symmetric tensors of stress **T** and damage **D**. It is necessary to specify the constitutive regarding strain E:

$$\mathbf{E} = \mathbf{E}(\mathbf{T}, \mathbf{D}) \tag{6.1}$$

as well as an evolution equation for the damage:

$$\dot{\mathbf{D}} = \dot{\mathbf{D}}(\mathbf{T}, \mathbf{D}). \tag{6.2}$$

The tensor-valued tensor functions (6.1) and (6.2) can be written in their polynomial representations involving both a specified number of the tensor generators $\mathbf{G}_i$ and the associate number of scalar-valued functional multipliers, expressed in terms of the basic invariants $\mathbf{I}_i$ of the independent tensors.

A simple anisotropic-damage theory will involve a limited number of the basic invariants and generators. When considering failure criteria we shall specify the form

$$f(\text{tr}\,\mathbf{T}, \text{tr}\,\mathbf{T}^2, \text{tr}\,\mathbf{TD}, \mu) = 0 \tag{6.3}$$

involving three invariant $\mathbf{I}_i$ only, as well as μ.

A constitutive relation, applicable until a failure criterion is attained, will be assumed:

$$\mathbf{T} = \alpha_1 \mathbf{I} + \alpha_2 \mathbf{E} + \alpha_3 (\mathbf{DE} + \mathbf{ED}), \tag{6.4}$$

whereas the damage growth, given directly in terms of the overall deformation $\mathbf{E}$, will supposedly, be of the form

$$\mathbf{D} = \beta_1 \mathbf{I} + \beta_2 \mathbf{E} \tag{6.5}$$

The scalar functions α_i, β_i depends on the basic invariants of the independent tensor variables. A power series expansion can be used to specify these functions. The simplest forms will likely retain only tr $\mathbf{E}$, tr $\mathbf{ED}$, as well as λ and μ introduced earlier.

Since the damage variation is expressed in (6.5) directly in terms of strain, the evolution equation is simplified in the sense that it is no longer a differential equation of time. Eventually the constitutive equation (6.4) will involve strains only as the independent variables. It will thus be a generalization of three-dimensional cases of the non-linear stress—strain relation proposed by Janson and Jult [21] for the uniaxial case within the Kachanov damage theory. Still, however, the overall response (6.4) might appear anisotropic as the principal directions of the damage tensor do not necessarily coincide with those of stress. The mutual orientation of the principal directions will enter (6.4) via the reference systems employed.

Prior, however, to advancing any specific form of the constitutive law our attention is concentrated on simulation of the oriented damage growth [12].

7. Conclusions

Preliminary experiments on modelling of continuous damage are reported. It was intended to visualize directional properties of internal continuous damage as well as the overall anisotropic response of a homogenized damaged solid. The tests on aluminium alloy sheets concerned both elastic and plastic response as well as failure loads and failure modes. Differences between elastic and plastic anisotropy due to fissuration are shown. It is to be noted that different failure modes are observed depending on the mutual orientation of the principal directions of stress and those of the oriented damage. The tests allowed the specification of the elastic properties of an equivalent orthotropic material and the disclosure of the variation of the yield point stress in tension with crack orientation. The differences between elastic, plastic and failure anisotropies are visualized.

Experiments, in addition to giving information on directional properties of internal damage, allowed specification of an experimental program concerning behaviour of damaged solids under combined stress, as well as an appropriate program for an oriented-damage growth simulation. A mathematical model for anisotropic continuous damage is proposed within the tensor-function representations. The polynomial representations. are used both to specify the general form of the stress--strain relation as well as the equation describing the damage growth with the applied overall strain.

Additional tests regarding complex stress and deformation states and intended to furnish the experimental background for an anisotropic continuum damage theory, and further hypotheses will be needed when devising the mechanical laws. In particular Drucker's material stability postulate [22] in combination with the tensor functions approach will be employed [23].

References

1 L.M. Kachanov, Izv. Akad. Nauk SSSR, OTN, 8 (1958) 26—31.
2 J. Lemaître, 22nd Polish Solid Mechanics Conf., Golun, Sept. 3—11, 1980, Mech. Teoret. Stos., *20* (1982) (in press).
3 D. Krajcinovic and G.J. Fonseka, J. Appl. Mech., *48* (1981) 809--824.
4 M. Kachanov, Continuum model of medium with cracks, Brown Univ. Rep. 16, Providence, RI, Sept. 1978.
5 S. Murakami and N. Ohno, 3nd IUTAM Symp. Creep in Structures, Leicester, Sept. 8—12, 1980, Springer, Berlin, 1981, pp. 422—443.
6 A. Sawczuk, Proc. 8th Int. Congr. Ship Struct., Palaiseau, Aug. 31—Sept. 3, 1982 (in press).
7 J.P. Boehler, J. Méc., *17* (1978) 153—190.
8 S. Murakami and A. Sawczuk, Nucl. Engng. Design, *65* (1981) 33—47.
9 J. Kubik and A. Sawczuk, Ing. Archiv, *53* (1983) (in press).

10 A. Litewka and A. Sawczuk, Ing. Archiv, *50* (1981) 242—250.
11 S. Murakami and T. Imaizumi, J. Méc. Théor. Appl., 1 (1982) 743—761.
12 A. Litewka and A. Sawczuk, Euromech Colloquium 152, Wuppertal, Sept. 21—24, 1982.
13 S.G. Lecknitskij, Theory of Elasticity for Anisotropic Solids (in Russian), Nauka, Moscow, 1977.
14 D. Gross, Ing. Archiv, *51* (1982) 301—310.
15 D. Gross, 1982, personal communication.
16 W. Olszak and W. Urbanowski, Arch. Mech. Stos., *8* (1956) 671—694.
17 R. Hill, The Mathematical Theory of Plasticity, Clarendon Press, Oxford, 1950, ch. 12, p. 318.
18 A. Litewka, Nucl. Engng. Design, *57* (1980) 417—425.
19 J.P. Boehler and A. Sawczuk, Acta. Mech., *27* (1977) 185—206.
20 D. Allirot, J.P. Boehler and A. Sawczuk Res. Mechanica, *4* (1982) 97—113.
21 J. Janson and J. Hult, J. Méc. Appl., *1* (1977) 69—84.
22 D.C. Drucker, Proc. 1st U.S. National Congr. Appl. Mech., ASME, New York, 1951, 487—491.
23 M. Basista and A. Sawczuk, CNRS Colloquium, Villard-de-Lans, June 20—24, 1983.

On the Unified Framework Provided by Mathematical Programming to Plasticity

G. MAIER and A. NAPPI*

Department of Structural Engineering, Technical University (Politecnico), Milan (Italy)

Abstract

In the traditional description of plastic behaviour, which rests on the notions of time-independence, yield surface, association (normality) or nonassociation, hardening or softening, a typical mathematical structure is the complementarity relation (in each pair of corresponding sign-constrained variables at least one is required to vanish). As a consequence, a few simple notions in finite dimensional constrained optimization and linear complementarity theory, are shown to lead directly, in a unified manner, to several known and some new results at three levels of plasticity theory: constitutive laws in terms of rates, stress—strain relations in finite increments for computational purposes, and discretized boundary value problems.

1. Introduction

In the early fifties Drucker's postulate [1, 2] assumed a central role in the foundations of plasticity. It provided both a compact expression of mechanical features common to diverse material models and a simple basis for the derivation of many theorems for mechanical systems founded on those models, i.e. for the development of a unitary theory of plasticity. Outside its range of validity, plasticity theory may be said to have become less simple, rich and elegant.

The natural trend towards generalization and unification is probably one of the reasons for the recent growth of the literature on models of material behaviour. Constitutive laws with broader coverage, i.e. laws endowed with greater versatility and interpretative power, represent important advancements, although in this respect a warning by Drucker may be appropriate: "full generality is not the goal; full generality is complete chaos and contains no information. Unifying instructive classi-

* On leave from Rome University

fications of physical behaviour are required which contain the essence of the macroscopic problem" [3].

Another way of pursuing unification and, after all, economy of thought, is to focus on the formal structure of some basic governing relations and to utilize the relevant mathematics. The central and typical formal structure of traditional plasticity (i.e. resting on the notion of yield surface, hardening rule, etc.) is identified here in the "complementarity" relation. This requires at least one variable to vanish in each pair of corresponding sign-constrained variables. This relation is typical of "complementarity problems" which are the object of a growing mathematical literature of their own [4]. They are, however, under certain circumstances fundamental, as Lagrange multiplier rules or optimality conditions, in mathematical programming, i.e. in mathematical optimization under inequality constraints (see e.g. [5]). The purpose of this paper is to present some of the results which are directly attainable, based on the above identification, from a few simple mathematical notions in optimization theory (basically, Kuhn—Tucker theorem, duality, linear complementarity problems, all summarized with some of their peculiarities and specializations in the Appendix). In fact, by means of these notions it becomes possible to generate "automatically" many known aspects and some apparently novel results in plasticity theory, with or without the validity of Drucker's postulate. Because of space limitations, emphasis is on plastic constitutive laws. It is concisely shown how the discussion of the constitution can be easily transferred, on the same restricted mathematical basis, to discretized boundary value problems in the "small deformation" range.

The same conceptual tools may have a similar fruitful and unifying role for other material models exhibiting complementarity (e.g. fracturing and locking materials) and in other areas of nonlinear structural analysis.

Finite dimensional optimization and complementarity theory, of which just a few basic concepts are used in what follows, are closely related to a large, still growing number of computational methods, not of concern here. Their potential in engineering plasticity has long been recognized, but probably not yet fully exploited.

The recent developments of infinite dimensional constrained optimization are currently the source of broader unifying frameworks for continuum mechanics of unilateral problems, including classical plasticity, though with a well expected burden of complications and abstractions; see e.g. [4, 6]. Numerical solutions, however, require to consider some discrete model and, hence, lead back to complementarity in finite dimensional spaces and to mathematical programming anyway.

2. Basic features of plastic behaviour: Hill's principle; Drucker's postulate "in the large"

A classical description of the manifestations of plastic strains in the behaviour of metallic materials (e.g. [7]) consists of the following relations (bold symbols denote vectors and matrices, a tilde means transpose):

$$\phi_i(\boldsymbol{\sigma}) \leqslant 0 \quad (i = 1, \ldots, m) \tag{1a}$$

$$\dot{\boldsymbol{\varepsilon}}^p = \sum_{i=1}^{m} \frac{\partial \phi_i}{\partial \boldsymbol{\sigma}} \dot{\lambda}_i, \qquad \dot{\lambda}_i \geqslant 0 \tag{1b, c}$$

$$\phi_i(\boldsymbol{\sigma})\dot{\lambda}_i = 0 \quad (\text{or } \tilde{\boldsymbol{\phi}}\,\dot{\boldsymbol{\lambda}} = 0) \tag{1d}$$

The (regular, differentiable) yield functions ϕ_i define through inequalities (1a) the current elastic domain and its boundary or yield surface Γ in the space of the independent stress components $\tilde{\boldsymbol{\sigma}} \equiv (\sigma_{11}\ \sigma_{22}\ \sigma_{33}\ \sigma_{12}\ \sigma_{23}\ \sigma_{31})$. Equation (1b, c) expresses the outward "normality" (in generalized sense at the intersection of two or more surfaces $\phi_i = 0$) of the plastic strain rate vector to Γ in the current stress point $\boldsymbol{\sigma}$, being understood that strains $\boldsymbol{\varepsilon} \equiv (\epsilon_{11}\ \epsilon_{22}\ \epsilon_{33}\ 2\epsilon_{12}\ 2\epsilon_{23}\ 2\epsilon_{31})$ will be referred to the same axes used for $\boldsymbol{\sigma}$. Prager's consistency rule stipulates that the ith plastic multiplier rate $\dot{\lambda}_i$ must be zero if the ith "yield mode" is "inactive", i.e. if $\phi_i(\boldsymbol{\sigma}) < 0$. Equation (1d) gives two alternative nontraditional representations of this rule: the former is expressed componentwise, the latter as orthogonality of the vectors $\tilde{\boldsymbol{\phi}} \equiv (\phi_i \ldots \phi_m)$, $\dot{\boldsymbol{\lambda}} \equiv (\dot{\lambda}_1 \ldots \dot{\lambda}_m)$; their equivalence obviously flows from the sign constraints.

A comparison of the relation set (1) with (A.1) in the Appendix reveals their strict analogy. The interpretation of (1) as Kuhn—Tucker conditions implies that $\dot{\boldsymbol{\varepsilon}}^p$ is stress-gradient of the function

$$\dot{D}(\boldsymbol{\sigma}) = \tilde{\boldsymbol{\sigma}}\,\dot{\boldsymbol{\varepsilon}}^p \tag{2}$$

and allows writing the relevant nonlinear program (NLP) straightforwardly:

$$\max_{\boldsymbol{\sigma}} \tilde{\boldsymbol{\sigma}}\,\dot{\boldsymbol{\varepsilon}}^p \tag{3a}$$

subject to:

$$\phi_i(\boldsymbol{\sigma}) \leqslant 0 \quad (i = 1, \ldots, m). \tag{3b}$$

If the yield functions ϕ_i and, hence, the elastic domain are convex, problems (1) and (3) are equivalent by the Kuhn—Tucker theorem (see Appendix) and one recognizes in (2) the dissipation density and in (3) Hill's principle. This reads: for given plastic strain rates $\dot{\boldsymbol{\varepsilon}}^p$, the actual stresses $\boldsymbol{\sigma}^0$ among all $\boldsymbol{\sigma}$ complying with the yield inequalities ("plastically admissible"), maximize the dissipation (see, e.g.[7]).

The analogy between (1) and (A.2) allows writing directly the dual of the maximum dissipation principle (3) as the analogue to (A.3), but its mechanical meaning and use are doubtful primarily because of the presence of the stresses among the dual variables. However, it is not so with piecewise-linear or linearized (PWL) yield surface Γ (i.e. polyhedrical elastic domain Ω). In fact, assume *linear* yield functions ϕ_i and form matrix $\mathbf{N} \equiv [\partial\phi_1/\partial\boldsymbol{\sigma} \ldots \partial\phi_m/\partial\boldsymbol{\sigma}]$ with their (constant) gradients and vector $\mathbf{R}$ with their constants (R_i is the distance of the ith yield plane from the origin, if the ith stress gradient is normalized). Then (1) and (3) in matrix notation reduce to the analogues of (A.4) for $\mathbf{M} = \mathbf{0}$ (LP) and (A.8), respectively:

$$\boldsymbol{\phi} = \tilde{\mathbf{N}}\boldsymbol{\sigma} - \mathbf{R} \leqslant \mathbf{0}, \dot{\boldsymbol{\varepsilon}}^{\mathrm{p}} = \mathbf{N}\dot{\boldsymbol{\lambda}}, \dot{\boldsymbol{\lambda}} \geqslant \mathbf{0}, \ \tilde{\boldsymbol{\phi}}\dot{\boldsymbol{\lambda}} = 0 \tag{4}$$

$$D(\dot{\boldsymbol{\varepsilon}}^{\mathrm{p}}) = \max_{\boldsymbol{\sigma}} \tilde{\boldsymbol{\sigma}}\dot{\boldsymbol{\varepsilon}}^{\mathrm{p}}, \text{ subject to: } \tilde{\mathbf{N}}\boldsymbol{\sigma} \leqslant \mathbf{R} \tag{5}$$

(5) is a linear program (LP) whose dual, analogue to (A.8), reads:

$$\dot{D}(\dot{\boldsymbol{\varepsilon}}^{\mathrm{p}}) = \min_{\boldsymbol{\lambda}} \tilde{\mathbf{R}}\dot{\boldsymbol{\lambda}}, \text{ subject to: } \mathbf{N}\dot{\boldsymbol{\lambda}} = \dot{\boldsymbol{\varepsilon}}^{\mathrm{p}}, \dot{\boldsymbol{\lambda}} \geqslant \mathbf{0} \tag{6}$$

Equation (6) formulates a *dual* principle which can be stated in the following mechanical terms: for given plastic strain rates $\dot{\boldsymbol{\varepsilon}}^{\mathrm{p}}$, the actual plastic multiplier rates $\dot{\boldsymbol{\lambda}}^0$, among all those complying with the normality rule, minimize the dissipation density. This dual of Hill's principle, with validity confined to piecewise-linear plasticity, turned out to be useful in shape optimization problems, as pointed out in Ref. [8].

Sometimes lack of normality is admitted, as, e.g., required for a realistic description of soils, rocks and other frictional materials. Accordingly the plastic strain rates are expressed in the "non associative" form:

$$\dot{\boldsymbol{\varepsilon}}^{\mathrm{p}} = \sum_{i=1}^{m} \frac{\partial\psi_i(\boldsymbol{\sigma})}{\partial\boldsymbol{\sigma}} \dot{\lambda}_i \text{ or } \dot{\boldsymbol{\varepsilon}}^{\mathrm{p}} = \mathbf{V}\dot{\boldsymbol{\lambda}} \tag{7}$$

assuming m "plastic potentials" ψ_i and setting $\mathbf{V} \equiv [\partial\psi_i/\partial\boldsymbol{\sigma}]$ as the matrix of their stress-gradients. After such modifications, the constitutive laws (1) and their PWL specialization (4) cannot be regarded as Kuhn—Tucker conditions; therefore the extremum dissipation properties are no longer valid.

Finally, if the convexity assumption on $\phi_i(\boldsymbol{\sigma})$ is relaxed in the associative laws (1), these are no longer sufficient (but only necessary) optimality conditions of problem (3) by the Kuhn—Tucker theorem (see Appendix) and, hence, Hill's principle is invalidated.

Convexity *and* normality in the above sense are known to be the implications of Drucker's polstulate "in the large" [2]. Basic mathematical programming notions are seen here to provide further evidence and re-interpretation of the central role of this mechanical concept in the foundations of the theory of plasticity: basically, this postulate requires that the yield inequalities and the flow rule are optimality conditions *equivalent* to an optimization problem in the stresses.

3. Elastoplastic constitutive laws in terms of rates; Drucker's postulate "in the small"

A fairly general nonassociative relationship between stress and strain rates (or to within δt, infinitesimal increments), is represented by the following relations [9] to be supplemented by (1a), (1d):

$$\dot{\boldsymbol{\varepsilon}} = \dot{\boldsymbol{\varepsilon}}^{\mathrm{e}} + \dot{\boldsymbol{\varepsilon}}^{\mathrm{p}} = \mathbf{C}\,\dot{\boldsymbol{\sigma}} + \sum_{r=1}^{p} \frac{\partial \psi_r}{\partial \boldsymbol{\sigma}}\,\dot{\lambda}_r, \qquad \dot{\lambda}_r \geqslant 0 \tag{8a}$$

$$\dot{\phi}_r = \frac{\partial \phi_r}{\partial \boldsymbol{\sigma}}\,\dot{\boldsymbol{\sigma}} - \sum_{s=1}^{p} H_{rs}\dot{\lambda}_s \tag{8b}$$

$$\dot{\phi}_r \leqslant 0, \qquad \dot{\phi}_r\dot{\lambda}_r = 0 (r=1, \ldots, p) \tag{8c, d}$$

where $\mathbf{C}$ is the (6×6 symmetric, positive definite) matrix representation of the elastic tensor; indices r and s run over the $p(\leqslant m)$ "*active*" yield modes, i.e. those which contain the stress point $\boldsymbol{\sigma}$ at the beginning of the incremental process considered: $\phi_r(\boldsymbol{\sigma}) = 0$ for $r = 1, \ldots, p$ ($p > 1$ at "corners"). For these p modes $\dot{\phi}_r$ cannot be positive and Prager's consistency requires $\dot{\lambda}_r = 0$ if $\dot{\phi}_r < 0$: this is expressed by eqn. 8(c, d). H_{rs} are "hardening coefficients" which reflect, at the rate level, the hardening phenomenon, i.e. the dependence of the yield functions on the plastic deformation history (e.g. through the plastic work), a dependence that did not need to be made explicit in eqns. (1). $H_{rs} = 0$ for $r \neq s$ reflects Koiter's hypothesis of "noninteractive modes". A thorough discussion of the general constitutive laws (8) is due to Mandel [9]. Let us define the hardening matrix $\mathbf{H} \equiv [H_{rs}]$ of order p and the $6 \times p$-matrices $\mathbf{N} \equiv [\partial \phi_r / \partial \boldsymbol{\sigma}]$ and $\mathbf{V} \equiv [\partial \psi_r / \partial \boldsymbol{\sigma}]$ containing as columns the stress-gradients of the p yield functions and plastic potentials, respectively; thus eqns. (8) acquire the form:

$$\dot{\boldsymbol{\varepsilon}} = \mathbf{C}\,\dot{\boldsymbol{\sigma}} + \mathbf{V}\dot{\lambda} \tag{9a}$$

$$\dot{\boldsymbol{\phi}} = \tilde{\mathbf{N}}\,\dot{\boldsymbol{\sigma}} - \mathbf{H}\dot{\boldsymbol{\lambda}} \leqslant 0, \quad \dot{\boldsymbol{\lambda}} \geqslant 0, \quad \dot{\boldsymbol{\phi}}\,\dot{\boldsymbol{\lambda}} = 0 \tag{9b}$$

By solving (9a) with respect to $\dot{\boldsymbol{\sigma}}$ and substituting into (9b), we obtain (if $\mathbf{D}$ denotes the elastic stiffness matrix of the material):

$$\dot{\boldsymbol{\sigma}} = \mathbf{D}\,\dot{\boldsymbol{\varepsilon}} - \mathbf{V}\dot{\boldsymbol{\lambda}} \tag{10a}$$

$$\dot{\phi} = \tilde{N}'\dot{\varepsilon} - H'\dot{\lambda} \leqslant 0, \ \dot{\lambda} \geqslant 0, \ \tilde{\dot{\phi}}\dot{\lambda} = 0 \tag{10b}$$

having set:

$$V' = DV, N' \equiv DN, H' \equiv H + \tilde{N}DV = H + \tilde{N}'CV'. \tag{11}$$

Equations (9) define $\dot{\varepsilon}$ for given $\dot{\sigma}$, eqns (10) $\dot{\sigma}$ for given $\dot{\varepsilon}$; hence, they will be referred to as *direct* and *inverse* elasto-plastic rate laws, respectively. Both include in (a) a linear equation, in (b) a nonlinear part in form of a linear complementarity problem (LCP) in the variables $\dot{\lambda}_r, \dot{\phi}_r$. If the hardening matrix H is nonsingular, then $\dot{\lambda}$ can be expressed in terms of $\dot{\phi}$ through the first of eqns. (9b) and substituted in (9a), which becomes:

$$\dot{\varepsilon} = C'\dot{\sigma} - VH^{-1}\dot{\phi}, \text{ with: } C' \equiv C + VH^{-1}\tilde{N} \tag{12}$$

If H' (which might be called "elastoplastic hardening" matrix) is nonsingular, then the same manipulations in eqns. (10) lead to:

$$\dot{\sigma} = D'\dot{\varepsilon} + V'(H')^{-1}\dot{\phi}, \text{ with: } D' \equiv D - V'(H')^{-1}\tilde{N}' \tag{13}$$

Equations (12), (9b) and eqns. (13), (10b) provide alternative formulations of the direct and inverse elastoplastic flow rules, respectively. Setting $\dot{\phi} = 0$ in both, these formulations show that C' in (12) and D' in (13) represent the elastoplastic (or "tangent") compliance and stiffness matrix, respectively, when none of the p active plastic modes undergoes "unloading" ($\dot{\phi} < 0$). By means of a well known matrix indentity, or some algebra, it may be easily seen that $C'D' = I$.

We will consider below two categories of situations separately, the latter being a generalization of the former.

(A) NORMALITY AND SYMMETRIC HARDENING

We will assume first *symmetry* (or "reciprocity") of hardening $\tilde{H} = H$ and *normality* $V = N$, which together imply that $\tilde{H}' = H'$ as well. The additional symmetric term $\tilde{N}DN$ in H' is positive definite if N is full-column rank, i.e., if the stress gradients $\partial\phi_r/\partial\sigma$ are linearly independent, as we will assume; otherwise, it is at least positive semidefinite.

Using the mathematical concepts expressed in the Appendix by (A.5)–(A.7) and (A.9), after having identified (9b) and (10b) with (A.6), we can write [10]:

$$\min_{\dot{\lambda}} \{\tfrac{1}{2}\tilde{\dot{\lambda}}H\dot{\lambda} - \dot{\sigma}N\dot{\lambda}\}, \text{ subject to: } \dot{\lambda} \geqslant 0 \tag{14a}$$

$$\min_{\dot{\lambda}} \ \tfrac{1}{2}\tilde{\dot{\lambda}}H\dot{\lambda}, \text{ subject to: } H\dot{\lambda} \geqslant \tilde{N}\dot{\sigma} \tag{14b}$$

$$\min_{\dot{\lambda}} \{\tfrac{1}{2}\, \tilde{\dot{\lambda}} H' \dot{\lambda} - \tilde{\dot{\varepsilon}} N' \dot{\lambda}\}, \text{ subject to: } \dot{\lambda} \geqslant 0 \tag{15a}$$

$$\min_{\lambda} \tfrac{1}{2}\, \tilde{\lambda}\, \mathbf{H}' \lambda, \text{ subject to: } \mathbf{H}' \lambda \geqslant \tilde{\mathbf{N}}' \dot{\varepsilon} \tag{15b}$$

$$\min_{\dot{\lambda}} \{\tilde{\dot{\lambda}} \mathbf{H} \dot{\lambda} - \dot{\sigma} \mathbf{N} \dot{\lambda}\}, \text{ subject to: } \dot{\lambda} \geqslant \mathbf{0}, \mathbf{H} \dot{\lambda} \geqslant \tilde{\mathbf{N}} \dot{\sigma} \tag{16}$$

$$\min_{\dot{\lambda}} \{\tilde{\dot{\lambda}} \mathbf{H}' \dot{\lambda} - \tilde{\dot{\varepsilon}} \mathbf{N}' \dot{\lambda}\}, \text{ subject to: } \dot{\lambda} \geqslant \mathbf{0}, \mathbf{H}' \dot{\lambda} \geqslant \tilde{\mathbf{N}}' \dot{\varepsilon} \tag{17}$$

Clearly, if det $\mathbf{H} \neq \mathbf{0}$ and/or det $\mathbf{H}' \neq \mathbf{0}$, the substitution $\dot{\lambda}(\dot{\boldsymbol{\phi}})$ from (9b) and/or (10b) would transform (14), (16) and (15), (17) into QP problems in $\dot{\boldsymbol{\phi}}$ only.

When $p = 1$, i.e. for flow rules at a "*smooth*" point of the yield surface, with outward normal vector $\mathbf{N} \equiv \{\partial\phi/\partial\boldsymbol{\sigma}\}$, the situation has been examined in detail from different standpoints in [11, 12]. Some conclusions, which are made almost self-evident by the above formulations, are summarized below.

If $H > 0$ (hardening): $\tilde{\mathbf{N}}\dot{\boldsymbol{\sigma}} > 0, = 0, < 0$ in the direct law correspond to yielding, "neutral", "unloading path", respectively; the same correspondence holds for $\tilde{\mathbf{N}}'\dot{\boldsymbol{\varepsilon}} > 0, = 0, < 0$ in the inverse law, where $H' > 0$ as a consequence of $H \geqslant 0$. Referred to a mode, yielding process or "path" means $\dot{\lambda} > 0$, neutral loading $\dot{\lambda} = \dot{\phi} = 0$, unloading $\dot{\phi} < 0$.

If $H = 0$ (nonhardening): the same as above for the inverse law; $\tilde{\mathbf{N}}\dot{\boldsymbol{\sigma}} > 0$ is not admissible, $\tilde{\mathbf{N}}\dot{\boldsymbol{\sigma}} = 0$ corresponds to yielding or neutral path, $\tilde{\mathbf{N}}\dot{\boldsymbol{\sigma}} < 0$ to unloading; $\mathbf{C}'$ is meaningless, while $\mathbf{D}' = \mathbf{0}$ as can be formally proved by easy matrix algebra. If $H < 0$ but $H' > 0$, i.e. $H > -\tilde{\mathbf{N}}\mathbf{D}\mathbf{N}$ (softening): nothing new for the inverse law; $\tilde{\mathbf{N}}\dot{\boldsymbol{\sigma}} > 0$ not admissible, $\tilde{\mathbf{N}}\dot{\boldsymbol{\sigma}} = 0$ corresponds to neutral path, $\tilde{\mathbf{N}}\dot{\boldsymbol{\sigma}} < 0$ to yielding or unloading as alternatives. The QP formulation (14a) is no longer equivalent to (9b).

If $H = -\tilde{\mathbf{N}}\mathbf{D}\mathbf{V}$, so that $H' = 0$ ("critical softening"): the remarks of the preceding case on the direct law still apply; clearly, the same remarks on the direct law for $H = 0$ hold here for the inverse one, considering $\tilde{\mathbf{N}}'\dot{\boldsymbol{\varepsilon}}$ instead of $\tilde{\mathbf{N}}\dot{\boldsymbol{\sigma}}$; $\mathbf{C}' = \mathbf{0}$, $\mathbf{D}'$ is meaningless ("stress jumps").

Finally, if $H < -\tilde{\mathbf{N}}\mathbf{D}\mathbf{V}$, so that $H' < 0$ ("subcritical softening"): the remarks on the direct law for $H < 0$ still hold and apply also to the inverse law replacing $\tilde{\mathbf{N}}\dot{\boldsymbol{\sigma}}$ by $\tilde{\mathbf{N}}'\dot{\boldsymbol{\varepsilon}}$. Also the QP formulations (15) are invalidated, whereas (16), (17) are valid in all cases. The mechanical meaning of "critical" and "subcritical" softening has been discussed in [11, 13].

At a *singular point* or "corner" of the yield surface ($p > 1$) the situation is more complex. Therefore, corners, as in the Tresca yield criterion, are often rounded-off in numerical procedures. On the other hand, the interest in examining this situation is also motivated by its analogy with some classes of finite-increment laws (see next Section). To this aim, and in order to generalize the above results for $p = 1$, the LCP and QP formulations in eqns. (9) to (17) appear useful both theoretically and computationally.

In fact, let the hardening matrix **H** be positive definite. Then the QP problems (14) and (15) are equivalent, by Kuhn—Tucker theory, to the LCP's (9b) and (10b), respectively. All become strictly convex and, hence, every $\dot{\boldsymbol{\sigma}}$ defines uniquely $\dot{\boldsymbol{\lambda}}$ through (14) and $\dot{\boldsymbol{\varepsilon}}$ through (9a); every $\dot{\boldsymbol{\varepsilon}}$ defines uniquely $\dot{\boldsymbol{\lambda}}$ through (15) and $\dot{\boldsymbol{\sigma}}$ through (10a). In other words, the flow laws represent, in this case, a one-to-one correspondence between stess and strain rates, as in the hardening case at smooth points.

We now assume **H** to be positive-semidefinite, which includes the perfectly plastic case **H** = 0. Then matrix **H**$'$ is positive definite, since its second addend is so. Hence, the inverse $\dot{\boldsymbol{\varepsilon}} \rightarrow \dot{\boldsymbol{\sigma}}$ relation still exhibits uniqueness for *every* $\dot{\boldsymbol{\varepsilon}}$. The direct relation for some $\dot{\boldsymbol{\sigma}}$ leads to an infinite number of $\dot{\boldsymbol{\varepsilon}}$, by virtue of a theorem on the solution multiplicity stated in the Appendix (since det **H** = 0, at least one principal minor of **H** is zero).

Equations (9b) and (10b) still can be regarded as sufficient and necessary optimality conditions of QP (LP if **H** = **0**). The QP problems (14), (15) are convex, the former ones no longer strictly convex. The second-order plastic work (up to δt^2) through eqns. (9) can be expressed in the form of the objective function of (14b):

$$\dot{W}^{\mathrm{p}} \equiv \tfrac{1}{2}\tilde{\dot{\boldsymbol{\sigma}}}\,\dot{\boldsymbol{\varepsilon}}^{\mathrm{p}} = \tfrac{1}{2}\tilde{\dot{\boldsymbol{\sigma}}}\mathbf{N}\dot{\boldsymbol{\lambda}} = \tfrac{1}{2}\tilde{\dot{\boldsymbol{\lambda}}}\mathbf{H}\dot{\boldsymbol{\lambda}} \tag{18}$$

Thus, for **H** positive semidefinite, Drucker's postulate "in the small", stating the nonnegativeness of $\dot{W}^{\mathrm{p}}$, is seen to be fulfilled, as it is also in its so-called "extended form" [1, 2]. In fact, this form postulates the nonnegativeness of the following quantity (a and b marking two infinitesimal processes starting from the same static situation):

$$2\dot{W}^{\mathrm{p}}_{ab} \equiv (\tilde{\dot{\boldsymbol{\sigma}}}_a - \tilde{\dot{\boldsymbol{\sigma}}}_b)(\dot{\boldsymbol{\varepsilon}}^{\mathrm{p}}_a - \dot{\boldsymbol{\varepsilon}}^{\mathrm{p}}_b) = (\tilde{\dot{\boldsymbol{\lambda}}}_a - \tilde{\dot{\boldsymbol{\lambda}}}_b)\mathbf{H}(\dot{\boldsymbol{\lambda}}_a - \dot{\boldsymbol{\lambda}}_b) - \tilde{\dot{\boldsymbol{\phi}}}_a\dot{\boldsymbol{\lambda}}_b - \tilde{\dot{\boldsymbol{\phi}}}_b\dot{\boldsymbol{\lambda}}_a \tag{19}$$

where the last expression is obtained by using the first of (9b) and, clearly, cannot be negative in any of its addends. In eqn. (18) $\dot{W}^{\mathrm{p}} \geqslant 0$ is guaranteed by a somewhat weaker condition on **H**, i.e. its "copositiveness" ($\tilde{\dot{\boldsymbol{\lambda}}}\mathbf{H}\dot{\boldsymbol{\lambda}} \geqslant \mathbf{0}$ for any $\dot{\boldsymbol{\lambda}} \geqslant \mathbf{0}$).

When **H** is indefinite in sign (i.e. some of its principal minors and eigenvalues become negative), the *stability* in the sense $\dot{W}^{\mathrm{p}} \geqslant 0$ for any $\dot{\boldsymbol{\varepsilon}}$ is no longer guaranteed and also the sufficiency of (9b) for (14a) is invalidated. In correspondence with the above mentioned ranges of H and H' for the smooth case ($p = 1$), one might distinguish the following categories of situations: **H** indefinite, **H**$'$ positive definite (softening); **H** indefinite, **H**$'$ positive semidefinite (critical softening); **H** and **H**$'$ indefinite (subcritical softening). The names adopted and the main mechanical aspects of these ranges at corners are suggested by the previous remarks on the counterparts in the $p = 1$ case. The quantitative study and the implementation of the rate laws at corners imply recourse to more detailed aspects of LCP theory and to relevant algorithms, not considered here because of space limitations.

(B) NONNORMALITY OR UNSYMMETRIC HARDENING

Lack of normality ($\mathbf{N} \neq \mathbf{V}$, which entails $\tilde{\mathbf{H}}' \neq \mathbf{H}'$) makes the QP (15) related to the inverse law meaningless, while *unsymmetric hardening* ($\tilde{\mathbf{H}} \neq \mathbf{H}$) invalidates all QP's (14) and (15), in the sense that a LCP with nonsymmetric matrix cannot be interpreted as optimality conditions. However, the QP problems (16) (17) are still equivalent to LCP (9b) and (10b), respectively (see the Appendix).

It is easy to realize that the quantities $\dot{W}^{\mathrm{p}}$ and $\dot{W}^{\mathrm{p}}_{ab}$ can be given the expressions (18) and (19), respectively, even if $\tilde{\mathbf{H}} \neq \mathbf{H}$; they can not if $\mathbf{N} \neq \mathbf{V}$. Hence, the positive semidefiniteness of the hardening matrix suffices for Drucker's postulate "in the small" to be valid (both in the narrower and in the extended form) in the presence of normality; it is no longer sufficient in its absence.

Other LCP properties which still hold whatever the matrix may be, concern the number of solutions (see Appendix). Namely, for given $\dot{\boldsymbol{\sigma}}$, (9b) or (16) define a unique $\dot{\boldsymbol{\varepsilon}}$, if and only if $\mathbf{H}$ is a "*P*-matrix", i.e. has all principal minors positive; a finite number of $\dot{\boldsymbol{\varepsilon}}$, if and only if it has all principal minors different from zero. The same can be stated for the inverse law, by replacing $\dot{\boldsymbol{\sigma}}$ with $\dot{\boldsymbol{\varepsilon}}$, $\dot{\boldsymbol{\varepsilon}}$ with $\dot{\boldsymbol{\sigma}}$ and $\mathbf{H}$ with $\mathbf{H}'$.

In view of the small size of the matrices involved ($\mathbf{H}$ or $\mathbf{H}'$), the above characterizations (of definite matrices, *P*-matrices, etc.) have a straightforward operative value and any LCP or QP algorithm is efficiently applicable in order to implement the above rate laws at any level of generality. However, for unsymmetric nonassociative laws, some difficulties arise in a qualitative study capable of providing an insight into their mechanical aspects, such as the locus of "unstable paths" ($\dot{\boldsymbol{\varepsilon}}$ such that $\dot{\boldsymbol{\sigma}}\,\dot{\boldsymbol{\varepsilon}} < 0$). This study has been carried out for $p = 1$ [13]; for $p > 1$ (corners) it does not appear to be available in the literature, but the above developments and LCP and QP theory are believed to provide a suitable basis. Ref. [14] discussed the role of geometric effects when the material models embodied in the kind of constitution considered here are generated by an assembly of structural components.

4. Elastic-plastic laws in finite increments

Elastic-plastic boundary value problems require the time-integration of the governing relations in terms of rates along the history of the external actions. In practice, finite steps Δt and approximations are clearly inevitable. Solving the rate problem and taking its solutions as constant over Δt is known to lead to significant error accumulation, (see e.g. [15]).

Various corrective procedures are used in computational plasticity in order to enhance the fulfilment of the constitutive law by the finite stress and strain increments. An attractive approach is to construct from the chosen material model, as an approximation of it, a relationship between finite increments. In this line of thought, we will consider here the following constitutive relations for a step $t_0 \rightarrow t_0 + \Delta t$ (t being physical time or some increasing function of it):

$$\Delta \boldsymbol{\varepsilon} = \mathbf{C}\Delta \boldsymbol{\sigma} + \sum_{r=1}^{q} \frac{\partial \psi_r}{\partial \boldsymbol{\sigma}} \Delta\lambda_r, \qquad \Delta\lambda_r \geqslant 0 \tag{20a}$$

$$\phi_r = \frac{\partial \phi_r}{\partial \boldsymbol{\sigma}} [\boldsymbol{\sigma}(t_0) + \Delta \boldsymbol{\sigma}] - \sum_{s=1}^{q} H_{rs}\Delta\lambda_s - R_r \leqslant 0 \tag{20b}$$

$$\phi_r \Delta\lambda_r = 0 \qquad (r = 1, \ldots, q) \tag{20c}$$

In (20) the stress-gradients are constant, i.e. the yield functions ϕ_r and the plastic potentials ψ_r are assumed linear in the stresses; H_{rs} and R_r are also constant. In other terms, (20b) defines a polyhedrical (*piecewise linear*, PWL) approximation of the current elastic domain, either locally in the vicinity of the starting stress point $\boldsymbol{\sigma}(t_0)$, or globally [16]. Each yield plane (or mode) $\phi_r = 0$, when activated, translates if $H_{rr} \neq 0$ and gives a plastic strain contribution with fixed direction, normal to it in the associative case $\psi_r = \phi_r$. The q modes considered are potentially active, but not necessarily active in t_0, so that $\Delta\phi_r$ are not sign-constrained; Prager's consistency is expressed by (20c). Let us set, for convenience:

$$\phi_r^0 \equiv \frac{\partial \phi_r}{\partial \boldsymbol{\sigma}} \boldsymbol{\sigma}(t_0) - R_r \leqslant 0 \tag{21}$$

The meaning and interest of eqns. (20) are further clarified by the following remarks.

(a) For $\Delta t = \delta t \rightarrow 0$, eqns. (20) reduce (to within the common factor δt) to the rate relations considered in Sec. 3. In fact, in the limit $\phi_r^0 = 0$ since only the modes active in t_0 are to be considered; hence, $\phi_r = \delta\phi_r = \dot{\phi}_r \delta t$ ($r = 1, \ldots, q$, with $q = p$).

(b) With $\mathbf{H} = \mathbf{0}$ (perfect plasticity) and $\psi_r = \phi_r$ (normality), eqns. (20) cover the PWL approximation of yield surfaces frequently used in limit analysis, and the local "a posteriori" PWL approximation proposed for elastoplastic analysis in order to control and reduce the violation of yield criteria [15].

(c) When $\boldsymbol{\sigma}(t_0) = \mathbf{0}$ (hence: $\boldsymbol{\phi}^0 = \mathbf{R}$, $\Delta \boldsymbol{\sigma} = \boldsymbol{\sigma}$, $\Delta \boldsymbol{\varepsilon} = \boldsymbol{\varepsilon}$), eqns (20) describe a behaviour which is easily recognized as nonlinear-elastic (specifically PWL) or "holonomic" elastoplastic, in the spirit of the so-called "deformation theory" of plasticity [7].

(d) If $\boldsymbol{\sigma}$ and $\boldsymbol{\varepsilon}$ become scalars ($n = 1$) and are interpreted as axial force (or moment) and elongation (or curvature), eqns. (20) represent the familiar elastic-linearly hardening model assumed in plastic analysis of trusses and beams (typically, PWL hardening for reinforced concrete beams).

(e) The hardening matrix $\mathrm{H} = [H_{rs}]$ governs the evolution of the elastic domain $\phi_r \leqslant 0$ at yielding. Hence, when basic features of this evolution are prescribed (e.g. isotropic, Prager's kinematic hardening), it acquires a special structure with few parameters (or a single one) available for matching the original constitutive law or fitting to the experimental data. Some of these special structures have been determined in [16].

For the present purpose, the essential remark is that eqns. (20) are fully *analogous* to the rate relations of the preceding Section. In fact, with symbols similar to those adopted previously, we can write:

$$\Delta \boldsymbol{\varepsilon} = \mathrm{C}\Delta \boldsymbol{\sigma} + \mathrm{V}\Delta \boldsymbol{\lambda} \tag{22a}$$

$$\boldsymbol{\phi} = \boldsymbol{\phi}^0 + \tilde{N}\Delta\boldsymbol{\sigma} - H\Delta\boldsymbol{\lambda} \leqslant \mathbf{0}, \qquad \Delta\boldsymbol{\lambda} \geqslant \mathbf{0}, \qquad \tilde{\boldsymbol{\phi}}\Delta\boldsymbol{\lambda} = 0 \tag{22b}$$

The analogy between eqns. (22) and (9) allows transferring immediately to the former in finite increments, all developments of Sec. 3 concerning the latter relation set in terms of rates.

5. Elastoplastic analysis of discretized systems: extremum theorems, overall stability conditions

The finite element discretization with compatible modelling of the displacement field, is a familiar versatile approach to the numerical approximate solution of boundary value problems.

Constant strain triangular and tetrahedrical homogeneous elements in two- and three-dimensional problems, respectively, led rather straightforwardly to the notion of "element behaviour" described by means of natural generalized stresses and strains ("natural" means selfequilibrated and unaffected by rigid body motions, respectively) analogous to the preceding description of material behaviour [16]. Formulations of the discretized elastoplastic analysis as linear complementarity or quadratic programming problems were derived both in terms of rates and in terms of finite increments with piecewise linearized (globally or for the load step) laws of local deformability for materials or elements [16]. It has been shown recently [17, 18] that separate modeling of displacements and plastic multipliers permits such formulations and, by some "consistency" provisions, still leads to the notion of element behaviour, for *any* order of interpolation functions. We will adopt for materials the finite increment relations (20), since they cover the rate relations (9) as a special case and are endowed with computational, operative value as noted in Sec. 3. Following [18], first we model the displacement and plastic multiplier fields over the eth finite element volume Ω_e ($e = 1, \ldots, E$) by interpolation matrices $\mathbf{U}(\mathbf{x})$, $\mathbf{\Lambda}(\mathbf{x})$, respectively:

$$\Delta\mathbf{u}_e(\mathbf{x}) = \mathbf{U}(\mathbf{x})\Delta\mathbf{u}_e, \qquad \Delta\boldsymbol{\lambda}_e(\mathbf{x}) = \mathbf{\Lambda}(\mathbf{x})\Delta\boldsymbol{\lambda}_e \tag{23,a b}$$

$\Delta\boldsymbol{\lambda}_e$ being a vector of parameters (e.g. identified with the material $\Delta\boldsymbol{\lambda}$ in some suitably chosen "strain points" in Ω_e). Then:

$$\Delta\boldsymbol{\varepsilon}_e(\mathbf{x}) = \mathbf{B}(\mathbf{x})\Delta\mathbf{u}_e, \qquad \boldsymbol{\phi}_e \equiv \int_{\Omega_e} \mathbf{\Lambda}(\mathbf{x})\boldsymbol{\phi}(\mathbf{x})d\Omega \tag{24a, b}$$

Equation (24a) flows from (23a) through compatibility, as usual; eqn. (24b) is a definition of vector $\boldsymbol{\phi}_e$, as weighted average over Ω_e of the material yield functions. Equivalence, in the usual sense of the virtual work principle, defines the element nodal force vector as:

$$\Delta\mathbf{Q}_e = \int_{\Omega_e} \tilde{\mathbf{B}}(\mathbf{x})\Delta\boldsymbol{\sigma}(\mathbf{x})d\Omega - \Delta\mathbf{P}_e \tag{25a}$$

where:

$$\Delta\mathbf{P}_e \equiv \int_{\Omega_e} \mathbf{U}(\mathbf{x})\Delta\mathbf{X}(\mathbf{x})d\Omega + \int_{\Sigma_e} \tilde{\mathbf{U}}(\mathbf{x})\Delta\mathbf{T}(\mathbf{x})d\Sigma \tag{25b}$$

X denoting body forces and T tractions on the surface Σ_e, the element may have in common with the unconstrained boundary Σ_T of the domain Ω.

Let us substitute $\Delta\boldsymbol{\sigma}(\mathbf{x})$ in (25a) and in (22b) by (22a), $\boldsymbol{\phi}(\mathbf{x})$ in (24b) by (22b), thus making use of the inverse incremental laws (analogous to the inverse laws (10) in terms of rates). We obtain, through (23) and (24a) the following *elemental relations:*

$$\Delta\mathbf{Q}_e = \mathbf{K}^e_{uu}\Delta\mathbf{u}_e - \mathbf{K}^e_{u\lambda}\Delta\boldsymbol{\lambda}_e - \Delta\mathbf{P}_e \tag{26a}$$

$$\boldsymbol{\phi}_e = \mathbf{K}^e_{\lambda u}\Delta\mathbf{u}_e - \mathbf{K}^e_{\lambda\lambda}\Delta\boldsymbol{\lambda}_e + \boldsymbol{\phi}^0_e \tag{26b}$$

having set:

$$\mathbf{K}^e_{uu} \equiv \int_{\Omega_e} \tilde{\mathbf{B}}(\mathbf{x})\mathbf{D}(\mathbf{x})\mathbf{B}(\mathbf{x})d\Omega, \qquad \mathbf{K}^e_{\lambda\lambda} \equiv \int_{\Omega_e} \tilde{\mathbf{\Lambda}}(\mathbf{x})\mathbf{H}'(\mathbf{x})\mathbf{\Lambda}(\mathbf{x})d\Omega \tag{27a, b}$$

$$\mathbf{K}^e_{\lambda u} \equiv \int_{\Omega_e} \tilde{\mathbf{\Lambda}}(\mathbf{x})\tilde{\mathbf{N}}'(\mathbf{x})\mathbf{B}(\mathbf{x})d\Omega, \qquad \mathbf{K}^e_{u\lambda} \equiv \int_{\Omega_e} \tilde{\mathbf{B}}(\mathbf{x})\mathbf{V}'(\mathbf{x})\mathbf{\Lambda}(\mathbf{x})d\Omega \tag{27c, d}$$

$$\boldsymbol{\phi}_e^0 \equiv \int_{\Omega_e} \Lambda(\mathbf{x})\boldsymbol{\phi}^0(\mathbf{x})d\Omega \tag{27e}$$

Equations governing the step response of the whole discretized system are derived by assembling eqns. (26) in the usual way:

$$\Delta\mathbf{u}_e = \mathbf{L}_e\Delta\mathbf{u} \qquad \sum_{e=1}^{E} \tilde{\mathbf{L}}_e\Delta\mathbf{Q}_e = \mathbf{0} \tag{28a, b}$$

where: $\mathbf{L}_e$ is a connectivity (or "localization") Boolean matrix, $\mathbf{u}$ the vector of the unconstrained nodal displacements of the element aggregate (imposed displacements and strains and nodal external loads are not considered for simplicity.)

By setting $\Delta\tilde{\bar{\boldsymbol{\lambda}}} \equiv \{\ldots\Delta\tilde{\boldsymbol{\lambda}}_e\ldots\}$, $\tilde{\bar{\boldsymbol{\phi}}} \equiv \{\ldots\tilde{\boldsymbol{\phi}}_e\ldots\}$, the following relations hold:

$$\Delta\boldsymbol{\lambda}_e = \mathbf{M}_e\Delta\bar{\boldsymbol{\lambda}}, \bar{\boldsymbol{\phi}} = \sum_{e=1}^{E} \tilde{\mathbf{M}}_e\boldsymbol{\phi}_e \tag{29}$$

where $\mathbf{M}_e$ is a block-row matrix which contains the identity matrix as eth block and zeros elsewhere. Hence, we obtain:

$$\Delta\mathbf{P} = \mathbf{K}_{uu}\Delta\mathbf{u} - \mathbf{K}_{u\lambda}\Delta\bar{\boldsymbol{\lambda}} \tag{30a}$$

$$\bar{\boldsymbol{\phi}} = \mathbf{K}_{\lambda u}\Delta\mathbf{u} - \mathbf{K}_{\lambda\lambda}\Delta\bar{\boldsymbol{\lambda}} + \bar{\boldsymbol{\phi}}^0 \tag{30b}$$

where

$$\mathbf{K}_{uu} \equiv \sum_{e=1}^{E} \tilde{\mathbf{L}}_e\mathbf{K}_{uu}^e\mathbf{L}_e, \qquad \mathbf{K}_{\lambda\lambda} \equiv \sum_{e=1}^{E} \tilde{\mathbf{M}}_e\mathbf{K}_{\lambda\lambda}^e\mathbf{M}_e \tag{31a, b}$$

$$\mathbf{K}_{\lambda u} \equiv \sum_{e=1}^{E} \tilde{\mathbf{M}}_e\mathbf{K}_{\lambda u}^e\mathbf{L}_e, \qquad \mathbf{K}_{u\lambda} \equiv \sum_{e=1}^{E} \mathbf{L}_e\mathbf{K}_{u\lambda}^e\mathbf{M}_e \tag{31c, d}$$

$$\Delta\mathbf{P} \equiv \sum_{e=1}^{E} \tilde{\mathbf{L}}_e\Delta\mathbf{P}_e, \qquad \boldsymbol{\phi}^0 \equiv \sum_{e=1}^{E} \tilde{\mathbf{M}}_e\boldsymbol{\phi}_e^0 \tag{31e, f}$$

Equations (30) will be supplemented by the complementarity rule:

$$\bar{\boldsymbol{\phi}} \leqslant \mathbf{0}, \qquad \Delta\bar{\boldsymbol{\lambda}} \geqslant \mathbf{0}, \qquad \bar{\boldsymbol{\phi}}\Delta\bar{\boldsymbol{\lambda}} = 0 \tag{32}$$

Cleary, this implies the fulfilment of complementarity at the finite element level (in $\Delta\boldsymbol{\lambda}_e, \boldsymbol{\phi}_e$), but only in a weighted average sense, and not everywhere, at material level (in $\Delta\boldsymbol{\lambda}(\mathbf{x}), \boldsymbol{\phi}(\mathbf{x})$).

If no rigid-body motion is permitted, $\mathbf{K}_{uu}$ (the usual elastic stiffness matrix of the aggregate) is positive definite. Then (30a) can be substituted into (30b), which becomes:

$$\bar{\phi} = \mathbf{K}_{\lambda u}\Delta\mathbf{u}^E + \bar{\phi}^0 - \mathbf{K}\Delta\bar{\lambda} \tag{33}$$

where:

$$\Delta\mathbf{u}^E \equiv \mathbf{K}_{uu}^{-1}\,\Delta\mathbf{P}, \qquad \mathbf{K} \equiv \mathbf{K}_{\lambda\lambda} - \mathbf{K}_{\lambda u}\mathbf{K}_{uu}^{-1}\mathbf{K}_{u\lambda} \tag{34}$$

$\Delta\mathbf{u}^E$ being the fictitious elastic displacement response to the external action step. Equations (32), (33) represent a governing relation set in LCP form for a mechanical system. It is analogous to the material description provided by eqn. (22b) in finite increments and by eqn. (9b) or (10b) in terms of rates. The analogy allows transferring directly, merely by re-interpretation of symbols, all considerations developed in Sec. 3 for stress—strain rate relations; in particular, the QP formulations and the solution uniqueness and overall stability criteria. Only some conclusions will be concisely expounded below.

For normality and symmetric hardening, case (A), matrix $\mathbf{K}$ turns out to be symmetric (and $\mathbf{K}_{\lambda u} = \tilde{\mathbf{K}}_{u\lambda}$) through (27), (31) and (34). Hence, eqns. (32) and (33) can be interpreted as Kuhn—Tucker conditions of a QP problem, namely:

$$\min_{\Delta\bar{\lambda}} \{\tfrac{1}{2}\,\Delta\tilde{\bar{\lambda}}\mathbf{K}\Delta\bar{\lambda} - \Delta\tilde{\bar{\lambda}}(\mathbf{K}_{\lambda u}\Delta\mathbf{u}^E + \bar{\phi}^0)\} \text{ subject to } \Delta\bar{\lambda} \geqslant \mathbf{0} \tag{35}$$

The dual of (35) and the QP problem unconditionally equivalent to (32), (33) read, respectively:

$$\min_{\Delta\bar{\lambda}} \tfrac{1}{2}\,\Delta\tilde{\bar{\lambda}}\mathbf{K}\Delta\bar{\lambda}, \text{ subject to: } \mathbf{K}\Delta\bar{\lambda} \geqslant \mathbf{K}_{\lambda u}\Delta\mathbf{u}^E + \bar{\phi}^0 \tag{36}$$

$$\min_{\Delta\bar{\lambda}} \{\Delta\tilde{\bar{\lambda}}\mathbf{K}\Delta\bar{\lambda} - \Delta\tilde{\bar{\lambda}}(\mathbf{K}_{\lambda u}\Delta\mathbf{u}^E + \bar{\phi}^0)\} \tag{37a}$$

$$\text{subject to: } \mathbf{K}\Delta\bar{\lambda} \geqslant \mathbf{K}_{\lambda u}\Delta\mathbf{u}^E + \bar{\phi}^0,\ \Delta\bar{\lambda} \geqslant \mathbf{0} \tag{37b}$$

The total plastic work $\Delta\bar{W}^p$ performed by any external agency acting on the system over Δt can be expressed in a form similar to (18):

$$\Delta\bar{W}^p = \tfrac{1}{2}\,\Delta\tilde{\bar{\lambda}}\mathbf{K}\Delta\bar{\lambda} \tag{38}$$

so that for $\Delta t \to 0$ (i.e. in terms of rates) positive semidefiniteness of $\mathbf{K}$ ensures overall stability "in the small" in Drucker's sense; it is worth noting, however, that this is a condition less restrictive than the analogous one on the material hardening matrix $\mathbf{H}$, as it does not require material stability. When this sufficient (not necessary) overall stability condition holds, then (35), (36) become extremum properties of the elasto-plastic solution for the current loading step.

In the absence of normality or hardening reciprocity, case (B), the extremum theorem (37b) still holds, while (35), (36) do not.

In contrast with the constitutive law context, now the LCP formulation is computationally expensive to generate and to solve, primarily because of the large size of the stiffness matrix $\mathbf{K}_{uu}$ to invert and because of the large size and high density of matrix $\mathbf{K}$. Therefore it is useful to cast the formulation (30), (32) into the LCP format (A.6) without substituting $\Delta\mathbf{u}$ and computing $\mathbf{K}_{uu}^{-1}$. In order to achieve this, we simply substitute the free variables by sign-constrained ones ($\Delta\mathbf{u} = \Delta\mathbf{u}^+ - \Delta\mathbf{u}^-$, $\Delta\mathbf{u}^+ \geqslant \mathbf{0}$, $\Delta\mathbf{u}^- \geqslant \mathbf{0}$). Then, by expressing eqn. (30a) as a double inequality with auxiliary variables $\mathbf{v}^+$, $\mathbf{v}^-$, eqns. (30), (32) can be rewritten as follows:

$$\begin{Bmatrix} -\Delta\mathbf{P} \\ \Delta\mathbf{P} \\ -\bar{\boldsymbol{\phi}}^0 \end{Bmatrix} + \begin{bmatrix} \mathbf{K}_{uu} & -\mathbf{K}_{uu} & -\mathbf{K}_{u\lambda} \\ -\mathbf{K}_{uu} & \mathbf{K}_{uu} & \mathbf{K}_{u\lambda} \\ -\mathbf{K}_{\lambda u} & \mathbf{K}_{\lambda u} & \mathbf{K}_{\lambda\lambda} \end{bmatrix} \begin{Bmatrix} \Delta\mathbf{u}^+ \\ \Delta\mathbf{u}^- \\ \Delta\bar{\boldsymbol{\lambda}} \end{Bmatrix} = \begin{Bmatrix} \mathbf{v}^+ \\ \mathbf{v}^- \\ -\bar{\boldsymbol{\phi}} \end{Bmatrix} \geqslant \mathbf{0},$$

$$\begin{Bmatrix} \Delta\mathbf{u}^+ \\ \Delta\mathbf{u}^- \\ \Delta\bar{\boldsymbol{\lambda}} \end{Bmatrix} \geqslant \mathbf{0}, \qquad \begin{aligned} \Delta\tilde{\mathbf{u}}^+\mathbf{v}^+ &= 0 \\ \Delta\tilde{\mathbf{u}}^-\mathbf{v}^- &= 0 \\ \tilde{\bar{\boldsymbol{\phi}}}\Delta\bar{\boldsymbol{\lambda}} &= 0 \end{aligned} \tag{39}$$

In case (A) of normality and symmetric hardening the matrix of the LCP (39) becomes symmetric and eqn. (39) analogue to (A.6) generates the QP problems in forms (A.5), (A.7), (A.8) once again via Kuhn—Tucker and duality. These become, by obvious identifications of $\mathbf{x}$, $\mathbf{M}$, $\mathbf{c}$ with the quantities in (39) and by some trivial algebra:

$$\min_{\Delta\bar{\boldsymbol{\lambda}} \geqslant \mathbf{0}, \Delta\mathbf{u}} \{\tfrac{1}{2}\Delta\tilde{\mathbf{u}}\mathbf{K}_{uu}\Delta\mathbf{u} + \tfrac{1}{2}\Delta\tilde{\bar{\boldsymbol{\lambda}}}\mathbf{K}_{\lambda\lambda}\Delta\bar{\boldsymbol{\lambda}} - \Delta\tilde{\bar{\boldsymbol{\lambda}}}\mathbf{K}_{\lambda u}\Delta\mathbf{u} - \Delta\tilde{\mathbf{P}}\Delta\mathbf{u} - \tilde{\bar{\boldsymbol{\phi}}}^0\Delta\bar{\boldsymbol{\lambda}}\} \tag{40}$$

$$\min_{\Delta\bar{\boldsymbol{\lambda}}, \Delta\mathbf{u}} \{\tfrac{1}{2}\Delta\tilde{\mathbf{u}}\mathbf{K}_{uu}\Delta\mathbf{u} + \tfrac{1}{2}\Delta\tilde{\bar{\boldsymbol{\lambda}}}\mathbf{K}_{\lambda\lambda}\Delta\bar{\boldsymbol{\lambda}} - \Delta\tilde{\bar{\boldsymbol{\lambda}}}\mathbf{K}_{\lambda u}\Delta\mathbf{u}\} \tag{41a}$$

subject to: $\mathbf{K}_{uu}\Delta\mathbf{u} - \mathbf{K}_{u\lambda}\Delta\bar{\boldsymbol{\lambda}} = \Delta\mathbf{P}, \qquad \mathbf{K}_{\lambda u}\Delta\mathbf{u} - \mathbf{K}_{\lambda\lambda}\Delta\bar{\boldsymbol{\lambda}} \leqslant -\bar{\boldsymbol{\phi}}^0$

(41b, c)

$$\min_{\Delta\bar{\boldsymbol{\lambda}}, \Delta\mathbf{u}} \{\Delta\tilde{\mathbf{u}}\mathbf{K}_{uu}\Delta\mathbf{u} + \Delta\tilde{\bar{\boldsymbol{\lambda}}}\mathbf{K}_{\lambda\lambda}\Delta\boldsymbol{\lambda} - \Delta\tilde{\bar{\boldsymbol{\lambda}}}(\mathbf{K}_{\lambda u} + \tilde{\mathbf{K}}_{u\lambda})\Delta\mathbf{u} - \Delta\tilde{\mathbf{P}}\Delta\mathbf{u} - \tilde{\bar{\boldsymbol{\phi}}}^0\Delta\bar{\boldsymbol{\lambda}}\} \tag{42a}$$

subject to: $\Delta\bar{\boldsymbol{\lambda}} \geqslant \mathbf{0}$ and to (41b, c) (42b)

All the preceding QP problems turn out to be convex when $\mathbf{H}$ is positive semidefinite. The constraints of (41) represent approximate formulations of equilibrium and yield conditions; if parameters governing the stress field were suitably defined, they could replace $\Delta\mathbf{u}$ in (41), thus making explicit its nature of a counterpart to the complementarity principle of elasticity in the present context of *compatible* discrete models.

In case (B) of nonassociative or asymmetrically hardening materials, $\mathbf{K}_{\lambda u} \neq \tilde{\mathbf{K}}_{u\lambda}$ and, once again, the only extremum characterization valid is (42).

In what precedes the use of finite element discretization, besides its computational merits, has allowed the treatment of both material models and boundary value problems (or structural analysis) in a single manner, by applying the same simple finite-dimensional, mathematical notions. Parallel developments might be based on equilibrium finite element models. Of course, compatible finite element modelling implies local violations of equilibrium and constitutive laws. However, this is hardly detrimental to the generality of the results of this section. In fact, the discretization adopted, without limitations on mesh size, could be considered as a description of a structural system alternative to the tensor field description of continuum mechanics (see e.g. [19]). It is easy to transfer every conclusion achieved in the first description to the second one, and re-demonstrate it by the classical path of reasoning of the latter (e.g. [7, 20]). The extremum theorems for continua, in turn, can be used for generating discrete approximations (see e.g. [18]).

The extent of the unification and generalization achieved here can be clarified by mentioning below previous separate results covered by the above extremum properties of solutions. For $\Delta t \to 0$ (rates), eqns. (35) and (36) become formulations of the theorems established in Refs. [21] and [22], respectively; for $\boldsymbol{\sigma}(t_0) = \mathbf{0}$ (holonomic piecewise linear plasticity) they reduce to those presented in Ref. [22]. For $\Delta t \to 0$, eqn. (37) was discussed in Ref. [23], where continua have been dealt with. Theorems (40) and (41), the latter in terms of stresses and $\boldsymbol{\lambda}$, were first proposed for $\boldsymbol{\sigma}(t_0) = \mathbf{0}$ in [24] with reference to perfectly plastic trusses and in [25] with reference to continua with H diagonal; in [24] the limit analysis theorems were derived from the solvability conditions of the QP problems (40) and (41); the latter, in total stresses, was interpreted there as a formulation of the Haar—Kármán principle [7] and eqn (40) as its dual. For $\Delta t \to 0$, eqns. (40) and (41) are counterparts to the statements given in [26] and [27], respectively; Ref. [27] pointed out their links with the classical theorems of Greenberg and Prager—Hodge [7, 20]. Comprehensive treatments from other standpoints and approaches of extremum and variational theorems in plasticity can be found in Refs. [20], [28], [29] and [30].

6. Conclusions

In what precedes some central aspects of plasticity have been revisited in the light of a few notions of mathematical programming and related complementarity theory.

Focusing on the complementarity relations implied by the concept of yield locus, mathematically analogous formulations have been given to rate constitutive laws, piecewise linear laws in terms of finite increments and discretized boundary value problems. In all these contexts, extremum theorems and statements on the solution multiplicity and stability (in terms of second order work) have been derived as direct formal consequences of the governing relations; their validity has been found subordinate to conditions related to those postulated by Drucker.

The same path of reasoning adopted in this paper could be easily applied, and partly has already been applied, to other categories of material models, where complementarity is implicitly or explicitly involved: e.g. locking materials [31, 32]; fracturing materials [33, 34]; materials with "coupling" between elastic properties and plastic deformations [13].

Acknowledgement

A grant from M.P.I. is gratefully acknowledged.

References

1 D.C. Drucker, Plasticity, in: N. Goodier and N.J. Hoff (Eds.), Structural Mechanics, Pergamon Press, 1960.
2 D.C. Drucker, On the postulate of stability of material in mechanics of continua, J. de Mécanique, 3 (1964) 235—249.
3 D.C. Drucker, Closing comments by Session Chairman — session II, in: U.S. Lindholm (Ed.). Mechanical Behaviour of Materials under Dynamic Loads, Springer—Verlag, 1968, pp. 405—409.
4 R.W. Cottle, F. Giannessi and J.L. Lions (Eds.), Variational Inequalities and Complementarity Problems — Theory and Applications, John Wiley and Sons, 1980.
5 M.R. Hestenes, Optimization Theory — The Finite Dimensional Case, John Wiley and Sons, 1975.
6 P.D. Panagiotopoulos, Strong physical nonlinearities, variational inequalities and applications, Annual of the School of Technology, Aristotelian University, 1981, pp. 345—371.
7 J.B. Martin, Plasticity, MIT Press, 1975.
8 G. Maier, Limit design in the absence of a given layout: a finite element zero—one programming problem, J. of Struct. Mech., 1 (1972) 213—230.
9 J. Mandel, Generalisation de la théorie de W.T. Koiter, Int. J. Solids and Structures, 1 (1965) 273.

10 G. Maier, 'Linear' flow-laws of elastoplasticity: a unified general approach, Rend. Acc. Naz. Lincei, Serie VIII, Vol. XLVII, fasc. 5, nov. 1969.
11 G. Maier, Sui legami associati tra sforzi e deformazioni incrementali in elastoplasticità, Rend. Ist. Lomb. di Scienze e Lettere, 100 (1966) 809—838.
12 Y. Yamada, N. Yoshimura and T. Sakurai, Plastic stress—strain matrix and its application for the solution of elastic-plastic problems by the finite element method, Int. J. Mech. Sci., 10 (1968) 343—354.
13 G. Maier and T. Hueckel, Nonassociated and coupled flow rules of elastoplasticity for rock-like materials, Int. J. Rock Mech. Min. Sci., 16 (1979) 77—92.
14 D.C. Drucker and G. Maier, Effects of geometry change on essential features of inelastic behaviour, J. Eng. Mech. Div., Proc. ASCE, 99 (EM4) (Aug. 1973) 819—834.
15 P.G. Hodge Jr., Automatic piecewise linearization in ideal plasticity, Comp. Meth. Appl. Mech. and Engineering, 10 (1977) 249—272.
16 G. Maier, A matrix structural theory of piecewise-linear plasticity with interacting yield planes, Meccanica, 5 (1) (1970) 55—66.
17 L. Corradi, On compatible finite element models for elastic plastic analysis, Meccanica, 13 (1978) 133.
18 L. Corradi, A displacement formulation for the finite element elastic-plastic problem, Meccanica, to appear.
19 J.F. Besseling, The complete analogy between the matrix equations and the continuous field equations of structural analysis, Int. Symp. on Analogue and Digital Techniques Applied to Aeronautics, Liège, 1963, pp. 223—242.
20 D.C. Drucker, Variational principles in the mathematical theory of plasticity, Proc. 1956 Symp. in Appl. Math., Vol. 8, McGraw-Hill, 1958, pp. 7—22.
21 G. Ceradini, A maximum principle for the analysis of elastic plastic systems, Meccanica, 1 (1966) 77—86.
22 G. Maier, Some theorems for plastic strain rates and plastic strains, Journal de Mécanique, 8 (1) (March 1969) 5—19.
23 G. Maier, A minimum principle for incremental elastoplasticity with non-associated flow laws, J. of the Mech. and Phys. of Solids, 18 (1970) 319—330.
24 G. Maier, Quadratic Programming and theory of elastic-perfectly plastic structures, Meccanica, 3 (4) (1968) 1—9.
25 G. Maier, Teoremi di minimo in termini finiti per continui elastoplastici con leggi constitutive linearizzate a tratti, Rend. Ist. Lomb. di Scienze e Lettere, 103 (1969) 1066—1080.
26 M. Capurso, Principi di minimo per la soluzione incrementale di problemi elastoplastici, Acc. Lincei — Rend. Sc. Fis. Mat. Nat., 46 (1969) 417 and 552.
27 M. Capurso and G. Maier, Incremental elastoplastic analysis and quadratic optimization, Meccanica, 5 (1970) 107—120.
28 P.G. Hodge, Numerical applications of minimum principles in plasticity, in: Engineering Plasticity, Cambridge Univ. Press, 1968, pp. 237—256.
29 M.J. Sewell, The governing equations and extremum principles of elasticity and plasticity generated from a single functional, J. Struct. Mech., 2 (1973) 1—32 and 135—158.
30 N. Dang Hung, Sur les principes variationnels en plasticité, Collection des Publications de la Faculté des Sciences Appliquées de l'Université de Liège, 84, 1980.
31 W. Prager, On Elastic, Perfectly Locking Materials, IBM Zürich Research Lab. Res. Paper RZ, 122, 1964.
32 L. Corradi and G. Maier, A matrix theory of elastic-locking structures, Meccanica, 3 (4) (1969) 289—313.
33 S.W. Dougill, On stable progressively fracturing solids, J. Appl. Math. Phys. (ZAMP), 27 (1976) 423.

34 M. Capurso, Extremum theorems for the solution of the rate-problem in elastic-plastic-fracturing structures, J. Struct. Mech., 7 (4) (1979) 411—434.
35 H.P. Künzi and W. Krelle, Nonlinear Programming, Blaisdell Publishing Company, 1966.
36 R.W. Cottle, Complementarity and variational problems, Symposia Mathematica, 19 (1975) 177—208.
37 K.G. Murty, On the number of solutions of the complementarity problem and spanning properties of complementarity cones, Linear Algebra and its Applications, 5 (1972) 65—108.

Appendix — some basic notions in mathematical programming

The few mathematical concepts employed in the paper are concisely summarized here. Not only proofs are omitted, but also various definitions and specifications, necessary for mathematical rigour, will be implicitly taken for granted or ignored altogether. This simplified survey seems justified by several reasons, besides space limitations: the present, growing popularity of these notions in the mechanics community; the availability of systematic, easily accessible presentations, such as [5, 35]; the fact that some mathematically meaningful restrictions (e.g., typically, the so-called "constraint qualification" for the Kuhn—Tucker theorem) are clearly satisfied in the mechanical applications in point; finally, our intention to emphasize the intrinsic simplicity and compactness of the ideas underlaying the basic theoretical framework that mathematical programming (MP) has in common with plasticity, as this paper is meant to show.

A *nonlinear programming* problem (NLP) can be formulated as follows:

$$\min_{\mathbf{x}} F(\mathbf{x}) \text{ subject to: } g_i(\mathbf{x}) \leqslant 0 \ (i = 1, \ldots, m) \tag{A.1}$$

where: $\mathbf{x}$ is a vector of n components (a point in an n-dimensional Euclidean space), F and g_i are "objective" and "constraint" functions, respectively (real-valued; differentiable). The constraints in (A.1) define the "feasible domain".

Consider the following relations or Kuhn—Tucker optimality conditions of problem (A.1):

$$g_i(\mathbf{x}) \leqslant 0 \ (i = 1, \ldots, m) \tag{A.2a}$$

$$\frac{\partial F(\mathbf{x})}{\partial \mathbf{x}} = \sum_{i=1}^{m} \frac{\partial g_i(\mathbf{x})}{\partial \mathbf{x}} y_i, \qquad y_i \geqslant 0 \tag{A.2b, c}$$

$$y_i g_i(\mathbf{x}) = 0 \ (\text{or } \tilde{\mathbf{y}}\mathbf{g} = 0) \tag{A.2d}$$

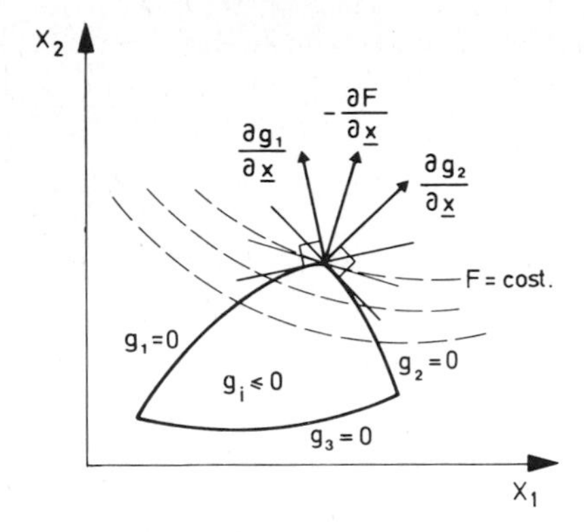

Fig. 1. Schematic illustration of Kuhn—Tucker optimality conditions.

where the new variables y_i are Lagrange multipliers. Note that the "complementarity" equation (A.2d) can be equivalently expressed as orthogonality of the two vectors in view of their sign constraints; it makes 0 the y_i corresponding to "inactive" constraints $g_i(\mathbf{x}) < 0$. Equation (A. 2b) expresses the gradient of the objective function as positive linear combination of the gradients of the active constraint functions, so that $\partial F/\partial \mathbf{x}$ will belong to the cone defined by the outward normals to the constraints active in $\mathbf{x}$.

Figure 1 visualizes problem (A.1) and illustrates geometrically (and makes intuitive) the following circumstances (*Kuhn—Tucker theorem*): the conditions (A.2) are fulfilled by any solution $\mathbf{x}^0$ of problem (A.1), i.e. they are necessary for optimality; if F and g_i are convex (and if a suitable "constraint qualification" condition is fulfilled), then conditions (A.3) are also sufficient, i.e. characterize the solution to (A.1), so that (A.1) and (A.2) are equivalent problems. The solution (or "optimal vector") $\mathbf{x}^0$ is unique if F and g_i are strictly convex, as Fig. 1 suggests.

Consider the problem:

$$\max_{\mathbf{x},\mathbf{y}} \{\hat{F}(\mathbf{x}) \equiv F(\mathbf{x}) + \sum_{i=1}^{m} y_i g_i(\mathbf{x})\} \tag{A.3a}$$

$$\text{subject to:} \quad \frac{\partial F(\mathbf{x})}{\partial \mathbf{x}} + \sum_{i=1}^{m} y_i \frac{\partial g_i(\mathbf{x})}{\partial \mathbf{x}} = 0, \qquad y_i \geqslant 0 \tag{A.3b}$$

Problem (A.3), constructed with the data of (A.2), is its "dual"; (A.1) is then referred to as "primal". The two problems are interrelated by various links, among which the following *duality* theorems: when F and g_i are all convex, if vectors $\mathbf{x}'$ and $(\mathbf{x}, \mathbf{y})$ are feasible for the primal and the dual respectively, then $F(\mathbf{x}') \geqslant \hat{F}(\mathbf{x}, \mathbf{y})$, and if $\mathbf{x}^0$ is primal-optimal then there is a $\mathbf{y}^0$ such that $(\mathbf{x}^0, \mathbf{y}^0)$ is dual-optimal and $F(\mathbf{x}^0) = \hat{F}(\mathbf{x}^0, \mathbf{y}^0)$.

When $F(\mathbf{x})$ is quadratic and g_i are linear, problem (A.1) becomes a *quadratic programming* problem (QP):

$$\min \{F(\mathbf{x}) \equiv \tfrac{1}{2}\tilde{\mathbf{x}}\mathbf{M}\mathbf{x} + \tilde{\mathbf{c}}\mathbf{x}\} \text{ subject to: } \mathbf{A}\mathbf{x} \leqslant \mathbf{b} \tag{A.4}$$

where matrices **M**, **A** and vectors **c**, **b** are data and **M** is symmetric.

Of interest here is a further specialization for $\mathbf{b} = \mathbf{0}$ and $\mathbf{A} = -\mathbf{I}$. In this case the primal problem (A.1) = (A.4), the optimality conditions (A.2) and the dual (A.3) become, respectively:

$$\min_{\mathbf{x}} \{\tfrac{1}{2}\tilde{\mathbf{x}}\mathbf{M}\mathbf{x} + \tilde{\mathbf{c}}\mathbf{x}\} \text{ subject to: } \mathbf{x} \geqslant \mathbf{0} \tag{A.5}$$

$$\mathbf{y} = \mathbf{M}\mathbf{x} + \mathbf{c} \leqslant \mathbf{0}, \ \mathbf{x} \geqslant \mathbf{0}, \ \tilde{\mathbf{x}}\mathbf{y} = 0 \tag{A.6}$$

$$\max_{\mathbf{x}} \{-\tfrac{1}{2}\tilde{\mathbf{x}}\mathbf{M}\mathbf{x}\} \text{ subject to: } \mathbf{M}\mathbf{x} \leqslant -\mathbf{c} \tag{A.7}$$

If **M** is positive semidefinite (A.5) is convex and equivalent to (A.6). Every solution of (A.5) solves (A.7); the solutions of (A.7), which are feasible also for (A.5), solve (A.5). For positive definite **M**, (A.5), (A.6) and (A.7) have a unique, common solution.

The specialization $\mathbf{M} = \mathbf{0}$ reduces (A.4) to *linear programming* (LP), and the dual (A.3) turns out to acquire the form:

$$\max_{\mathbf{y}} \{-\tilde{\mathbf{b}}\mathbf{y}\} \text{ subject to: } \tilde{\mathbf{A}}\mathbf{y} + \mathbf{c} = \mathbf{0}, \qquad \mathbf{y} \geqslant \mathbf{0} \tag{A.8}$$

A linear relation between two sign-constrained and orthogonal vectors is called *linear complementarity* (LCP) and is represented by (A.6), matrix **M** being now *not* necessarily symmetric. Among many results in LCP theory (e.g. [36, 37]), the following ones are of use in this paper:

(a) a LCP has a unique solution for any **c** if and only if matrix **M** has all principal minors positive; has a finite number of solutions for any **c**, if and only if **M** has all principal minors nonzero;

(b) the following QP problem can be associated to the LCP (A.6)

$$\min_{\mathbf{x}} \{\tilde{\mathbf{x}}\mathbf{M}\mathbf{x} + \tilde{\mathbf{c}}\mathbf{x}\} \text{ subject to: } \mathbf{M}\mathbf{x} \leqslant -\mathbf{c}, \qquad \mathbf{x} \geqslant \mathbf{0} \tag{A.9}$$

in the sense that, for any **M**, if (A.6) has a solution, this solves also (A.9); any solution of (A.9), if the corresponding minimum is zero, solves (A.6).

Estimates for the Constitutive Moduli of Ductile Materials Containing a Small Volume Fraction of Voids or Inclusions

ROBERT M. McMEEKING

Department of Theoretical and Applied Mechanics, University of Illinois at Urbana-Champaign, Urbana, IL 61801 (U.S.A.)

Summary

A technique for estimating some of the constitutive moduli of a ductile material containing a small volume fraction of voids or inclusions is proposed. The estimates are obtained from a volume average of the strain-rate in the matrix and strain-rate in the voids or inclusions. To first order in the volume fraction of holes or inclusions, the result for the response of a single hole or inclusion in an infinite or very large block of the matrix can be used to determine the average strain-rate in the voids or inclusions. The estimates obtained are shown to be in agreement with the first order estimates derived from a self-consistent averaging method. Results are derived for an elastic-plastic matrix containing voids, where the matrix material is assumed to obey J_2 flow theory, and for a power law creeping material containing voids growing by plasticity. Comparisons are made with other available estimates.

Introduction

Drucker [1] has often advocated the application of the methods of mechanics to problems in which the length scales involved are conventionally considered to be microscopic. The resulting insight into the nature of the behavior of the material is often substantial, especially in circumstances where critical experiments are difficult or expensive. A case in point is the work of McClintock [2] followed by the contributions of Rice and Tracey [3] and Tracey [4] on the growth of voids in ductile materials. Together, these researchers have established the importance of triaxial stress in determining the rate of void growth and have used the results in a model for ductile rupture due to the coalescence of holes. (More recently, Budiansky, Hutchinson and Slutsky [5] have discussed extensively the growth of voids in viscous solids including the special case of the perfectly plastic material for which they provide improved results.) Prior to final rupture but after the nucleation of the cavities, the macroscopic behavior of the material involves a plastic dilation due to the growth of the voids. Stress—strain relations for the macroscopic behavior of this porous elastic-plastic solid are useful for improved predictions of ductile rupture and would also be valuable for the analysis of non-fully dense sintered metal alloys and for metal forming processes involving an unwanted cavitation. Such laws have been developed by Gurson [6, 7] for a matrix of J_2 flow theory material containing spherical or long cylindrical cavities. Noting that in the absence of further void nucleation the porous solid is governed by a flow law in which the strain rate is normal to a yield surface, Gurson estimated the constitutive law by calculating the yield surfaces for cells of material containing a single void. These yield surfaces are dependent on the mean normal macroscopic stress and the void volume fraction as well as the usual second invariant of the macroscopic stress deviator. The resulting stress rate—strain rate relationships have been used to model ductile rupture by the localization of flow [8].

A related problem is the growth of voids on the grain boundaries in a ductile material at high temperature. Edward and Ashby [9] and Needleman and Rice [10] have shown that there is significant coupling with diffusion in some stress states at some temperatures. However, in other stress states or temperatures or after a grain boundary is completely cavitated to form a larger void, the void growth can be by plasticity alone. Budiansky et al. [5] have analyzed void growth in secondary creep. Recently there has been some preliminary work on developing macroscopic constitutive laws for secondary creep with void growth [10, 11]. These estimates are based on the response of a cell of material containing a single growing void.

In this paper, estimates for the constitutive moduli for a ductile material containing voids or inclusions which do not crack or separate are obtained to first order when the volume fraction f of voids or inclusions is small. The estimates are obtained by letting the macroscopic strain rate be the volume average of the strain rates in the individual components. Averaging to obtain composite properties in ductile materials is a well established technique, having been used or suggested by Bishop and Hill [12], Hill [13] and Hutchinson [14]. However, the technique has been directed mainly toward establishing composite moduli for a polycrystal. Of course, averaging is implicit when properties are estimated from the behavior of a finite unit cell containing a single void or inclusion as in refs. 6, 7, 10 and 11. In the analysis to follow, the averaging is carried out explicitly based on estimates for the average strain rates in individual components. These estimates are derived from the behavior of single voids in infinite bodies [3, 5] or very large bodies. The results are not as far reaching as those based on finite unit cells containing single voids [6, 7, 10, 11] in that the resulting constitutive laws are definitely limited to small or very small volume fractions of voids or inclusions. This can be a very extreme restriction as was noted by Budiansky et al. [5]. However, moduli for a material containing a non-spherical or non-round cylindrical void or inclusions can be determined if results for the same void or inclusion shape in an infinite matrix are available. This is perhaps a simpler problem than that of the given void or inclusion in a suitably shaped finite unit cell. Furthermore, it can be shown that the averages that result from the technique are also the first order estimates resulting from a self-consistent averaging method [13] of a more general nature. The constitutive law would be useful for the early stages of void growth in ductile rupture especially with the evolving shape of the voids taken into account. If instability leading to rupture occurs in a given material in some state of stress while f is quite small, the derivation of laws for higher void fractions in that case could be avoided.

Estimates of moduli by averaging

Consider a large cell of material with enough voids or inclusions to endow the cell with the average properties of the material. Henceforth, inclusions will be understood to mean also voids. It must be assumed that the voids are not arranged in any regular way to cause nearest neighbor interactions to dominate the material deformation, e.g. by causing shear bands. We wish to determine the macroscopic uniform strain rate due to a macroscopic uniform stress applied to a unit cell. Such unit cells have been used by Bishop and Hill [12] and Gurson [6] to determine the overall nature of composite material behavior. In particular, Gurson shows that the macroscopic deformation rate D of the composite is given by

$$\mathbf{D} = \frac{1}{V}\int_V \mathbf{d}\, dV = (1-f)\mathbf{d}^{\mathrm{M}} + f\mathbf{d}^{\mathrm{I}} \tag{1}$$

where $\mathbf{d}$ is the symmetric part of the local velocity gradient, V is the volume of the unit cell, f is the volume fraction of inclusions, $\mathbf{d}^{\mathrm{M}}$ is the volume average over the matrix of the local deformation rate and $\mathbf{d}^{\mathrm{I}}$ is the volume average over the inclusions. In the case of voids, the average deformation rate is to be understood as the average of the deformation rates compatible with the surface velocities. In contrast to Bishop and Hill [12] and Gurson [6] but following Hill [13] we will take the macroscopic stress $\boldsymbol{\sigma}$ to be a volume average rather than a surface average:

$$\boldsymbol{\sigma} = (1-f)\boldsymbol{\sigma}^{\mathrm{M}} + f\boldsymbol{\sigma}^{\mathrm{I}} \tag{2}$$

where again M and I indicate volume averages over matrix and inclusion respectively. Of course the stress in a void would be zero.

The matrix material will be assumed to obey the relationship

$$\mathbf{d}^{\mathrm{M}} = \mathbf{M}^{\mathrm{M}}(\boldsymbol{\sigma}^{\mathrm{M}}, \mathscr{K}^{\mathrm{M}})\overset{*}{\boldsymbol{\sigma}}{}^{\mathrm{M}} \tag{3}$$

where $\overset{*}{\boldsymbol{\sigma}} = \dot{\boldsymbol{\sigma}} - \boldsymbol{\Omega}\boldsymbol{\sigma} + \boldsymbol{\sigma}\boldsymbol{\Omega}$ is the Jaumann rate of change of the stress and is used to assure objectivity to superimposed spin of the constitutive law. The tensor $\boldsymbol{\Omega}$ is the spin rate, that is the antisymmetric part of the velocity gradient. The fourth order tensor of inverse moduli $\mathbf{M}^{\mathrm{M}}$ will be in general a function of the stress and the history of deformation. A set of internal variables $\mathscr{K}^{\mathrm{M}}$ is included to account for history and other effects. For creeping materials, the Jaumann stress rate in eqn. (3) is replaced by the stress which leads to a law in which elastic effects are omitted. In some cases, the tensor $\mathbf{M}^{\mathrm{M}}$ will be multivalued as in the case of elastic-perfect plasticity when plastic flow is occurring. In these cases the stress-rate will be unique and it will be convenient to proceed as if $\mathbf{M}^{\mathrm{M}}$ and its inverse $\mathbf{L}^{\mathrm{M}}$ both exist and obtain limiting cases later, e.g. by finally allowing the hardening modulus to disappear. In special cases the resulting law may be non-invertible and only the form giving stress rate from strain rate will be valid.

At this stage we introduce the relationship

$$\mathbf{d}^{\mathrm{I}} = \mathbf{A}(\boldsymbol{\sigma}, \mathscr{K}, S)\mathbf{D} \tag{4}$$

where the parameter S indicates a dependence on current inclusion shape. Obviously, the fourth order tensor $\mathbf{A}$ relates the average strain rate in inclusions to the average strain rate in the composite material. We will assume from now on that the average strain rate in inclusions can be approximated by the average strain rate in a single typical inclusion embedded in an infinite body of the composite material subject to macroscopically uniform stresses at infinity. Self-consistent averaging techniques [13] are based on this assumption and eqn. (4) is used in that method. As Hill notes, the problem of finding the strain rate in a typical inclusion embedded in the composite material is tractable when the matrix is hypoelastic and homogeneous. In that case the inclusion results of Eshelby [15] can be used if the inclusion is ellipsoidal or can be approximated as ellipsoidal. With a fully nonlinear matrix as in flow theory plasticity or creep, general results for $\mathbf{A}$ will be difficult. However, there are results for voids embedded in fully nonlinear homogeneous infinite or very large bodies with standard constitutive laws [2—6, 16]. These results will be used to estimate $\mathbf{A}$ when the volume fraction of inclusions in the composite material is very low. Certainly, the composite constitutive equations with f very small should recover the behavior of large or infinite homogeneous bodies containing a single inclusion as long as the inclusions in the composite being modelled are distributed so any representative volume has a dilute dispersion, i.e. no clustering.

The response on the surface of an infinite body with a single inclusion to the uniform stress applied at infinity will be the same as the response in the absence of an inclusion. Defining $\mathbf{M}^0 = \mathbf{M}^{\mathrm{M}}(\boldsymbol{\sigma}, \mathscr{K})$, we can combine eqns. (1, 2, 3 and 4) to give

$$\mathbf{D} = \mathbf{M}^{\mathrm{M}}(\boldsymbol{\sigma}^{\mathrm{M}}, \mathscr{K}^{\mathrm{M}})[\overset{*}{\boldsymbol{\sigma}} - f\overset{*}{\boldsymbol{\sigma}}{}^{\mathrm{I}} - \dot{f}(\boldsymbol{\sigma}^{\mathrm{I}} - \boldsymbol{\sigma}^{\mathrm{M}})] + f\mathbf{A}\mathbf{M}^0\overset{*}{\boldsymbol{\sigma}} \tag{5}$$

where we assume that all components experience the same spin rate to leading order. Strictly, this will exclude some inclusions such as slender ones oriented to the principal shear direction in uniaxial stress. When it is noted that $\dot{f}/f = \text{trace (AD)}$ and the relationship $\overset{*}{\boldsymbol{\sigma}}{}^{I} = \mathbf{L}^{I}(\boldsymbol{\sigma}^{I}, \mathscr{K}^{I})\mathbf{d}^{I}$ is introduced, there follows from (5)

$$\mathbf{D} = \mathbf{M}(\boldsymbol{\sigma}, \mathscr{K})\overset{*}{\boldsymbol{\sigma}}$$

where

$$\mathbf{M} = \mathbf{M}^{M} + f[\mathbf{A}\mathbf{M}^{0} + \mathbf{M}^{M}(\boldsymbol{\sigma}^{M} - \boldsymbol{\sigma}^{I}) \otimes \text{trace}(\mathbf{A})\mathbf{M}^{0} - \mathbf{M}^{M}\mathbf{L}^{I}\mathbf{A}\mathbf{M}^{0}] \qquad (6)$$

The trace operation will always be carried out on the leading indices of a fourth order tensor, i.e. trace $(\mathbf{A})_{kl} = \mathrm{A}_{iikl}$. Since the estimate (6) is only valid to first order in f and some quantities will be only slightly changed from the inclusionless case, we can replace $f\mathbf{M}^{M}$ by $f\mathbf{M}^{0}$ and $\boldsymbol{\sigma}^{M}$ by $\boldsymbol{\sigma}$. The form of (6) for creep ($\overset{*}{\boldsymbol{\sigma}}$ replaced by $\boldsymbol{\sigma}$ etc. in eqns. (3), (6)) would be somewhat simpler and can be easily derived. The time integration of (6) would be accomplished by keeping track of the volume fraction of inclusions and the separate stress averages for the matrix and inclusions. Note that the character of the inclusion is restricted; e.g. only creeping inclusions in creeping matrices and elastic or elastic-plastic inclusions in elastic-plastic matrices. Voids are always permissible.

In the elastic-plastic material, some consideration must be given to when plastic flow would first commence. If both the matrix and the inclusions have an average stress which is below their first yield stress, the methods of analysis for elastic composites could be used to determine the stresses in each component. A survey of this topic has recently been given by Willis [17]. If the inclusions yield before the matrix on average and are approximated as ellipsoids, the analysis of Eshelby can be used to form the tensor **A** since the plastic strain in the inclusion can be treated as a transformation strain [18]. Equation (6) would then be used with $\mathbf{M}^{M}$ taken as the elastic compliance tensor for the matrix material. Once the matrix has reached its yield strength, the Eshelby form of **A** must be abandoned in favor of the fully nonlinear version. However, a possibility that will not be taken up here is a suggestion of Hill [13] that the yielded matrix material be approximated by a homogeneous hypoelastic material. The hypoelastic moduli everywhere would be set equal to the moduli of the elastic-plastic (or creeping) material at the macroscopic stress state. This would allow the continued use of the Eshelby form for **A**, based of course on the hypoelastic moduli rather than the initial elastic response. A comparison with more accurate estimates for the growth of isolated inclusions would allow an assessment of the accuracy of this approximation.

Self consistent estimates of composite moduli for material with voids

With the tensor A chosen to represent the behavior of a small void in an infinite or very large body the effect of long range interactions between voids as well as particular nearest neighbor interactions are omitted. These interactions must become important at larger volume fractions of voids than those for which eqn. (6) (with $L_I = 0$) is suitable. One method by which interactions can be taken into account in an approximate way is the estimation of composite properties from the behavior of finite cells containing single voids. The voids occupy a fraction f of the volume of the cell. The cell is subjected to boundary conditions designed to model the macroscopic state and the cell response dictates the macroscopic behavior. This is the approach adopted by Gurson [6, 7], Needleman and Rice [10] and Cocks and Ashby [11] and it has proved to be very useful. However, another technique for finding approximate composite properties is the self-consistent method of Hill [13, 19]. In this method the averages in eqns. (1) and (2) are adopted. The deformation rate averaged over the voids is estimated from the behavior of a single hole in an infinite body with the as yet unknown composite properties. The body is subject to appropriate boundary conditions at infinity. Long range interactions, but not nearest neighbor effects, are approximated by this technique.

Following Hill [13, 19], we will write the relationship between the void deformation rate and macroscopic rates as

$$\mathbf{D} - \mathbf{d}^{\mathrm{I}} = -\mathbf{M}^* \overset{*}{\boldsymbol{\sigma}} \tag{7}$$

The fourth order tensor $\mathbf{M}^*$ will depend on the current stress state at infinity, the history at infinity and the current void shape. The average relationships (1, 2) and eqn. (7) together imply that

$$\mathbf{D} - \mathbf{d}^{\mathrm{M}} = -\mathbf{M}^* (\overset{*}{\boldsymbol{\sigma}} - \overset{*}{\boldsymbol{\sigma}}{}^{\mathrm{M}}) - \mathbf{M}^* \boldsymbol{\sigma}^{\mathrm{M}}\, \dot{f}/(1-f) \tag{8}$$

Hill [19] found the same equation, but for $\dot{f} = 0$ and $\boldsymbol{\sigma}^{\mathrm{I}} \neq \mathbf{0}$. Following Hill's use of equations equivalent to (7, 8), we find that the inverse composite moduli estimates are given by

$$\begin{aligned}\mathbf{M} = {} & (1-f)\mathbf{M}^{\mathrm{M}}(\mathbf{M}^{\mathrm{M}} + \mathbf{M}^*)^{-1}(\mathbf{M} + \mathbf{M}^*) + f(\mathbf{M} + \mathbf{M}^*) \\ & + f\mathbf{M}^{\mathrm{M}}(\mathbf{M}^{\mathrm{M}} + \mathbf{M}^*)^{-1}\mathbf{M}^*\boldsymbol{\sigma}^{\mathrm{M}} \otimes \mathrm{trace}\,(\mathbf{M} + \mathbf{M}^*)\end{aligned} \tag{9}$$

where we have assumed that all necessary tensor inversions are possible. If that situation does not prevail, the inverse form of (9), i.e. an expression for $\mathbf{L}$ where $\overset{*}{\boldsymbol{\sigma}} = \mathbf{LD}$, can be developed, again using Hill's procedures. Non-singularity can be artificially introduced such as by a finite bulk modulus instead of incompressibility and the limit taken later. We shall be interested in the case where f is small. The following expansions can be introduced:

$$\mathbf{M} = \mathbf{M}^0 + f\mathbf{M}^{(1)} + \ldots$$
$$\mathbf{M}^{\mathrm{M}} = \mathbf{M}^0 + f\mathbf{M}^{\mathrm{M}(1)} + \ldots$$
$$\boldsymbol{\sigma}^{\mathrm{M}} = \boldsymbol{\sigma} + f\boldsymbol{\sigma}^{(1)} + \ldots$$
$$\mathbf{M}^* = \mathbf{M}^{*(0)} + f\mathbf{M}^{*(1)} + \ldots \tag{10}$$

The tensor $\mathbf{M}^{*(0)}$ governs the behavior of a single void in an infinite body of the matrix material. In some cases such as the incompressible material subject to purely hydrostatic stressing, the meaning of $\mathbf{M}^{*(0)}$ will not be obvious. However, a consideration of the appropriate behavior of large bodies with a single void would clarify this. After the introduction of the expansions into eqn. (9) and the collection of first order terms, we find that

$$[\mathscr{I} - \mathbf{M}^0(\mathbf{M}^0 + \mathbf{M}^{*(0)})^{-1}](\mathbf{M}^{(1)} - \mathbf{M}^{\mathrm{M}(1)})$$
$$= \mathbf{M}^{*(0)} + \mathbf{M}^0(\mathbf{M}^0 + \mathbf{M}^{*(0)})^{-1}\mathbf{M}^{*(0)}\boldsymbol{\sigma} \otimes \mathrm{trace}\,(\mathbf{M}^0 + \mathbf{M}^{*(0)}), \tag{11}$$

where $\mathscr{I}$ is the fourth order unit tensor. From (7) we can see that $\mathrm{A}\mathbf{M}^0 = \mathbf{M}^{*(0)} + \mathbf{M}^0$. A premultiplication of eqn. (11) by A gives

$$(\mathrm{A} - \mathscr{I})(\mathbf{M}^{(1)} - \mathbf{M}^{\mathrm{M}(1)}) = (\mathrm{A} - \mathscr{I})\mathrm{A}\mathbf{M}^0$$
$$+ (\mathrm{A} - \mathscr{I})\mathbf{M}^0\boldsymbol{\sigma} \otimes \mathrm{trace}\,(\mathrm{A})\mathbf{M}^0 \tag{12}$$

and finally reforming **M** to first order we find

$$\mathbf{M} = \mathbf{M}^0 + f\mathbf{M}^{(1)} = \mathbf{M}^{\mathrm{M}} + f[\mathrm{A}\mathbf{M}^0 + \mathbf{M}^0\boldsymbol{\sigma} \otimes \mathrm{trace}\,(\mathrm{A})\mathbf{M}^0] \tag{13}$$

in agreement with eqn. (6) with $\mathbf{M}^{\mathrm{M}} = \mathbf{M}^0 + f\mathbf{M}^{\mathrm{M}(1)}$ used consistently only to within first order in f. We see now that the basic result required is the strain rate in the isolated void due to a stress-rate at infinity

$$\mathrm{d}^{\mathrm{I}} = \mathrm{A}\mathbf{M}^0\overset{*}{\boldsymbol{\sigma}} \tag{14}$$

or in the case of power law creep

$$\mathrm{d}^{\mathrm{I}} = \mathrm{A}\mathbf{M}^0\boldsymbol{\sigma} \tag{15}$$

in which case the term involving trace (A) in eqn. (13) is omitted. In elastic-perfect plasticity, eqn. (14) will be unsuitable. However, A will be meaningful where $\mathrm{A}\mathbf{M}^0\overset{*}{\boldsymbol{\sigma}}$ is not and if preferred an inverse law $\overset{*}{\boldsymbol{\sigma}} = \mathrm{LD}$ can be developed from eqn. (13) before we take the limit of perfect plasticity.

In the next sections we give some examples of estimates for a few situations. We will confine our consideration to the cases of a material containing voids.

Elastic-plastic materials with spherical voids

The Prandtl—Reuss equations of J_2 flow theory [16] will be used as the plasticity law for the matrix material. The estimates need not be confined to J_2 flow theory, but we consider that case as the most commonly used law. The equations are

$$d_{ij}^{\mathrm{M}} = \frac{1}{2G}\overset{*}{\sigma}{}_{ij}^{\mathrm{M}} + \frac{1}{3}\left(\frac{1}{3K} - \frac{1}{2G}\right)\delta_{ij}\overset{*}{\sigma}{}_{kk}^{\mathrm{M}} \tag{16}$$

when the material is responding elastically and

$$d_{ij}^{\mathrm{M}} = \frac{1}{2G}\overset{*}{\sigma}{}_{ij}^{\mathrm{M}} + \frac{1}{3}\left(\frac{1}{3K} - \frac{1}{2G}\right)\delta_{ij}\overset{*}{\sigma}{}_{kk}^{\mathrm{M}} + \frac{9}{4h}\frac{\sigma_{ij}^{\prime\mathrm{M}}\sigma_{kl}^{\prime\mathrm{M}}}{(\bar{\sigma}^{\mathrm{M}})^2}\overset{*}{\sigma}{}_{kl}^{\mathrm{M}} \tag{17}$$

when plastic flow is occurring. This occurs when $\bar{\sigma} = \sigma_{\mathrm{y}}$ the yield stress and as long as $\sigma_{ij}\mathrm{d}_{\mathrm{p}ij} > 0$. G is the elastic shear modulus, K is the elastic bulk modulus, δ_{ij} is the Kronecker delta, $\boldsymbol{\sigma}'$ is the stress deviator, $(\bar{\sigma})^2 = \frac{3}{2}\sigma_{ij}'\sigma_{ij}'$ is the square of the tensile equivalent stress and h is the hardening rate such that $\dot{\bar{\sigma}}^{\mathrm{M}} = h\dot{\bar{\epsilon}}_{\mathrm{p}}^{\mathrm{M}}$ where $\dot{\bar{\epsilon}}_{\mathrm{p}}^{\mathrm{M}} = (\frac{2}{3}d_{\mathrm{p}ij}^{\mathrm{M}}d_{\mathrm{p}ij}^{\mathrm{M}})^{1/2}$ is the rate of increase of the tensile equivalent plastic strain. The plastic deformation rates $\boldsymbol{d}_{\mathrm{p}}^{\mathrm{M}}$ are the total deformation rates given by eqn. (17) less the elastic deformation rates given by eqn. (16). The hardening rate is usually taken to be a function of $\bar{\epsilon}^{\mathrm{p}}$, but here we shall be concerned only with the case where h is a non-zero constant. In that case $\boldsymbol{M}^{\mathrm{M}} = \boldsymbol{M}^0$.

The results of analyses of void growth in rigid-perfectly plastic materials [3, 5] suggest that the deformation rate of a single spherical void in a stress state close to that of uniaxial tension with a superimposed hydrostatic stress can be approximated by

$$A_{ijkl} = \alpha_1\delta_{ik}\delta_{jl} + \alpha_2\delta_{ij}\delta_{kl} + \\ + \alpha_3\delta_{ij}\frac{\sigma_{kl}'}{\bar{\sigma}} + \alpha_4\frac{\sigma_{ij}'}{\bar{\sigma}}\delta_{kl} + \alpha_5\frac{\sigma_{ij}'}{\bar{\sigma}}\frac{\sigma_{kl}'}{\bar{\sigma}} \tag{18}$$

where α_1 and a contribution to α_2 would arise from elastic deformation and the remaining coefficients α_i represent ductile growth. The plasticity effects in α_i, $i = 2, \ldots, 5$ would cause these terms to be functions of the state of stress at infinity. For example, Rice and Tracey [3] give estimates for D (our α_3) and $1 + E$ (our $2\alpha_5/3$) when the state of stress at infinity is an hydrostatic tension superimposed on a uniaxial stress in a rigid perfectly plastic material. At high ratios of hydrostatic to flow stress they find that $\alpha_3 = 0.283 \exp(\sigma_{kk}/2\sigma_{\mathrm{y}})$. Further results for the same problem are given by Budiansky et al. [5] in plots (marked ∞) of $\dot{V}/\epsilon V$ (our $3\alpha_3$) versus $\sigma_{\mathrm{m}}/\sigma$ (our $\sigma_{kk}/3\sigma_{\mathrm{y}}$) in their Figs. 5.1(a) and (b). These last authors also give a plot which can be interpreted to give values for α_5. Approximating their strain rate field by one which would cause the void to grow into an ellipsoid, we can equate their $\dot{a}/\dot{b}$ to our $(\alpha_3 + 2\alpha_5/3)/(\alpha_3 - \alpha_5/3)$ and use their Fig. 5.2 to give values for α_5 as a function of our $\sigma_{kk}/\bar{\sigma}$.

We proceed now to give the form of eqn. (13) when eqn. (17) is used to define $\boldsymbol{M}^0$ and $\boldsymbol{A}$ is given by eqn. (18). The results would pertain to continued plastic flow after yielding has already taken place and because the elastic contribution to A would be minor in these circumstances we neglect α_1 and elastic contributions to α_2. The result is

$$M_{ijkl} = \frac{1}{2G}\left(\delta_{ik}\delta_{jl} - \frac{1}{3}\delta_{ij}\delta_{kl}\right) + \frac{1}{9K}\delta_{ij}\delta_{kl}$$

$$+ \frac{9}{4H}\frac{\sigma'_{ij}}{\bar{\sigma}}\frac{\sigma'_{kl}}{\bar{\sigma}} + f\left[\alpha_3\left(\frac{1}{2G} + \frac{3}{2h}\right)\left(1 + \frac{\sigma_{kk}}{3K}\right)\delta_{ij}\frac{\sigma'_{kl}}{\bar{\sigma}}\right. \tag{19}$$

$$\left. + \left(\frac{\alpha_4}{3K} + \alpha_2\frac{\bar{\sigma}}{K}\left(\frac{1}{2G} + \frac{3}{2h}\right)\right)\frac{\sigma'_{ij}}{\bar{\sigma}}\delta_{kl} + \frac{\alpha_2}{3K}\left(1 + \frac{\sigma_{kk}}{3K}\right)\delta_{ij}\delta_{kl}\right]$$

where

$$\frac{1}{H} = \frac{1}{h} + \frac{4}{9}f\left[\alpha_5\left(\frac{1}{2G} + \frac{3}{2h}\right) + 3\alpha_3\left(\frac{1}{2G} + \frac{3}{2h}\right)^2\bar{\sigma}\right] \tag{20}$$

Some simplification results from the fact that the plastic flow law should be such that the plastic potential surface coincides with the yield surface for the macroscopic sample of material. Gurson [6], using results established by Berg [20] from an argument of Bishop and Hill [12], shows that a flow law for a voided material will obey a normality rule to the macroscopic yield surface as long as the matrix material obeys a normality rule itself. The Bishop and Hill deduction is based on volume averaged strain rates and surface averaged stresses. However, the argument can be extended to cover the case of volume averaged stresses as follows. Consider a macroscopic sample of material subject to surface tractions in equilibrium with an homogeneous macroscopic stress state. If it is assumed that the stress on any set of parallel planes within the sample is the same on average, then volume averaged stresses are entirely equivalent to area averaged stresses.

The constitutive law (17) involves a plastic flow normal to the yield surface. Thus the moduli given in eqn. (19) should be a good approximation to a flow law associated by normality to a yield surface for the porous material. As discussed by Rudnicki and Rice [21], such a form would be given by

$$M_{ijkl} = \frac{1}{2G}\left(\delta_{ik}\delta_{jl} - \frac{1}{3}\delta_{ij}\delta_{kl}\right) + \frac{1}{9K}\delta_{ij}\delta_{kl}$$

$$+ \frac{1}{H}\left(\frac{3\sigma'_{ij}}{2\bar{\sigma}} + \frac{\beta}{\sqrt{3}}\delta_{ij}\right)\left(\frac{3\sigma'_{kl}}{2\bar{\sigma}} + \frac{\beta}{\sqrt{3}}\delta_{kl}\right) \tag{21}$$

By equating coefficients of $\delta_{ij}\sigma'_{kl}/\bar{\sigma}$ in (19) and (21) and retaining only leading order terms in f we can establish that

$$\beta = \sqrt{3}\, f\alpha_3 \left(1 + \frac{h}{3G}\right)\left(1 + \frac{\sigma_{kk}}{3K}\right) \tag{22}$$

It will be assumed that α_2 and α_4 can be approximated in such a way that the symmetry on interchange of subscripts ij with kl in (21) is retained. Finally, as an example, we shall use the Rice and Tracey [3] result for perfect plasticity, where $\alpha_3 = 0.283 \exp(\sigma_{kk}/2\bar{\sigma})$. This implies that $\beta/\sqrt{3} = 0.283\, f \exp(\sigma_{kk}/2\bar{\sigma})$ where $\sigma_{kk}/3K$ has been neglected compared to 1. In addition $H = -0.85\, f\bar{\sigma} \exp(\sigma_{kk}/2\bar{\sigma})$. These results can be compared with the equivalent terms in Gurson's macroscopic constitutive law for spherical holes in an elastic-plastic matrix [6]. Gurson's laws have the same form as eqn. (21) and for perfect plasticity to first order in f Gurson finds that $\beta/\sqrt{3} = \frac{1}{2} f \sinh(\sigma_{kk}/2\bar{\sigma})$ and $H = -1.5\, f\bar{\sigma} \sinh(\sigma_{kk}/2\bar{\sigma}) \cosh(\sigma_{kk}/2\bar{\sigma})$. At high triaxialities, for which the Rice—Tracey estimate of α_3 is appropriate, $\sinh(\sigma_{kk}/2\bar{\sigma}) \approx \frac{1}{2} \exp(\sigma_{kk}/2\bar{\sigma})$ and the estimates for β are within 12% of each other. The expressions for H are somewhat different, but would be very close when $\sigma_{kk}/\bar{\sigma} = 1.64$. The similarity of the two results is not surprising, since Gurson's laws were inferred from the behavior of spherical cells containing a single spherical hole. When the cells are very large compared to the hole, they will behave in a manner similar to the response of a single hole in an infinite matrix. Of course, the Gurson laws were designed for a larger range of void volume fractions, and are not necessarily optimal for very low volume fractions.

Power law creeping materials with spherical voids

In this section the matrix material will be assumed to obey a power law creep relationship such that

$$d_{ij}^{\mathrm{M}} = B(\bar{\sigma}^{\mathrm{M}})^{n-1} \sigma_{ij}^{\mathrm{M}\prime} \tag{23}$$

This means that

$$M_{ijkl}^{\mathrm{M}} = [1 + (n-1)f]\, B\bar{\sigma}^{n-1} (\delta_{ik}\delta_{jl} - \tfrac{1}{3}\delta_{ij}\delta_{kl}) \tag{24}$$

where only terms up to order f have been retained. B is a material parameter which is usually temperature dependent. The form of eqn. (13) suitable for creep is

$$\mathbf{D} = \mathbf{M}\boldsymbol{\sigma} \qquad \mathbf{M} = \mathbf{M}^{\mathrm{M}} + f\mathbf{A}\mathbf{M}^0 \tag{25}$$

The void growth tensor $\mathbf{A}$ represents the void growth due to plasticity effects alone rather than that coupled with diffusion effects. It will be assumed for spherical voids in stress states close to uniaxial stress with a superimposed hydrostatic stress that, at least approximately,

$$A_{ijkl} = \alpha_1 \delta_{ik}\delta_{jl} + \alpha_2 \delta_{ij}\delta_{kl} + \alpha_3 \delta_{ij} \frac{\sigma'_{kl}}{\bar{\sigma}} + \alpha_4 \frac{\sigma'_{ij}}{\bar{\sigma}} \delta_{kl} \tag{26}$$

From eqns. (25) and (26) it follows that

$$M_{ijkl} = B\bar{\sigma}^{n-1}\{[1 + f(n-1+\alpha_1)]\,(\delta_{ik}\delta_{jl} - \tfrac{1}{3}\delta_{ij}\delta_{kl}) + f\alpha_3\delta_{ij}\sigma'_{kl}/\bar{\sigma}\} \tag{27}$$

in such stress states.

Budiansky et al. [5] give estimates from which α_3 and α_1 can be inferred. Taking their high triaxial stress approximation for the dilatational part of the growth rate, we can find from their eqn. (6.15) that

$$\alpha_3 = \frac{1}{2}\left[\frac{1}{2n}\,\frac{\sigma_{kk}}{\bar{\sigma}} + g(n)\right]^n \tag{28}$$

for $\sigma_{kk} > 0$. The function g is given by Budiansky et al., but we will approximate it by $g = 1 - 1/n$. The term α_1 will be a function of n and $(\sigma_{kk}/\bar{\sigma})$ and can be deduced by regarding $(\alpha_1 + \alpha_3)/(\alpha_3 - \frac{1}{2}\alpha_1)$ as an approximation to $\dot{a}/\dot{b}$ in Fig. 5.2 of ref. [5]. It would be useful to develop simple approximations to the graphs of $\dot{a}/\dot{b}$ given in [5]. However, we will confine ourselves to very high triaxialities such that $|\sigma_{kk}|/3\bar{\sigma} \geqslant 3$ and the range of indices $5 \leqslant n \leqslant 10$. A rough estimate in that case gives

$$\alpha_1 = -6\bar{\sigma}\alpha_3/(\sigma_{kk} + 3\bar{\sigma}) \tag{29}$$

and eqn. (27) then gives

$$D_{ij} = \left\{1 + f\left[n - 1 - 3\bar{\sigma}\,\frac{\left(\frac{1}{2n}\,\frac{\sigma_{kk}}{\bar{\sigma}} + \frac{n-1}{n}\right)^n}{(\sigma_{kk} + 3\bar{\sigma})}\right]\right\} B\bar{\sigma}^{n-1}\sigma'_{ij}$$
$$+ f\left[\frac{1}{2n}\,\sigma_{kk} + \frac{n-1}{n}\,\bar{\sigma}\right]^n \frac{B}{3}\,\delta_{ij} \tag{30}$$

It should be noted that eqn. (29) gives a dilatational response to pure hydrostatic stress even when $\bar{\sigma} = 0$. According to Budiansky et al. the estimate for isolated void growth leading to our eqn. (28) is accurate in the limit of pure hydrostatic stress. Consequently we judge the response to hydrostatic stress inherent to eqn. (29) to be reliable at least for very small values of f. However, Budiansky et al. [5] have pointed out that the response of a finite sphere of material containing a spherical hole occupying a volume fraction f will strain much faster in hydrostatic stressing than an infinite block containing the same hole even at volume fractions as low as 0.1% when $n = 3$. Thus the hydrostatic response in eqn. (29) must become suspect at similar volume fractions and is likely to underestimate the dilatational deformation rate.

Discussion

First order corrections to the matrix constitutive laws have been found for an elastic-plastic material containing a small volume fraction of spherical voids and for a power law creeping material containing a small volume fraction of voids. Specific estimates for states of high triaxial stress superimposed on a uniaxial one have been given. In the case of elastoplasticity, this estimate was restricted to perfect plasticity. The negative value predicted for H in that case indicates that this material would be subject to destabilizing localizations of flow as soon as or shortly after substantial amounts of plastic flow begin to occur [21]. More useful estimates would be those that account for matrix strain hardening. Gurson [6, 7] accounted for strain hardening by assuming that his pressure sensitive yield surfaces were subject to isotropic hardening. He also equated the total macroscopic plastic work rate to the total matrix plastic work rate and used that result along with the consistency condition of plastic flow [16] to determine the contribution to H arising from matrix strain hardening. The implication is that void growth rates would be smaller than in the perfectly plastic case, which was predicted by Tracey [4]. Further work on the behavior of isolated voids in infinite blocks of strain hardening material like that in ref. [4] would be useful. The resulting void growth rates could be incorporated into expressions for $\mathbf{A}$ and so into a continuum description of the voided material.

The estimates given in this paper have all been for materials containing spherical voids. As a result the material behavior is isotropic. However, as the voids grow in any stress state but purely hydrostatic stress, they will develop non-spherical shapes. Budiansky et al. [5] have shown that the void does not always lengthen fastest in the tensile direction but in certain situations will broaden faster in the transverse direction. In fact, this effect at high triaxality in power law creep gives rise to the negative sign in eqn. (29). The shape changing effect will be substantially similar in almost all voids in a macroscopic sample of material. This must give rise to anisotropy in the expressions for $\mathbf{A}$ and a consequent anisotropy in the continuum descriptions for the material. Anisotropy could also be present because the inclusions from which the voids nucleated were not initially equiaxed, say due to rolling. It remains to be seen how significant this effect will be at substantial void volume fractions. However, it is obvious that at very small volume fractions, the effect can only enter at order f and so may not be very significant. This appears to be the case in some reverse torsion experiments in creep of Hayhurst et al. [22]. After torsion in one direction, substantial anisotropic creep damage was apparent and it had accelerated the creep rate beyond the steady state level. Reverse torsion was applied, and the creep rate continued at the same elevated level rather than showing any substantial change. The final rupture, however, was controlled by the anisotropic part of the damage. It will be interesting to see whether micromechanics models like the one initiated in this paper will predict a similar lack of a strong influence of the orientation of the creep voids on the strain rate. Certainly one would imagine that crack like voids formed by the complete cavitation of a grain boundary would have a substantial anisotropic influence.

The concept of creep damage has proved to be of considerable utility in explaining the creep rupture of structures [23]. Although of most value when used as a comparative tool for predicting what fraction of the creep life has elapsed, damage can also be thought of as an indication of the net section area which has been lost to creep voids. The true matrix stress is elevated by a factor of $1/(1-d)$ where d is the damage. In uniaxial stress σ, the creep deformation rate C would be such that

$$C = \frac{2}{3}B\left(\frac{\sigma}{1-d}\right)^n = \frac{2}{3}B(1+nd)\sigma^n \tag{31}$$

to first order in d where d is small. This can be compared with the result for C derived from eqn. (30)

$$C = \frac{2}{3}B\left[1+f\left(n-1-\frac{1}{4}\left(\frac{1}{2n}+\frac{n-1}{n}\right)^n\right)\right]\sigma^n \tag{32}$$

Strictly, eqn. (30) was derived for stress states with $\sigma_{kk}/\bar{\sigma} \gg 1$ but we shall use eqn. (32) anyway. Equation (32) predicts a somewhat lower creep rate than (31) when d and f are considered to be the same. Perhaps (31) would be better written as

$$C = \frac{2B\sigma^n}{3(1-d)^{n-1}} = \frac{2}{3}B(1+(n-1)d)\sigma^n \tag{33}$$

in recognition of the fact that C represents the creep rate of a fraction $(1-d)$ of the macroscopic volume. The difference between eqn. (33) and eqn. (32) represents the effect of void growth on the creep rate. That eqn. (32) underestimates eqn. (33) is due to the greater transverse growth of the voids modelled by eqn. (30) at high triaxial stress. This arises because the inappropriate high triaxiality result was used and the behavior is wrong for the purely uniaxial stress in that voids stretch out in the tensile direction in that case according to Budiansky et al. [5]. However, the comparison is interesting and is only meant to indicate a direction of research for making damage a more precise notion.

Acknowledgement

This work was carried out while the author was supported by the Office of Naval Research through Contract NR 064-N00014-81-K-0650 with the University of Illinois. Discussions with Professor F.A. Leckie have been helpful in the development of this work.

References

1 D.C. Drucker, The continuum theory of plasticity on the macroscale and the microscale, 1966 Marburg Lecture, J. Materials, ASTM, 1 (1966) 873—910; and Yielding, flow and failure, in: C.T. Herakovich (Ed.), Inelastic Behavior of Composite Materials, AMD-Vol. 13, ASME, New York, 1975, pp. 1—15.
2 F.A. McClintock, A criterion for ductile fracture by the growth of holes, J. Appl. Mech., 35 (1968) 363—371.
3 J.R. Rice and D.M. Tracey, On the ductile enlargement of voids in triaxial stress fields, J. Mech. Phys. Solids., 17 (1969) 201—217.
4 D.M. Tracey, Strain-hardening and interaction effects on the growth of voids in ductile fracture, Engrg. Fracture Mech., 3 (1971) 301—315.
5 B. Budiansky, J.W. Hutchinson and S. Slutsky, Void growth and collapse in viscous solids, in: H.G. Hopkins and M.J. Sewell (Eds.), Mechanics of Solids, The Rodney Hill 60th Anniversary Volume, Pergamon, Oxford 1982, pp. 13—45.
6 A.L. Gurson, Continuum theory of ductile rupture by void nucleation and growth, J. Engrg. Materials and Tech., 98 (1976) 2—15.
7 A.L. Gurson, Porous rigid-plastic materials containing rigid inclusions — yield function, plastic potential and void nucleation, in: D.M.R. Taplin (Ed.), Fracture 1977 (Proc. 4th Int. Conf. Fracture), Vol. 2, Pergamon, Oxford, 1977, p. 357.
8 H. Yamamoto, Conditions for shear localization in ductile fracture of void-containing materials, Int. J. Fracture, 14 (1978) 247—265.
9 G.H. Edward and M.F. Ashby, Intergranular fracture during power law creep, Acta Metallurgica, 27 (1979) 1505—1518.
10 A. Needleman and J.R. Rice, Plastic creep flow effects in the diffusive cavitation of grain boundaries, Acta Metallurgica, 28 (1980) 1315—1332.
11 A.C.F. Cocks and M.F. Ashby, On creep fracture by void growth, Prog. Materials Sci. (1982) 1—56.
12 J.F.W. Bishop and R. Hill, A theory of plastic distortion of a polycrystalline aggregate under combined stress, Phil. Mag., 42 (1951) 414—427.
13 R. Hill, Continuum micro-mechanics of elastoplastic polycrystals, J. Mech. Phys. Solids, 13 (1965) 89—101.
14 J.W. Hutchinson, Elastic-plastic behaviour of polycrystalline metals and composites, Proc. Roy. Soc. Lond., A319 (1970) 247—272.
15 J.D. Eshelby, The determination of the elastic field of an ellipsoidal inclusion and related problems, Proc. Roy. Soc. Lond., A241 (1957) 376—396.
16 R. Hill, The Mathematical Theory of Plasticity, Oxford Univ. Press, Oxford, 1950.
17 J.R. Willis, Elastic theory of composites, in: H.G. Hopkins and M.J. Sewell (Eds.), Mechanics of Solids, The Rodney Hill 60th Anniversary Volume, Pergamon, Oxford, 1982, pp. 653—686.
18 J.W. Rudnicki, The inception of faulting in a rock mass with a weakened zone, J. Geophys. Res., 82 (1977) 844—854.
19 R. Hill, A self-consistent mechanics of composite materials, J. Mech. Phys. Solids, 13 (1965) 213—222.
20 C.A. Berg, Plastic dilation and void interaction, in: M.F. Kanninen et al. (Eds.), Inelastic Behavior of Materials, McGraw-Hill, New York, 1970, pp. 171—209.
21 J.W. Rudnicki and J.R. Rice, Conditions for the localization of deformation in pressure-sensitive dilatant materials, J. Mech. Phys. Solids, 23 (1975) 371—394.
22 D.R. Hayhurst, W.A. Trampczynski and F.A. Leckie, Creep rupture under non-proportional loading, Acta. Met., 28 (1980) 1171—1183.
23 L.M. Kachanov, see Y.N. Rabotnov, Creep Problems in Structural Members, North-Holland, Amsterdam, 1969.

Some Remarks on Rate-Dependent Plasticity

P.M. NAGHDI

Department of Mechanical Engineering, University of California, Berkeley, CA 94720 (U.S.A.)

Summary

First, some motivation is provided for the development of a relatively simple rate-dependent theory of plasticity based on an idealized elastic—viscoplastic constitutive model. This idealization includes unloading and allows for suitable definition of plastic strain. Next, within the scope of a purely mechanical theory, a special constitutive response is used to discuss the nature of constitutive restrictions for a finitely deforming rate-dependent material, obtained from an appropriate work inequality. With the help of these restrictions, certain features of the theory are elaborated upon with particular reference to normality and convexity.

1. Introduction

This paper elaborates on certain aspects of rate-dependent plasticity in the presence of finite deformation and includes, as a special case, the corresponding results for a rate-independent theory. The discussion is carried out entirely within the scope of the purely mechanical theory, leaving aside relevant thermodynamical aspects of the subject.

By way of background, it should be recalled here that in the usual theory of rate-independent plasticity, the rate of plastic strain and the rate of work-hardening are expressed as linear functions of the rate of strain or the rate of stress. The coefficients of response functions in these expressions, as well as the loading functions and other constitutive response functions, are all independent of the rate of strain, the rate of stress and time derivatives of other kinematical ingredients. Indeed, by the very nature of the idealization of rate-independent theory, the time rates of quantities are independent of the time scale used to compute the rate of change so that, for example, the plastic strain rate is homogeneous of degree one in the rate of strain (or the rate of stress). In fact, in the rate-independent theory, the time variable may be interpreted as any parameter which is monotonically increasing with time during deformation.

By contrast, a rate-dependent theory of plasticity is intended to characterize "rate-dependent" behavior by including rate quantities in the constitutive response functions and possibly also in the yield or loading functions.

For definiteness, consider the response of a rate-dependent material (say that of a typical ductile metal) in a one-dimensional test — either in tension or compression — in which the strain may be moderately large or even very large. Let e and s stand, respectively, for the component e_{11} of the Lagrangian strain tensor and the component s_{11} of the symmetric Piola—Kirchhoff stress tensor. Figure 1 shows the familiar plot of stress versus strain of an elastic—viscoplastic material* at various strain rates $\dot{e}_1, \dot{e}_2, \ldots$, where a superposed dot designates differentiation with respect to time t. Also shown in Fig. 1 is the corresponding rate-independent idealization represented by the curve OAB, where in the linear range OA the material is elastic. The point A representing the elastic limit is assumed to be coincident with the initial yield.

Rate-dependent behavior of materials encompasses an important and difficult chapter in the theory of inelastic behavior of materials. Consider the enormous difficulties one encounters in attempting to include all relevant features in the development of a rate-independent theory of plasticity, where all time effects are ignored. Given this, the further difficulties of rate-dependent theory are evident. The terms "rate-dependent" or "rate-sensitive" behavior of materials, frequently used in the literature, refer loosely to various developments intended to reflect a variety of material behavior. For example, elastic—viscoplastic or visco-elastic—plastic theories, or still other rate-dependent theories of material behavior, although difficult to classify, include some dependence on the rate of strain. One particular category which has attracted considerable attention from time to time is one in which the theory attempts to describe material behavior during loading with the initial response prior to yield regarded as elastic, but usually such theories do not deal with the problem of unloading. In such rate-dependent theories an important feature which seems to have been dealt with from various points of view is the prediction of an increase in the linear range of the stress—strain response with increasing strain rate which, in turn, leads to the prediction of higher initial yield stress at higher strain rates. The interpretations of experimental results in this area seem inconclusive and appear to lack any discussion about the nature of unloading and a suitable definition for plastic strain in rate-dependent materials. For general background information on the subject, reference may be made to an article by Perzyna [1] and to a recent experimental paper by Klepaczko and Duffy [2], where a large number of additional references are cited.

*Although the stress—strain responses shown in Fig. 1 are plotted in the s—e plane, the corresponding plots of engineering stress versus engineering strain or true stress versus true strain exhibit similar features.

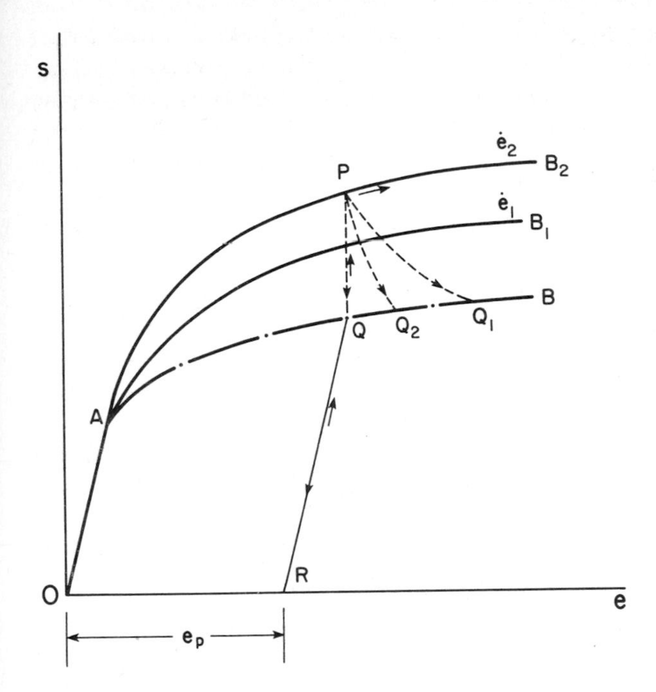

Fig. 1. Idealized mechanical response of an elastic—viscoplastic material at various strain rates. The curves OAB_1 and OAB_2 represent the material response at higher rates of strain associated with a "rate-dependent" model, while the curve OAB represents the "rate-independent" response and may be viewed as representing the material response when the strain rate approaches zero. The dashed line curves PQ_1, PQ_2, . . . , PQ represent unloading with various degrees of acceleration (the slowest being PQ_1) from a point P on curve OAB_2. The instantaneous unloading from P takes place along the rapid path PQ. Also shown are elastic unloading from a point Q and the identification of plastic strain by OR along the e-axis. The values of the total strain and the plastic strain are the same for every point along the dashed curve PQ.

An important aspect of the development of theories of the type under discussion involves obtaining realistic material response — in accordance with experimental observations — by imposing physically plausible restrictions on the constitutive equations. For example, such restrictions may be effected by some appropriate statement (or statements) of the Second Law of Thermodynamics, or by an appeal to certain stability criteria such as Hadamard's stability condition. In this connection, a fruitful idea involving the notion of non-negative work in a closed stress cycle was advanced by Drucker [3] in 1951 and, with the limitation to small deformation, it was expressed in the context of the purely mechanical, rate-independent theory of elastic—plastic materials. This idea, again in the presence of small deformation, was subsequently extended by Drucker [4] to rate-dependent theory and was referred to by him as a "stability postulate". A related postulate was introduced in 1961 by Il'iushin [5] in the context of the linearized theory of plasticity with small strain. The latter again involves a work inequality which is defined over a closed strain cycle and, as noted in [5], is less restrictive than Drucker's postulate [3]. Mention should also be made of related papers by Drucker [6, 7] and by Palmer et al. [8], which contain some discussion pertaining to the normality of plastic strain rate and convexity of the loading surfaces in the presence of small deformation.

More recently, in the context of finite deformation and in a strain space setting, Naghdi and Trapp [9] introduced a physically plausible work assumption and used this primitive assumption to derive a work inequality. The work inequality was then used to obtain constitutive restrictions in the rate-independent theory of elastic—plastic materials [9, 10]. The work inequality derived in [9], upon specialization to small deformation, may be compared with the postulates of Il'iushin and Drucker (see [9], Section 3). Moreover, it is clear from the derivation of the work inequality in [9] that its validity is not limited to rate-independent plasticity but, in fact, is valid for a fairly large class of materials including those which may be rate-dependent.

In the present paper, after some preliminary background information pertaining to both "rate-independent" and "rate-dependent" inelastic behavior of materials, the main features of a *model* for a class of "rate-dependent" materials introduced previously in [11] are recalled in Section 2. The constitutive response of this relatively simple idealized model, which characterizes an elastic—viscoplastic behavior, is capable of describing material response during both loading and unloading and accommodates a suitable definition for plastic strain. Next, using a special constitutive equation for the stress response (see equations (2.17a, b)) and the work inequality derived in [9], constitutive restrictions are derived in Section 3. A part of these results may be regarded as a special case of those obtained in [11] with the use of fairly general constitutive equations. Most of the mathematical details of the developments in Section 3 leading to the inequality (3.6) are placed in Appendix A at the end of the paper. Further, as discussed in Section 4, the restrictions derived in Section 3 bear on the normality of certain expressions involving plastic strain rate and on the convexity of loading surfaces in strain space. Also included in Section 4 is a geometrical interpretation of the one-dimensional version of the main work inequality over a strain cycle utilized in Section 3.

2. Preliminaries and general background

Throughout the paper, we use both the standard *direct* and the Cartesian tensor notations. Often, however, we write the various expressions in component forms and display vector and tensor fields in terms of their rectangular Cartesian components. All subscripts take the values 1, 2, 3, and the usual summation convention is employed over repeated subscripts.

Let the motion of a body be referred to a fixed system of rectangular Cartesian axes and let the position of a typical particle in the present configuration at time t be designated by $\mathbf{x}$ with rectangular Cartesian components x_i, where $x_i = \chi_i(X_A, t)$ and X_A is a reference position of the particle. Further, let E with rectangular Cartesian components e_{KL} be the symmetric Lagrangian strain defined by

$$e_{KL} = \tfrac{1}{2}(F_{iK}F_{iL} - \delta_{KL}), \quad F_{iK} = \partial\chi_i/\partial X_K, \tag{2.1}$$

where F_{iK} are the components of the deformation gradient relative to the reference position and δ_{KL} is the Kronecker symbol. We also recall the relationship between the nonsymmetric Piola–Kirchhoff stress tensor **P** and the symmetric Piola–Kirchhoff stress tensor S, namely

$$p_{iK} = F_{iL}s_{LK}, \quad s_{LK} = s_{KL}, \tag{2.2}$$

where p_{iK} and s_{KL} are, respectively, the rectangular Cartesian components of **P** and S.

For purposes of comparison with later developments, we now summarize the main ingredients of a rate-independent theory of a finitely deforming elastic–plastic solid and base our results on the purely mechanical aspects of the subject contained in the papers of Green and Naghdi [12, 13] and Naghdi and Trapp [14]. Thus, in addition to the strain tensor e_{KL}, we assume the existence of a symmetric second order tensor-valued function $\mathbf{E}^p$, with rectangular Cartesian components $e^p_{KL}(X_A, t)$, called the plastic strain at X_A and t, and a scalar-valued function $\kappa = \kappa(X_A, t)$ called a measure of work-hardening. It is assumed that the stress S is given by the constitutive equation

$$s_{KL} = \hat{s}_{KL}(\mathscr{U}), \quad \mathscr{U} = (e_{MN}, e^p_{MN}, \kappa) \tag{2.3}$$

and that for fixed values of $\mathbf{E}^p$ and κ, $(2.3)_1$ possesses an inverse of the form

$$e_{KL} = \hat{e}_{KL}(\mathscr{V}), \quad \mathscr{V} = (s_{MN}, e^p_{MN}, \kappa). \tag{2.4}$$

We use a strain space formulation of plasticity* and, as in the paper of Casey and Naghdi [16], we also regard the loading criteria of the strain space formulation as primary. The associated loading conditions in stress space can then be derived with the use of constitutive equations for stress. Thus, we admit the existence of a continuously differentiable scalar-valued yield (or loading) function $g(\mathscr{U})$ such that, for fixed values of $\mathbf{E}^p$ and κ, the equation

$$g(\mathscr{U}) = 0 \tag{2.5}$$

---

*Advantages of a strain space formulation of plasticity have been emphasized in [14, 15].

represents a closed orientable hypersurface $\partial\mathscr{E}$ of dimension five enclosing an open region $\mathscr{E}$ of strain space. The function g is chosen so that $g(\mathscr{U})<0$ for all points in the region $\mathscr{E}$. The hypersurface $\partial\mathscr{E}$ is called the yield (or loading) surface in strain space. The constitutive equations for $\mathbf{E}^p$ and $\dot{\kappa}$ are [14]:

$$\dot{e}^p_{KL} = \begin{cases} 0 & \text{if } g < 0, & \text{(a)} \\ 0 & \text{if } g = 0 \text{ and } \hat{g} < 0, & \text{(b)} \\ 0 & \text{if } g = 0 \text{ and } \hat{g} = 0, & \text{(c)} \\ \lambda\rho_{KL}\hat{g} & \text{if } g = 0 \text{ and } \hat{g} > 0, & \text{(d)} \end{cases} \tag{2.6}$$

and

$$\dot{\kappa} = \mathscr{C}_{KL}\dot{e}^p_{KL}, \tag{2.7}$$

where $\mathscr{C}_{KL}$ is a symmetric tensor-valued function of the variables $\mathscr{U}$, a superposed dot denotes material time differentiation,

$$\hat{g} = \frac{\partial g}{\partial e_{MN}}\dot{e}_{MN} \tag{2.8}$$

and where λ and ρ_{KL} are, respectively, a scalar-valued and a symmetric tensor-valued function of $\mathscr{U}$. The conditions involving g and $\hat{g}$ in (2.6) are the loading criteria of the strain space formulation. Using conventional terminology, these four conditions in the order listed correspond to (a) an elastic state (or point in strain space); (b) unloading from an elastic—plastic state, i.e., a point in strain space for which $g=0$; (c) neutral loading from an elastic—plastic state; and (d) loading from an elastic—plastic state. We assume that the coefficient of $\hat{g}$ in (2.6d) is nonzero on the yield surface and, without loss in generality, we then set $\rho_{KL}\neq 0$, $\lambda>0$. The so-called "consistency" condition, namely $\dot{g}=0$, yields the relationship

$$1+\lambda\rho_{KL}\left(\frac{\partial g}{\partial e^p_{KL}}+\frac{\partial g}{\partial\kappa}\mathscr{C}_{KL}\right) = 0 \tag{2.9}$$

at all points on the yield surface through which loading can occur.

For a given loading function $g(\mathscr{U})$, with the aid of $(2.4)_1$, we can obtain a corresponding function $f(\mathscr{V})$ by means of the formula

$$g(\mathscr{U}) = g(e_{KL}(\mathscr{V}), e^p_{KL}, \kappa) = f(\mathscr{V}). \tag{2.10}$$

Because of the assumed smoothness of $(2.4)_1$, for fixed values of $\mathbf{E}^p$ and κ, the equation

$$f(\mathscr{V}) = 0 \tag{2.11}$$

represents a hypersurface $\partial\mathscr{S}$ in stress space having the same geometrical properties as the hypersurface $\partial\mathscr{E}$ in strain space. The region enclosed by $\partial\mathscr{S}$ is denoted by $\mathscr{S}$. It is clear from (2.10) that a point in strain space belongs to the region $\mathscr{E}$ (i.e., $g(\mathscr{U})<0$) if and only if the corresponding point in stress space satisfies $f(\mathscr{V})<0$ and hence belongs to $\mathscr{S}$. With the use of the constitutive equation $(2.4)_1$ loading conditions for stress space can now be derived from the loading criteria of strain space in (2.6) as discussed by Casey and Naghdi [16]. We do not record these conditions as they are not needed for the particular development of the present paper.

We are concerned in the present paper with a discussion of certain features of a relatively simple rate-dependent theory of plasticity. In order to give reasons for the subsequent development, we begin by examining briefly the familiar *idealized* one-dimensional response of test results, especially since many concepts concerning mechanical behavior of materials are extensions of observations made in simple tension or simple compression or simple shear. Both Figs. 1 and 2 depict certain idealized responses of materials. In both figures the curve OAB represents the so-called "static" response corresponding to an idealized rate-dependent model. With reference to the curve OAB, we note that from the point O to the proportional limit A the material is linearly elastic and, since the deformation in this range is reversible, unloading takes place along AO.

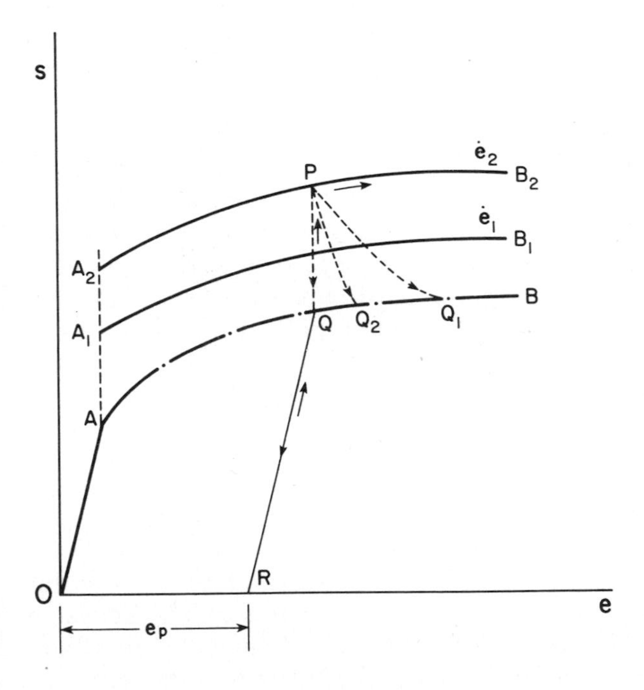

Fig. 2. An idealized one-dimensional response of a "rate-dependent" model with most of its features being similar to those depicted in Fig. 1, but with a different characteristic for the initial response at various rates of strain.

For loading above A (regarded to be coincident with the yield point), the deformation is irreversible and the rate-independent material strain hardens along AQB. Unloading from a point such as Q is assumed to take place elastically along QR, which is generally taken to be parallel to OA. There are other features associated with the rate-independent response such as the lower compressive yield limit, but these need not be discussed here.

It is reasonable to assume that for sufficiently low strain rate, i.e., as the strain rate $\dot{e}$ tends to zero (using the notation of Section 1), the rate-dependent response of the material approaches that of the rate-independent model represented by OAB in Fig. 1. At higher rates of strain, the rate-dependent response of the material would be represented by OAB_1 when the strain rate is $\dot{e}_1$ and by OAB_2 when the strain rate is $\dot{e}_2(>\dot{e}_1)$ and so on. Again, with reference to Fig. 1, consider a point P on the response curve $OAPB_2$ and through P draw a dashed line parallel to the s-axis intersecting the curve OAB at Q. Let e and e_p denote, respectively, the total strain and the plastic strain at Q on the rate-independent response curve OAB. Then, as indicated in Fig. 1, the segment OR represents the plastic strain at Q. In general, deceleration from a point P on the curve OAB_2 would proceed along the dashed curves such as PQ_1, PQ_2 to the right of PQ and such processes would be represented by dashed curves which end on the curve OAB representing the rate-independent response. In fact, as the process of deceleration takes place at a faster rate, in the limit one would approach the dashed line PQ. We may refer to PQ as a *rapid path* using an earlier terminology introduced in [17]. Clearly, every point along the line PQ experiences the same total strain e and the same amount of plastic strain e_p and this notion of rapid path enables one to define plastic strain for rate-dependent materials. Again, with reference to Fig. 1, reloading from a point along RQ at a given rate of strain (say for example at the rate $\dot{e}_2$) would proceed along RQ until one reaches Q and then, consistent with the idealization indicated, there has to be an allowance for jump in the value of the stress in order to reach the point P. Thereafter, as indicated in Fig. 1, further plastic deformation at the rate $\dot{e}_2$ will continue along PB_2.

Experimental results of a fairly large class of rate-dependent materials often exhibit the phenomenon that at higher rates of strain the value of yield stress, sometimes referred to as "dynamic yield" stress, occurs at a value that is higher than the corresponding "static yield" of the rate-independent theory. Such observations may be accommodated easily by a model whose one-dimensional response is shown in Fig. 2, where at higher rates of strain the value of the yield stress jumps to a higher value along the vertical of the line AA_1A_2. Remarks concerning unloading from a point P at a higher rate of strain, say $\dot{e}_2$, discussed with reference to Fig. 1, are also applicable to unloading from points such as P in Fig. 2. Again, every point along PQ experiences the same amount of total strain and the same amount of plastic strain.

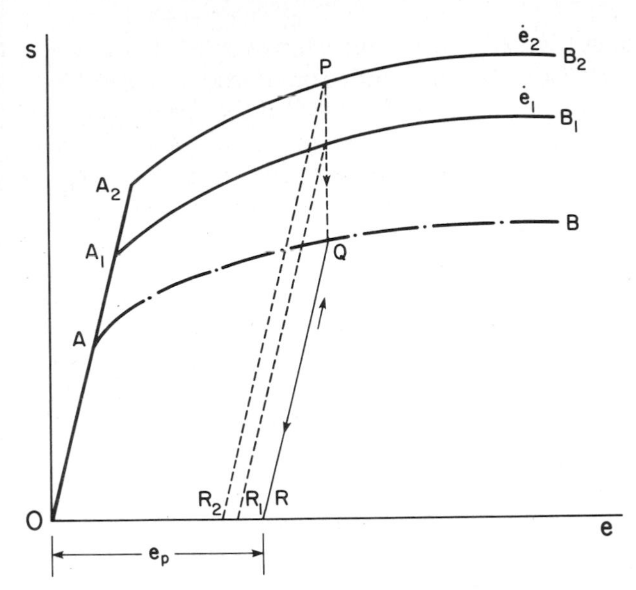

Fig. 3. An idealized one-dimensional response of a "rate-dependent" model displaying an increase in the linear range of the stress—strain response. Corresponding to different constant strain rates, the model exhibits "dynamic yield stress" at points such as A_1 and A_2. Also shown is the curve OAQB representing the "rate-independent" response.

The response in one dimension of the "rate-dependent" material behavior of the type discussed in the preceding two paragraphs is once more depicted in Fig. 3, where again the curve AOB represents the "rate-independent" idealization. This model displays an increase in the linear range of the stress—strain response and, corresponding to different constant strain rates $\dot{e}_1, \dot{e}_2$, admits the so-called "dynamic yield stress" at points such as A_1,A_2. It is important to observe that in contrast to the model associated with Fig. 1, the model associated with Fig. 3 also implies accumulation of plastic strain beyond the "static yield" stress above point A. However, this does not appear to have been reported experimentally and may even be difficult to observe. An unloading line from a point P parallel to OA leads to an unacceptable definition of plastic strain represented by the segment OR_2: compare OR_2 to the segment OR of Fig. 1, which for easy comparison is also reproduced in Fig. 3. A model corresponding to the material response indicated in Fig. 3 has been used during loading to predict higher "dynamic yield" stresses at higher strain rates by different procedures. In particular, this feature of the "rate-dependent" behavior has been discussed by Malvern [18] and by Wilkins and Guinan [19] using a rate-independent yield function. More recently, Rubin [20] has proposed a different approach for predicting "dynamic yield" stress such as A_1,A_2 in Fig. 3 with the use of a rate-dependent yield function along with a rate-independent constitutive equation for stress.

We now discuss constitutive equations for an elastic—viscoplastic material, which reflect properties for one-dimensional response discussed with reference to Fig. 1. Again, we adopt the strain space formulation of plasticity as primary, and admit the existence of a yield or loading function of the form (2.5). Keeping in mind the loading criteria of the strain space and the associated terminology in (2.6), we introduce constitutive equations for plastic strain and rate of work-hardening by the assumptions

$$\dot{e}^p_{KL} = \begin{cases} 0 & \text{, during unloading or neutral loading,} \quad (a) \\ \mathscr{M}_{KLMN}\dot{e}_{MN}, & \text{during loading,} \quad (b) \end{cases} \tag{2.12}$$

and an expression of the form (2.7) which, in view of $(2.12)_2$, may also be written as

$$\dot{\bar{\kappa}} = \mathscr{C}_{KL}\dot{e}_{KL}. \tag{2.13}$$

In (2.12) and (2.13), $\mathscr{M}_{KLMN}$ and $\mathscr{C}_{KL}$ are, respectively, the rectangular Cartesian components of a fourth-order tensor and symmetric second-order tensor functions $\mathscr{M}$ and $\mathscr{C}$ of the variables* $(2.3)_2$. As in the development of the rate-independent theory (see Green and Naghdi [12]), it can be shown that the response coefficient in (2.13), can be expressed in the form

$$\mathscr{M}_{KLMN} = \lambda\rho_{KL}\frac{\partial g}{\partial e_{MN}}, \tag{2.14}$$

or that, equivalently, $\dot{e}^p_{KL}$ is again given by (2.6d) during loading. Also, the consistency condition is again of the same form as (2.9).

Next, it is convenient to introduce the constitutive equation for the stress response in the form

$$\begin{cases} s_{KL} = {}_1s_{KL} + {}_2s_{KLMN}\dot{e}^p_{MN}, \text{ during loading when } g = 0, \hat{g} > 0, \quad (a) \\ s_{KL} = {}_1s_{KL}, \text{ during unloading and neutral loading when} \\ \qquad\qquad g = 0, \hat{g} \leqslant 0. \quad (b) \end{cases} \tag{2.15}$$

In (2.15), the second-order tensor ${}_1s_{KL}$ and the fourth-order tensor ${}_2s_{KLMN}$ satisfy obvious symmetries and each is regarded as depending on the variables** $\mathscr{U}$, i.e.,

$$s_1 = {}_1\hat{s}_{KL}(\mathscr{U}), \quad {}_2s_{KLMN} = {}_2\hat{s}_{KLMN}(\mathscr{U}). \tag{2.16}$$

The stress response characterized by (2.15) and (2.16) is linear in the

*The arguments of the response functions $\mathscr{M}$ and $\mathscr{C}$ could also include the strain rate $\dot{e}_{KL}$ but the forms of the response coefficient functions in $(2.12)_2$ and (2.13) suffice for our present discussion.

**Again the strain rate $\dot{e}_{KL}$ could be included in the argument of ${}_2s_{KLMN}$ but this is unnecessary in the present discussion.

strain rate through the constitutive equation for $\dot{e}^p_{MN}$ during loading. During unloading and neutral loading, the stress response has the same form as that of the rate-independent theory; hence, without loss in generality, we may identify ${}_1\hat{s}_{KL}$ in $(2.16)_1$ with $\hat{s}_{KL}$ of $(2.3)_1$. Also, the second part of the stress response during loading, namely ${}_2\hat{s}_{KLMN}\dot{e}^p_{MN}$, represents the jump in s_{KL} from a point such as Q (see Fig. 1) on the loading function and will assume different values depending on the rate of strain $\dot{e}_{MN}$.

In the next Section we restrict attention to a special form of the stress response (2.15) given by

$$\left\{\begin{array}{lll} s_{KL} = {}_1L_{KLMN}(e_{MN} - e^p_{MN}) + {}_2L_{KLMN}\dot{e}^p_{MN}, \text{ when } g = 0, \hat{g} > 0, & & \text{(a)} \\ s_{KL} = {}_1L_{KLMN}(e_{MN} - e^p_{MN}), \text{ when } g = 0, \hat{g} \leqslant 0, & & \text{(b)} \end{array}\right. \quad (2.17)$$

where ${}_1L_{KLMN}$ and ${}_2L_{KLMN}$ are the rectangular Cartesian components of constant fourth-order tensors ${}_1\mathbf{L}$ and ${}_2\mathbf{L}$, respectively.

3. Restrictions on constitutive equations for plastic strain rate and the stress response (2.17)

We first recall in this Section the primitive-work assumption introduced by Naghdi and Trapp [9] leading to the work inequality (3.2) and then we discuss the nature of the restrictions which can be placed on the constitutive equations for the plastic strain rate and the special response (2.17).

Consider a closed cycle of a spatially homogeneous motion in the closed time interval* $[t_1, t_2]$, $(t_1 < t_2)$. The cycle is said to be *smooth* if the time derivatives of displacement, strain and associated kinematical quantities are continuous in $[t_1, t_2]$; we assume the same values for each material point at times t_1 and t_2. We designate such a smooth spatially homogeneous closed cycle of deformation by $\mathscr{C}(t_1, t_2)$ and recall from [9] the following work assumption: the external work done on the body by surface tractions and by body forces in any smooth spatially homogeneous closed cycle is non-negative, i.e.,

$$\int_{t_1}^{t_2} \left[\int_{\partial \mathscr{R}_0} p_{iK} N_K v_i dA + \int_{\mathscr{R}_0} \rho_0 b_i v_i dV \right] dt \geqslant 0, \quad (3.1)$$

for all cycles $\mathscr{C}(t_1, t_2)$. In (3.1), $\mathscr{R}_0$ is the region of space occupied by the body in its reference configuration, $\partial \mathscr{R}_0$ is the closed boundary surface of

*Recall that a homogeneous motion is one whose deformation gradient is independent of the material coordinates so that, in a spatially homogeneous motion, the strain tensor $\mathbf{E}$ is a function of time only. For a closed spatially homogeneous cycle in the closed time interval $[t_1, t_2]$, the displacement $\mathbf{x}$ and the strain $\mathbf{E}$ assume the same values at times t_1 and t_2.

$\mathscr{R}_0$, ρ_0 is the mass density in the reference configuration, v_i are the components of the velocity, b_i are the components of the body force per unit mass, N_K are the components of the outward unit normal to $\partial\mathscr{R}_0$, $p_{iK}N_K$ represent the components of the stress vector measured per unit area in the reference configuration, and dA and dV refer to elements of area and volume in the reference configuration.

It is shown in [9] that for any cycle $\mathscr{C}(t_1, t_2)$ the assumption (3.1) leads to*

$$\int_{t_1}^{t_2} s_{KL}\dot{e}_{KL}\mathrm{d}t \geqslant 0. \tag{3.2}$$

As already noted in Section 1, the above work inequality is valid for rate-dependent materials even though originally it was used in [9] for rate-independent elastic—plastic solids.

Let $\mathbf{E}^0$ with rectangular Cartesian components e^0_{KL} refer to an existing state of strain inside a loading surface in the six-dimensional strain space such that $g(e^0_{KL}, e^p_{KL}, \kappa) < 0$ at time t_1. Let $\bar{t}$ designate the first occurrence of plastic strain at which time $g(e^y_{KL}, e^p_{KL}, \kappa) = 0$, where e^y_{KL} denotes the value of the strain at a point on the loading surface. We note that it is always possible to find a path inside $g(\mathbf{E}^Y, \mathbf{E}^p, \kappa) = 0$ from $\mathbf{E}^0$ to $\mathbf{E}^Y$ such that the strain rate is any desired value at** $\mathbf{E}^Y$. Consider now a closed spatially homogeneous strain cycle starting and ending at $\mathbf{E}^0$. From $\mathbf{E}^0$ to $\mathbf{E}^Y$ the loading takes place elastically during the interval $t_1 \leqslant t < \bar{t}$ and continues in the viscoplastic range in the interval $\bar{t} \leqslant t < \bar{t} + (1/n)$ at a constant strain rate $\mathbf{M}(= \dot{\mathbf{E}})$. At the end of this interval (corresponding to a point such as P in Fig. 1) the strain has a value $\mathbf{E}^Y + (1/n)\mathbf{M}$. This is followed by a constant deceleration process to a state of zero strain rate (corresponding to points such as Q_1, Q_2 on the rate-independent response curve in Fig. 1). Such deceleration processes are assumed to occur at a decreasing rate of strain specified by the constant rate $-l\mathbf{M}(= \ddot{\mathbf{E}})$ during the time interval $\bar{t} + (1/n) \leqslant t < \bar{t} + (1/n) + (1/l)$ while loading continues. At the end of this time interval, the strain rate vanishes and the strain has the value*** $\mathbf{E}^Y + [(1/n) + (1/2l)]\mathbf{M}$. A sequence of such decelerated processes (corresponding to paths PQ_1, PQ_2, . . . , in Fig. 1), in the limit as $l \to \infty$, results in a rapid path (corresponding to PQ in Fig. 1) to a state of zero strain rate with the value of strain being $\mathbf{E}^Y + (1/n)\mathbf{M}$ (corresponding to the value of strain at Q in Fig. 1). Thereafter, unloading takes place

*An inequality similar in form to (3.2), but with limitation to infinitesimal deformation, is the starting point of Il'iushin's discussion in [5] and is referred to by him as the "postulate of plasticity".

**Recall that if the path in strain space is parametrized with respect to time t, then the strain rate E is directed along the tangent to the path. If $\dot{\mathbf{E}} = $ const., then the path is a straight line.

***Note that if $\mathbf{E}^Y + [(1/n) + (1/2l)]\mathbf{M}$ is close to $\mathbf{E}^Y$, then the strain $\mathbf{E}^0$ will remain inside the yield surface since the loading function is continuous in E.

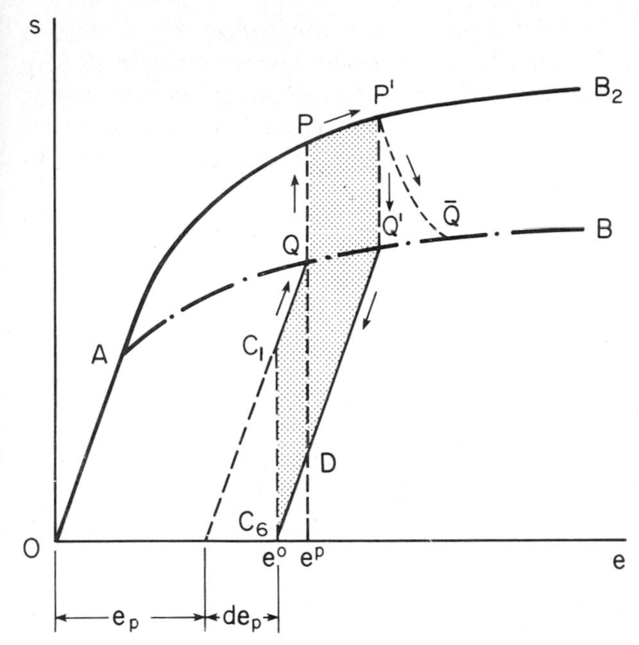

Fig. 4. One-dimensional version of a strain cycle $C_1QPP'Q'C_6$ corresponding to the cycle of homogeneous strain defined by (3.3) and used in the calculation of the work inequality (3.5). Also, the three integrals in (4.8) represent the (cross-hatched) areas enclosed by $C_1QPP'Q'C_6$.

during the time interval $\bar{t} + (1/n) + (1/l) \leqslant t < t_2$ with the strain returning to $\mathbf{E}(t) = \mathbf{E}^0$. The strain cycle just described may be summarized as follows:

$$
\mathbf{E}(t) = \begin{cases}
\mathbf{E}^0, & t < t_1, \quad \text{(a)} \\
\mathbf{E}^0 \to \mathbf{E}^Y, \quad \dot{\mathbf{E}} \to \mathbf{M} \text{ at } \mathbf{E}^Y, & t_1 \leqslant t < \bar{t}, \quad \text{(b)} \\
\mathbf{E}^Y + \mathbf{M}(t - \bar{t}), & \bar{t} \leqslant t < \bar{t} + \dfrac{1}{n}, \quad \text{(c)} \\
\mathbf{E}^Y + \left[\dfrac{1}{n} + \left(t - \bar{t} - \dfrac{1}{n}\right) - \dfrac{l}{2}\left(t - \bar{t} - \dfrac{1}{n}\right)^2\right] \mathbf{M}, & \\
\qquad \bar{t} + \dfrac{1}{n} \leqslant t < \bar{t} + \dfrac{1}{n} + \dfrac{1}{l}, & \text{(d)} \\
\mathbf{E}^Y \to \mathbf{E}^0, \ \dot{\mathbf{E}} = \mathbf{M} \text{ at } \mathbf{E}^Y, \ \bar{t} + \dfrac{1}{n} + \dfrac{1}{l} \leqslant t < t_2, & \text{(e)} \\
\mathbf{E}^0, \quad t_2 \leqslant t. & \text{(f)}
\end{cases}
\qquad (3.3)
$$

The one-dimensional version of the above strain cycle is indicated in Fig. 4

by the cycle $C_1QPP'Q'C_6$. The point C_1 at the strain e^0 in Fig. 4 corresponds to (3.3a); the elastic loading path C_1Q corresponds to (3.3b); after the jump in stress at Q, the path PP′ at a constant strain rate **M** corresponds to (3.3c); the path $P'\overline{Q}$, which in the limit of rapid deceleration becomes P′Q′, corresponds to (3.3d); the elastic unloading path $Q'C_6$ corresponds to (3.3e); and finally the point C_6 at the strain e^0 corresponds to (3.3f).

With the use of the special stress response (2.17), application of the work inequality (3.2) to the strain cycle (3.3) leads to

$$\int_{t_1}^{t_2} {}_1L_{KLMN}(e_{MN}-e^p_{MN})\dot{e}_{KL}\mathrm{d}t + \int_{\bar{t}}^{\bar{t}+(1/n+1/l)} {}_2L_{KLMN}\dot{e}^p_{MN}\dot{e}_{KL}\mathrm{d}t \geqslant 0, \tag{3.4}$$

where use has been made of the fact that $\dot{\mathbf{E}}^p$ is nonzero only during the time interval $[\bar{t}, \bar{t}+(1/n)+(1/l)]$. The strain cycle (3.3) begins and ends at the constant strain E^0. By writing $\dot{\mathbf{E}}$ as $(\overline{\mathbf{E}-\mathbf{E}^0})^{\cdot}$ and using the fact that $E(t)=E^0$ at t_1 and t_2, the first integral on the left-hand side of the above inequality can be integrated by parts and (3.4) reduces to

$$\int_{\bar{t}}^{\bar{t}+(1/n+1/l)} {}_1L_{KLMN}(e_{KL}-e^0_{KL})\dot{e}^p_{MN}\mathrm{d}t + \\ + \int_{\bar{t}}^{\bar{t}+(1/n+1/l)} {}_2L_{KLMN}\dot{e}_{KL}\dot{e}^p_{MN}\mathrm{d}t \geqslant 0. \tag{3.5}$$

After estimating the various integrals on the left-hand side of (3.5) by using the Taylor series expansion of the integrals about $t=\bar{t}+(1/n)$ and $t=\bar{t}$ and allowing in the limit $l\to\infty$, i.e., the "slowing-down period", to take place very fast and approach a rapid path (corresponding to P′Q′ in Fig. 4), the work inequality (3.5) yields (for details see Appendix A)

$$(\mathbf{E}^Y-\mathbf{E}^0)\cdot{}_1\mathbf{L}\mathscr{M}^Y\mathbf{M}+\mathbf{M}\cdot{}_2\mathbf{L}\mathscr{M}^Y\mathbf{M}\geqslant 0, \tag{3.6}$$

which in component form reads

$$(e^y_{KL}-e^0_{KL})_1L_{KLMN}\mathscr{M}^y_{MNRS}M_{RS}+M_{KL}\,{}_2L_{KLMN}\mathscr{M}^y_{MNRS}M_{RS}\geqslant 0, \tag{3.6a}$$

where for the inner product of any two tensors **A**, **B** we have used the notation $\mathbf{A}\cdot\mathbf{B}=\operatorname{trace}\mathbf{A}^T\mathbf{B}$, $\mathscr{M}^Y$ is the value at $\mathbf{E}^Y$ of the fourth-order tensor $\mathscr{M}$ whose components $\mathscr{M}_{MNRS}$ are given by (2.14) and M_{RS} are the components of the strain rate **M** which occurs in the strain cycle (3.3). The inequality (3.6) or (3.6a) is a necessary condition for the validity of (3.5) and hence also (3.2).

Now let $\mathbf{E}^0$ approach $\mathbf{E}^Y$ while maintaining the strain rate fixed at $\bar{t}$. It then follows from (3.6a) that

$${}_2L_{KLMN}\dot{e}_{KL}\dot{e}^p_{MN}\geqslant 0. \tag{3.7}$$

Since (3.6) must hold for all **M** and the coefficients of **M** in (3.6) are independent of rates, we may also deduce that

$$ {}_1L_{KLMN}(e^y_{KL} - e^0_{KL})\dot{e}^p_{MN} \geqslant 0. \tag{3.8} $$

The two inequalities (3.7) and (3.8) are both necessary and sufficient conditions for the validity of the coupled inequality (3.6). Further, by a special but arbitrary choice of $\boldsymbol{E}^0$ such that the path $\boldsymbol{E}^0$ to $\boldsymbol{E}^Y$ is traversed at a constant strain rate, (3.8) can be reduced to

$$ {}_1L_{KLMN}\dot{e}_{KL}\dot{e}^p_{MN} \geqslant 0. \tag{3.9} $$

The inequalities (3.7)–(3.9) are restrictions on the constitutive equations of a finitely deformed elastic–viscoplastic material with the special stress response (2.17). These restrictions hold in *all* motions, even though they have been deduced from consideration of homogeneous motions alone. We note that the results (3.7)–(3.9) include as a special case those relevant to the rate-independent theory of plasticity. In fact, (3.8) and (3.9) are identical to those derived previously by Naghdi and Trapp [10, equations (30) and (31)].

4. Further implications of the inequalities obtained in Section 3

We discuss now certain features of the restrictions (3.7)–(3.9). As shown in [11], it follows from (3.9) that

$$ \lambda\, {}_1L_{KLMN}\rho_{MN} = \gamma_1 \frac{\partial g}{\partial e_{KL}}, \qquad (\gamma_1 > 0), \tag{4.1} $$

where $\gamma_1 = \gamma_1(\mathscr{U})$ is a positive scalar independent of rates. Similarly, from (3.7), it can be deduced that

$$ \lambda\, {}_2L_{KLMN}\,\rho_{MN} = \gamma_2 \frac{\partial g}{\partial e_{KL}}, \qquad (\gamma_2 > 0), \tag{4.2} $$

where $\gamma_2 = \gamma_2(\mathscr{U})$ is a positive scalar independent of rates. The left-hand sides of both (4.1) and (4.2) involve rate of plastic strain (see equation (2.6a)). Moreover, these quantities (on the left-hand sides of (4.1) and (4.2)) are directed along the normal to the yield (or loading) surface in strain space. Next, by (3.8) and the fact that $\gamma_1 > 0$, we have

$$ (e^y_{KL} - e^0_{KL}) \frac{\partial g}{\partial e_{KL}} \geqslant 0 \tag{4.3} $$

and this implies convexity of the yield (or loading) surface in strain space.

We now examine the implication of the restrictions (4.1) and (4.2), which are obtained from (3.7) and (3.9), on the constitutive equation (2.17a). Since γ_1 and γ_2 are non-zero, by (4.1) and (4.2) we have

$$ \frac{\lambda}{\gamma_1}\, {}_1L_{KLMN}\,\rho_{MN} = \frac{\lambda}{\gamma_2}\, {}_2L_{KLMN}\,\rho_{MN}. \tag{4.4} $$

Now a part of the stress response in (2.17a) which involves ${}_2L$, with the help of (2.12b), (2.14) and (4.4), can be written as

$$\begin{aligned} {}_2L_{\mathrm{KLMN}}\dot{e}^p_{\mathrm{MN}} &= {}_2L_{\mathrm{KLMN}}\,\lambda\rho_{\mathrm{MN}}\hat{g} \\ &= \frac{\gamma_2}{\gamma_1}\,{}_1L_{\mathrm{KLMN}}\,\lambda\rho_{\mathrm{MN}}\hat{g} \\ &= \gamma\,{}_1L_{\mathrm{KLMN}}\dot{e}^p_{\mathrm{MN}}, \end{aligned} \tag{4.5}$$

where we have set

$$\gamma = \frac{\gamma_2}{\gamma_1}. \tag{4.6}$$

The conclusion $(4.5)_3$ enables us to rewrite the stress response (2.17a) in the form

$$s_{\mathrm{KL}} = {}_1L_{\mathrm{KLMN}}\left[(e_{\mathrm{MN}} - e^p_{\mathrm{MN}}) + \gamma\,\lambda\rho_{\mathrm{KL}}\hat{g}\right] \tag{4.7}$$

and this has the advantage of containing only the constitutive coefficient ${}_1\mathrm{L}$ and the scalar $\gamma = \gamma(\mathscr{U})$.

The inequality (3.5) when appropriately specialized to the strain cycle corresponding to $\mathrm{C_1QPP'Q'C_6}$ in Fig. 4 can be expressed as

$$\int_{t_1}^{t_2} (e^y_{\mathrm{KL}} - e^0_{\mathrm{KL}})_1 L_{\mathrm{KLMN}}\dot{e}^p_{\mathrm{MN}}\mathrm{d}t + \int_{t_1}^{t_2} (e_{\mathrm{KL}} - e^y_{\mathrm{KL}})_1 L_{\mathrm{KLMN}}\dot{e}^p_{\mathrm{MN}}\mathrm{d}t + \int_{t_1}^{t_2} ({}_2L_{\mathrm{KLMN}}\dot{e}^p_{\mathrm{MN}})\dot{e}_{\mathrm{KL}}\mathrm{d}t \geqslant 0, \tag{4.8}$$

and in one dimension (using the notations of Sections 1 and 2) has the form

$$\int_{\mathscr{C}(e)} (e^y - e^0)_1 L\mathrm{d}e_p + \int_{\mathscr{C}(e)} (e - e^Y)_1 L\mathrm{d}e_p + \int_{\mathscr{C}(e)} ({}_2L\dot{e}^p)\mathrm{d}e \geqslant 0. \tag{4.9}$$

In (4.9), the relevant material coefficients have, for convenience, been replaced by the constants ${}_1L$ and ${}_2L$, the quantity $({}_2L\,\dot{e}^p)$ in the integrand of the third integral represents the jump in stress from Q to P and $\mathscr{C}(e)$ refers to a closed strain cycle such as $\mathrm{C_1QPP'Q'C_6}$ in Fig. 4. The shaded areas enclosed by $\mathrm{C_1QPP'C_6}$ consist of: (i) the area of the parallelogram $\mathrm{C_6C_1QD}$, (ii) the area of the triangle $\mathrm{QQ'D}$ and (iii) the area enclosed by $\mathrm{PP'QQ'}$. It is easily seen that these areas represent the three integrals (from left to right) on the left-hand side of (4.9).

Acknowledgement

The results reported here were obtained in the course of research supported by the U.S. Office of Naval Research under Contract N00014-75-C-0148, Project NR 064-436 with the University of California, Berkeley.

References

1 P. Perzyna, in: C.S. Yih (Ed.), Advances in Applied Mechanics, Vol. 9, Academic Press, New York, 1966, pp. 243—377.
2 J. Klepaczko and J. Duffy, in: Mechanical Testing for Deformation Model Development, Special Technical Publication 765, Amer. Soc. Testing and Materials, Philadelphia, 1982, pp. 251—268.
3 D.C. Drucker, in: E. Sternberg (Ed.), Proc. First U.S. National Congr. Appl. Mech., Chicago, June 1951, Amer. Soc. Mechanical Engineers, New York, 1952, pp. 487—491.
4 D.C. Drucker, J. Appl. Mech., 26 (1959) 101—106.
5 A.A. Il'iushin, J. Appl. Math. Mech. [Transl. of PMM], 25 (1961) 746—752.
6 D.C. Drucker, in: N.J. Hoff and J.N. Goodier (Eds.), Proc. First Symp. Naval Structural Mechanics, Stanford, 1958, Pergamon, 1960, pp. 407—455.
7 D.C. Drucker, J. Mécanique, 3 (1964) 235—249.
8 A.C. Palmer, G. Maier and D.C. Drucker, J. Appl. Mech., 34 (1967) 464—470.
9 P.M. Naghdi and J.A. Trapp, Quart. J. Mech. and Appl. Math., 28 (1975) 25—46.
10 P.M. Naghdi and J.A. Trapp, J. Appl. Mech., 42 (1975) 61—66.
11 P.M. Naghdi, Constitutive Restrictions for Idealized Elastic-Viscoplastic Materials, UCB/AM-83-4, University of California, Berkeley, 1983. To appear in J. Appl. Mech.
12 A.E. Green and P.M. Naghdi, Arch. Rational Mech. Anal., 18 (1965) 251—281.
13 A.E. Green and P.M. Naghdi, in: H. Parkus and L.I. Sedov (Eds.), Proc. IUTAM Symp. on Irreversible Aspects of Continuum Mechanics and Transfer of Physical Characteristics in Moving Fluids, Springer-Verlag, 1966, pp. 117—131.
14 P.M. Naghdi and J.A. Trapp, Int. J. Engng. Sci., 13 (1975) 785—797.
15 P.M. Naghdi, in: Constitutive Equations in Viscoplasticity: Computational and Engineering Aspects, AMD Vol. 20, Amer. Soc. Mech. Engineers, New York, 1976, pp. 79—93.
16 J. Casey and P.M. Naghdi, J. Appl. Mech., 48 (1981) 285—296.
17 P.M. Naghdi and S.A. Murch, J. Appl. Mech., 30 (1963) 321—328.
18 L.E. Malvern, J. Appl. Mech., 18 (1951) 203—208.
19 M.L. Wilkins and M.W. Guinan, J. Appl. Phys., 44 (1973) 1200—1206.
20 M.B. Rubin, J. Appl. Mech., 49 (1982) 305—311.

Appendix

The purpose of this Appendix is to provide some of the mathematical details used in obtaining the estimates for the integrals in the work inequality (3.5). Most of the calculations are similar to corresponding developments carried out in [11] with the use of more general stress constitutive equations, but there are also some differences in the results obtained here. Previously in [11], the closed strain cycle in $\mathscr{E}$ was so constructed that the path from $\mathbf{E}^0$ to $\mathbf{E}^Y$, as well as the reverse path from $\mathbf{E}^Y$ to $\mathbf{E}^0$, are both traversed at an arbitrary, but at the same constant strain rate. This stipulation was necessary in [11] in order to effect explicit estimates for certain integrals. In the present paper, the path from $\mathbf{E}^0$ to $\mathbf{E}^Y$ and the reverse path are not necessarily traversed at constant strain rates. Nevertheless, explicit estimates for the integrals over these paths are possible because the response coefficient ${}_1\mathbf{L}$ in the special constitutive equation (2.17) is a constant tensor.

Let $G(s)$, defined by

$$G(s) = \int_{s_0}^{s} h(t)\mathrm{d}t, \tag{A1}$$

be continuous and at least twice differentiable in the interval $s_0 \leqslant t \leqslant s$ and, for later convenience, put

$$h(t) = f(t)g(t). \tag{A2}$$

Then, the Taylor series expansion of (A1) about $t = s_0$ is

$$\begin{aligned} G(s) = {} & (s-s_0)f(s_0)g(s_0) + \frac{1}{2!}(s-s_0)^2[f'(s_0)g(s_0) + f(s_0)g'(s_0)] \\ & + \frac{1}{3!}(s-s_0)^3[f''(s_0)g(s_0) + 2f'(s_0)g'(s_0) + f'(s_0)g''(s_0)] \\ & + \mathrm{O}((s-s_0)^4), \end{aligned} \tag{A3}$$

where O is the usual order symbol and prime denotes derivative with respect to t.

In order to indicate the manner in which the integrals in (3.5) can be estimated about $\bar{t}$, consider first the integral

$$I_1 = \int_{\bar{t}}^{\bar{t}+1/n} (\mathbf{E}-\mathbf{E}^0)\cdot{}_1\mathbf{L}\dot{\mathbf{E}}^p\,\mathrm{d}t, \tag{A4}$$

where the integrand of (A4) also occurs in the first integral in (3.5) and the notation for the inner product of any two tensors is defined following (3.6a). After substituting the relevant value of $\mathbf{E}$ from (3.3) and applying the Taylor expansion (A3) of the integral (A1) to (A4) with $f(t) = [\mathbf{E}^Y - \mathbf{E}^0 + \mathbf{M}(t-\bar{t})]$ and $g(t) = {}_1\mathbf{L}\mathbf{E}^p$, (A4) becomes

$$\begin{aligned} I_1 & = \int_{\bar{t}}^{\bar{t}+1/n} [\mathbf{E}^Y - \mathbf{E}^0 + \mathbf{M}(t-\bar{t})]\cdot{}_1\mathbf{L}\dot{\mathbf{E}}^p\,\mathrm{d}t \\ & = \left[\frac{1}{n}(\mathbf{E}^Y-\mathbf{E}^0) + \frac{1}{2n^2}\mathbf{M}\right]\cdot{}_1\mathbf{L}\dot{\mathbf{E}}^{pY} + \left[\frac{1}{2n^2}(\mathbf{E}^Y-\mathbf{E}^0) + \frac{1}{3n^3}\mathbf{M}\right]\cdot{}_2\mathbf{L}\ddot{\mathbf{E}}^p \end{aligned}$$

$$= \frac{1}{n}(\mathbf{E}^{Y} - \mathbf{E}^{0}) \cdot {}_1\mathbf{L}\mathscr{M}^{Y}\mathbf{M}$$

$$+ \frac{1}{2n^2}[\mathbf{M} \cdot {}_1\mathbf{L}\mathscr{M}^{Y}\mathbf{M} + (\mathbf{E}^{Y} - \mathbf{E}^{0}) \cdot {}_1\mathbf{L}\mathscr{N}^{Y}\mathbf{M}] + \mathrm{O}\left(\frac{1}{n^3}\right), \tag{A5}$$

where in writing $(\mathrm{A5})_3$ use is made of the fourth-order tensor $\mathscr{M}$ defined following $(2.12)_2$, the superscript Y denotes the value of a function on the yield surface at $t = \bar{t}$ and where the fourth-order tensor $\mathscr{N}$ stands for the abbreviation

$$\mathscr{N} = \mathscr{N}(\mathscr{U}, \mathbf{M}) = \frac{\partial \mathscr{M}}{\partial \mathbf{E}} \cdot \mathbf{M} + \frac{\partial \mathscr{M}}{\partial \mathbf{E}^p} \cdot \mathscr{M}\mathbf{M} + \frac{\partial \mathscr{M}}{\partial \kappa} \mathscr{C} \cdot \mathscr{M}\mathbf{M}. \tag{A6}$$

Similarly, consider the integral

$$I_2 = \int_{\bar{t}}^{\bar{t}+1/n} \dot{\mathbf{E}} \cdot {}_2\mathbf{L}\dot{\mathbf{E}}^p\, dt \tag{A7}$$

whose integrand also occurs in the second integral in (3.5). Substituting for the strain rate from (3.3c) and applying the Taylor expansion (A3) of the integral (A1) to (A7), we obtain

$$I_2 = \frac{1}{n}\mathbf{M} \cdot \left[{}_2\mathbf{L}\mathscr{M}^{Y}\mathbf{M} + \frac{1}{2n}\,{}_2\mathbf{L}\mathscr{N}^{Y}\mathbf{M}\right] + \mathrm{O}\left(\frac{1}{n^3}\right). \tag{A8}$$

Now by a procedure similar to that discussed above, we may obtain an estimate for the two integrals in (3.5). Thus, the Taylor expansion of the first integral in (3.5) about $\bar{t} + 1/n$ yields

$$I_3 = \int_{\bar{t}+1/n}^{\bar{t}+1/n+1/l} (\mathbf{E} - \mathbf{E}^{0}) \cdot {}_1\mathbf{L}\dot{\mathbf{E}}^p\, dt$$

$$= \frac{1}{2l}(\mathbf{E}^{Y} - \mathbf{E}^{0}) \cdot {}_1\mathbf{L}(\mathscr{M})_{\bar{t}+1/n}\mathbf{M}$$

$$+ \left(\frac{1}{2nl} + \frac{1}{8l^3}\right)\mathbf{M} \cdot {}_1\mathbf{L}(\mathscr{M})_{\bar{t}+1/n}\mathbf{M}$$

$$- \frac{3}{8l^2}(\mathbf{E}^{Y} - \mathbf{E}^{0}) \cdot {}_1\mathbf{L}(\mathscr{N})_{\bar{t}+1/n}\mathbf{M}$$

$$+ \mathrm{O}\left(\frac{1}{nl^2}, \frac{1}{l^3}\right). \tag{A9}$$

When the right-hand side of $(\mathrm{A9})_2$ is further expanded about $\bar{t}$, we obtain

$$I_3 = \frac{1}{2l}(\mathbf{E}^{\mathrm{Y}} - \mathbf{E}^0)\cdot{}_1\mathbf{L}\mathscr{M}^{\mathrm{Y}}\mathbf{M} + \frac{1}{2nl}\,\mathbf{M}\cdot{}_1\mathbf{L}\mathscr{M}^{\mathrm{Y}}\mathbf{M}$$

$$+ \frac{1}{2nl}(\mathbf{E}^{\mathrm{Y}} - \mathbf{E}^0)\cdot{}_1\mathbf{L}\mathscr{N}^{\mathrm{Y}}\mathbf{M}$$

$$- \frac{3}{8l^2}(\mathbf{E}^{\mathrm{Y}} - \mathbf{E}^0)\cdot{}_1\mathbf{L}\mathscr{N}^{\mathrm{Y}}\mathbf{M}$$

$$+ \mathrm{O}\left(\frac{1}{n^2 l}, \frac{1}{nl^2}, \frac{1}{l^3}\right). \tag{A10}$$

The estimate for the second integral on the left-hand side of (3.5) can be effected similarly and leads to

$$I_4 = \int_{\bar{t}+1/n}^{\bar{t}+1/n+1/l} \dot{\mathbf{E}}\cdot{}_2\mathbf{L}\dot{\mathbf{E}}^p\,dt$$

$$= \frac{1}{3l}\,\mathbf{M}\cdot{}_2\mathbf{L}\mathscr{M}^{\mathrm{Y}}\mathbf{M}$$

$$+ \left(\frac{1}{3nl} + \frac{1}{15l^2}\right)\mathbf{M}\cdot{}_2\mathbf{L}\mathscr{N}^{\mathrm{Y}}\mathbf{M}$$

$$+ \mathrm{O}\left(\frac{1}{n^2 l}, \frac{1}{nl^2}, \frac{1}{l^3}\right). \tag{A11}$$

Substituting the estimates (A5), (A8), (A10) and (A11) into (3.5) we finally obtain

$$\left(\frac{1}{n} + \frac{1}{2l}\right)(\mathbf{E}^{\mathrm{Y}} - \mathbf{E}^0)\cdot{}_1\mathbf{L}\mathscr{M}^{\mathrm{Y}}\mathbf{M}$$

$$+ \left(\frac{1}{2n^2} - \frac{3}{8l^2}\right)(\mathbf{E}^{\mathrm{Y}} - \mathbf{E}^0)\cdot{}_1\mathbf{L}\mathscr{N}^{\mathrm{Y}}\mathbf{M}$$

$$+ \left(\frac{1}{2n^2} + \frac{1}{2nl} - \frac{3}{8l^2}\right)(\mathbf{E}^{\mathrm{Y}} - \mathbf{E}^0)\cdot{}_1\mathbf{L}\mathscr{N}^{\mathrm{Y}}\mathbf{M}$$

$$+ \left(\frac{1}{n} + \frac{1}{3l}\right)\mathbf{M}\cdot{}_2\mathbf{L}\mathscr{M}^{\mathrm{Y}}\mathbf{M} + \left(\frac{1}{2n^2} + \frac{1}{3nl} + \frac{1}{15l^2}\right)\mathbf{M}\cdot{}_2\mathbf{L}\mathscr{N}^{\mathrm{Y}}\mathbf{M}$$

$$+ \mathrm{O}\left(\frac{1}{n^2 l}, \frac{1}{nl^2}, \frac{1}{n^3}, \frac{1}{l^3}\right) \geqslant 0. \tag{A12}$$

The inequality (A12) must hold for arbitrary **M**, for all values of the parameters n, l and for all $\mathbf{E}^0$. As in [11], we consider now a sequence of strain cycles such that n, $\mathbf{M}$, $\mathbf{E}^0$ are fixed while the "slowing down" period takes place faster and faster as l becomes larger and larger. In the limit, the sequence approaches the cycle involving the *rapid path* during deceleration and, as $l \to \infty$, (A12) reduces to

$$\frac{1}{n}(\mathbf{E}^{\mathrm{Y}} - \mathbf{E}^0) \cdot {}_1\mathbf{L}\mathscr{M}^{\mathrm{Y}}\mathbf{M} + \frac{1}{n}\mathbf{M} \cdot {}_2\mathbf{L}\mathscr{M}^{\mathrm{Y}}\mathbf{M} + O\left(\frac{1}{n^2}\right) \geqslant 0. \tag{A13}$$

Next, consider a second sequence of strain cycles each member of which has a similar rapid path of deceleration. For fixed values of $\mathbf{M}$ and $\mathbf{E}^0$ but with progressively larger values of n, we may apply (A13) to this second sequence of cycles. Then, after multiplying (A13) by n and taking the limit as $n \to \infty$, we deduce the inequality (3.6).

Shear Flow of Kinematically Hardening Rigid-Plastic Materials

E.T. ONAT

Department of Mechanical Engineering, Yale University, New Haven, Connecticut 06520 (U.S.A.)

Summary

Several versions of kinematic hardening are considered in the presence of finite deformations. Recent work on shear flow of such materials is reviewed and clarified.

Introduction

The original aim of this paper was to show the importance and the value of geometrical and global thinking in the representation of mechanical behavior of inelastic solids. Since my interest in this subject has its origins in the days when I worked with W. Prager and D.C. Drucker at Brown University, I thought that it would be appropriate to mark the present occasion with a praise of geometry. However the present paper has a narrower scope than originally intended.

Here I consider the shear flow of certain rigid-plastic materials. I hope to summarize and clarify the recent developments in this subject by using geometrical and global means.

Finite deformations of kinematically hardening rigid-plastic materials. Shear flow.

It is well known that strain-hardening metals develop anisotropy during the course of plastic deformation. The importance of (deformation induced) anisotropy in certain technical problems (e.g. metal forming; cyclic loading in nuclear reactors) has led to the development of constitutive equations that attempt to represent, in an explicit and detailed way, the evolution of anisotropy and its effects on mechanical behavior. The earliest of such work is by Prager [1] who also introduced the term "kinematic hardening". In later work of Mroz [3] and others the yield surface or surfaces are allowed to move and change shape in the stress-space. All of the above work is concerned with small deformations and rotations.

Lee [4] has clarified several basic issues that arise when attempts are made to generalize the classical laws of plasticity to the case of finite deformations. However the material considered by Lee remains isotropic in its unstressed state even after the occurrence of plastic flow. The case where the unstressed material becomes anisotropic was considered in [5]. In this work the very special occurrence of the rate of rotation tensor $\mathbf{\Omega}$ in the constitutive equations was emphasized.

Recent work of Nagtegaal and de Jong [6] is concerned with *finite* deformations of a kinematically hardening solid. These authors show that the joint presence of anisotropy and finite deformations could have unusual consequences in shear flow. In the present paper we confirm the results of [6] by simple global considerations and we clarify certain issues that arise in [6] and in the related work of Lee, Mallett and Wertheimer [7].

In order to define the class of rigid-plastic materials of interest, we first let s denote the deviatoric stress tensor:

$$\mathrm{s} = \boldsymbol{\sigma} - \tfrac{1}{3}\mathbf{I}\,\mathrm{Tr}\,\boldsymbol{\sigma} \tag{1}$$

where $\boldsymbol{\sigma}$ is the (Cauchy) stress tensor.

Next we introduce a symmetric traceless tensor $\boldsymbol{\alpha}$. It is assumed that the plastic state and orientation of the material is defined solely by $\boldsymbol{\alpha}$ [8]. The material obeys the modified von Mises yield condition:

$$Y(\mathrm{s}, \boldsymbol{\alpha}) = \tfrac{1}{2}(\mathrm{s} - \boldsymbol{\alpha})\cdot(\mathrm{s} - \boldsymbol{\alpha}) - k^2 \leqslant 0 \tag{2}$$

where k is the yield strength of the virgin undeformed material ($\boldsymbol{\alpha} = \mathbf{0}$) in pure shear. Here the dot indicates the usual inner product in the space of second rank tensors. Thus

$$(\mathrm{s} - \boldsymbol{\alpha})\cdot(\mathrm{s} - \boldsymbol{\alpha}) = \mathrm{Tr}(\mathrm{s} - \boldsymbol{\alpha})^2 = (s_{ij} - \alpha_{ij})(s_{ji} - \alpha_{ji}) \; (i, j = 1, 2, 3). \tag{3}$$

We see that $\boldsymbol{\alpha}$ is the shift tensor of kinematic hardening. One can also interpret $\boldsymbol{\alpha}$ as the "back stress" of the metallurgists.

PLASTIC BEHAVIOR

When a material element, which is in the current plastic state $\boldsymbol{\alpha}$, deforms, the deviatoric stress s carried by the element and the current rate of deformation $\mathbf{D}$ satisfy the yield condition:

$$Y(\mathrm{s}, \boldsymbol{\alpha}) = \tfrac{1}{2}(\mathrm{s} - \boldsymbol{\alpha})\cdot(\mathrm{s} - \boldsymbol{\alpha}) - k^2 = 0 \tag{4}$$

and the flow rule:

$$\mathbf{D} = \lambda\, \partial Y/\partial \mathbf{s} = \lambda(\mathbf{s} - \boldsymbol{\alpha}) \text{ with } \lambda > 0.$$

It follows from (4) that

$$\mathrm{Tr}\,\mathbf{D} = 0 \text{ (plastic flow preservcs volume)} \tag{5}$$

and

$$\lambda = \sqrt{\mathbf{D}\cdot\mathbf{D}}/\sqrt{2}k. \tag{6}$$

Thus during plastic deformation we have

$$\mathrm{s} = \boldsymbol{\alpha} + \sqrt{2}k\, \mathbf{D}/\sqrt{\mathbf{D}\cdot\mathbf{D}} \tag{7}$$

which shows that the material behaves as a non-Newtonian fluid during plastic flow. This equation shows also that once $\mathbf{D}$ and $\boldsymbol{\alpha}$ are known the deviatoric stress follows. Hence the unique importance of the law of evolution for $\boldsymbol{\alpha}$ which we consider next.

The evolution law is composed of statements concerning the time rate of change of $\boldsymbol{\alpha}$

$$\dot{\boldsymbol{\alpha}} = \mathrm{d}\boldsymbol{\alpha}/\mathrm{d}t$$

and its dependence upon the applied deformation. At time t, $\dot{\boldsymbol{\alpha}}$ will depend upon the present values of $\boldsymbol{\alpha}$ and s and also upon the present tensors of rate of deformation and rate of rotation, $\mathbf{D}$ and $\boldsymbol{\Omega}$ respectively.

It is known from general considerations [5], [8] that the dependence on $\boldsymbol{\Omega}$ must be of special form. Thus, in general,

$$\dot{\boldsymbol{\alpha}} = \mathbf{f}(\boldsymbol{\alpha}, \mathrm{s}, \mathbf{D}) + \boldsymbol{\Omega}\boldsymbol{\alpha} - \boldsymbol{\alpha}\boldsymbol{\Omega}. \tag{8}$$

We see that when $\boldsymbol{\Omega}$ is very large, the last two terms will be the controlling ones in (8) and hence the $\boldsymbol{\alpha}$ tensor's principal directions will rotate with angular velocity $\boldsymbol{\Omega}$.

Furthermore, following the classical ideas, we insist that during plastic flow the dependence on D must be "linear" in a restricted but well defined sense. This leads to the following form for $\dot{\boldsymbol{\alpha}}$

$$\dot{\boldsymbol{\alpha}} = \lambda \mathbf{A}(\mathrm{s}, \boldsymbol{\alpha}) + \boldsymbol{\Omega}\boldsymbol{\alpha} - \boldsymbol{\alpha}\boldsymbol{\Omega} \tag{9}$$

during plastic flow, i.e., when

$$Y(\mathrm{s}, \boldsymbol{\alpha}) = 0 \text{ and } \mathbf{D} = \lambda(\mathrm{s} - \boldsymbol{\alpha}),\ \lambda > 0.$$

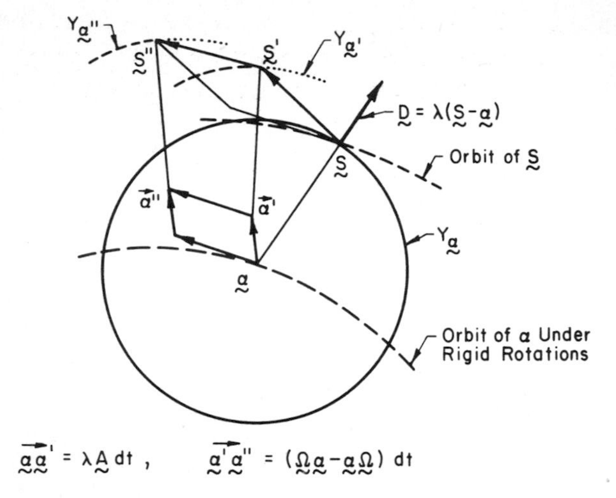

Fig. 1. Plastic deformation in the presence of spin.

A is a tensor valued (symmetric and traceless) function of the yield state (s, $\boldsymbol{\alpha}$). We know from general considerations that **A** must obey the invariance requirement.

$$\mathbf{QA}(\mathrm{s}, \boldsymbol{\alpha})\mathbf{Q}^{\mathrm{T}} = \mathbf{A}(\mathbf{Q}\mathrm{s}\mathbf{Q}^{\mathrm{T}}, \mathbf{Q}\boldsymbol{\alpha}\mathbf{Q}^{\mathrm{T}}) \tag{10}$$

where **Q** is a proper orthogonal transformation in R^3 and $\mathbf{Q}^{\mathrm{T}}$ is its inverse.

Special forms of **A** used in the paper satisfy this requirement. There is one more requirement that one places upon $\mathbf{A}(\mathrm{s}, \boldsymbol{\alpha})$ of (9). In order to motivate this requirement, let us consider the geometrical interpretation of growth laws, Fig. 1. In this figure $\overrightarrow{\boldsymbol{\alpha}\boldsymbol{\alpha}}'$ and $\overrightarrow{\mathrm{ss}}'$ denote the increments of $\boldsymbol{\alpha}$ and s during $(t, t + \mathrm{d}t)$ and when $\boldsymbol{\Omega} = \mathbf{0}$. What distinguishes this case from the one of unloading is that $\overrightarrow{\mathrm{ss}}'$ is directed towards the outside of the yield surface. This implies that

$$\mathbf{A}(\mathrm{s}, \boldsymbol{\alpha}) \cdot (\mathrm{s} - \boldsymbol{\alpha}) > 0. \tag{11}$$

It is easily seen that the requirement (11) implies that in uniaxial stressing the tangent modulus $\mathrm{d}\sigma/\mathrm{d}\epsilon > 0$, but the converse of this statement may not be true.

We are now ready to consider some special cases and the work reported in [6] and [7]. The earliest choice for $\mathbf{A}(\mathrm{s}, \boldsymbol{\alpha})$ is found in Prager's work. In a slightly modified version (cf. [2]) Prager's choice has the form:

$$\mathbf{A} = a(\mathrm{s} - \boldsymbol{\alpha}) \tag{12}$$

where a is a *scalar* which may be a constant or a function of the joint invariants of s and $\boldsymbol{\alpha}$ but often just a function of $\mathrm{Tr}\, \boldsymbol{\alpha}^2$.

For the sake of later developments, let us consider the response of this material to uniaxial tension. The rates of deformation and rotation associated with such a test performed with a virgin material are, in suitable units,

$$\mathbf{D} = \begin{bmatrix} 1 & 0 & 0 \\ 0 & -\frac{1}{2} & 0 \\ 0 & 0 & -\frac{1}{2} \end{bmatrix}, \qquad \mathbf{\Omega} = 0. \tag{13}$$

If ϵ denotes the logarithmic strain in the direction of extension, then according (13)

$$\dot{\epsilon} = 1 \tag{14}$$

We find from (6) and (7) that

$$\lambda = \frac{\sqrt{3}}{2k} \quad \text{and} \quad \mathbf{s} - \boldsymbol{\alpha} = \frac{2}{\sqrt{3}} k\mathbf{D} \tag{15}$$

and from (9)

$$\dot{\boldsymbol{\alpha}} = a\mathbf{D}$$

It follows from (14) and (15) that, σ denoting the only non-zero stress component, we have

$$d\sigma/d\epsilon = \tfrac{3}{2}a$$

which shows that when $a > 0$, the increasing deformation in a uniaxial tension test is accompanied by an increasing stress. This prediction is eminently reasonable for most materials of interest. But when one considers the response of the material to shear flow, then one meets with surprises, as was first pointed out by Nagtegaal and de Jong [6]. Let us therefore look at the case of shear flow with some care.

SHEAR FLOW

The flow is defined by the time independent velocity field

$$v_1 = 2x_2, \qquad v_2 = 0, \qquad v_3 = 0 \tag{16}$$

where v_i are the velocity components in a rectangular frame with coordinates x_i. Since the material considered here is "rate independent"

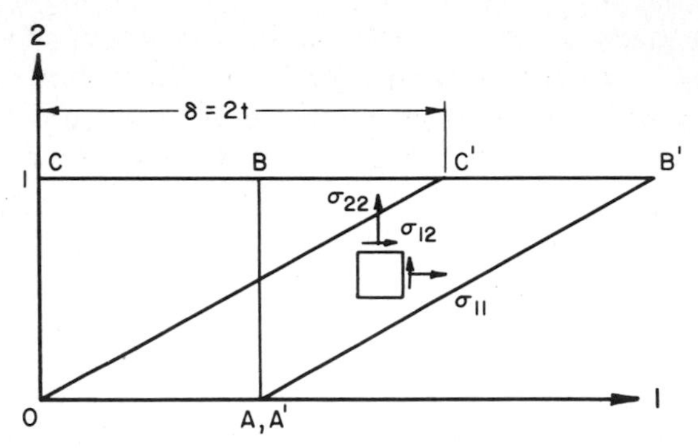

Fig. 2. Shear flow.

there is no need to specify the time units employed in (14). Under the action of this velocity field the unit material square $OABC$ at $t = 0$ becomes the parallelogram $OAB'C'$ at time t, Fig. 2. The rates of deformation and rotation for this flow are

$$\mathbf{D} = \begin{bmatrix} 0 & 1 & 0 \\ 1 & 0 & 0 \\ 0 & 0 & 0 \end{bmatrix}, \quad \mathbf{\Omega} = \begin{bmatrix} 0 & 1 & 0 \\ -1 & 0 & 0 \\ 0 & 0 & 0 \end{bmatrix}. \tag{17}$$

It is desired to find the stresses caused by this deformation in the material considered here. We see right away from (7) and (16) that

$$\boldsymbol{s} = \boldsymbol{\alpha} + k\mathbf{D} \tag{17'}$$

where $\mathbf{D}$ is given above. Thus the task of finding stresses reduces to finding $\boldsymbol{\alpha}$. This requires the integration of the system of differential equations (9) governing the evolution of $\boldsymbol{\alpha}$.

We write the first three equations of (9) without specifying the form of $\mathbf{A}$. Observing that $\lambda = 1/k$ for the present deformation (cf. (6)) we have from (9) and (17)

$$\begin{aligned} \dot{\alpha}_{12} &= A_{12}/k + (\alpha_{22} - \alpha_{11}) \\ \dot{\alpha}_{11} &= A_{11}/k + 2\alpha_{21} \\ \dot{\alpha}_{22} &= A_{22}/k - 2\alpha_{21} \end{aligned} \tag{18}$$

where the right-hand side terms containing $\boldsymbol{\alpha}$ are the relevant components of the "spin" tensor $\mathbf{\Omega}\,\boldsymbol{\alpha} - \mathbf{\Omega}\,\boldsymbol{\alpha}$. We do not write the remaining equations, because we can show that for the shear flow

$$\alpha_{13} = \alpha_{23} = 0 \tag{19}$$

and, of course, $\alpha_{33} = -(\alpha_{11} + \alpha_{22})$ always.

The assertion (19) follows from these observations: a coordinate rotation of 180° about the 3-axis leaves **D** and **Ω** in (17) invariant. This property is inherited by (**s**, **α**) and hence (19). Now we are ready to integrate (18) for various choices of **A**. In the case where **A** is given by (12), (18) becomes

$$\dot{\alpha}_{12} = a + \alpha_{22} - \alpha_{11}$$

$$\dot{\alpha}_{11} = 2\alpha_{21} \tag{20}$$

$$\dot{\alpha}_{22} = -2\alpha_{21}$$

It is convenient for the sake of later developments to introduce the following transformation

$$x = \alpha_{12}, \quad y = \alpha_{22} - \alpha_{11}, \quad z = \alpha_{22} + \alpha_{11} \tag{21}$$

and to write (20) in the following form:

$$\dot{x} = a + y$$

$$\dot{y} = -4x \tag{22}$$

$$\dot{z} = 0$$

and add the initial conditions

$$x = y = z = 0 \quad \text{at} \quad t = 0. \tag{22'}$$

Here we consider the case of a = constant. The solution of (22) is then found easily:

$$x = \alpha_{12} = \tfrac{1}{2} a \sin 2t$$

$$y = \alpha_{22} - \alpha_{11} = -a(1 - \cos 2t) \tag{23}$$

$$z = \alpha_{11} + \alpha_{22} = 0$$

which in view of (16) and (17′) leads to the following stress-shear displacement relations:

$$\sigma_{12} = k + \tfrac{1}{2} a \sin \delta$$

$$\sigma_{22} - \sigma_{11} = -a(1 - \cos \delta) \tag{24}$$

$$\sigma_{22} + \sigma_{11} = -2\sigma_{33}$$

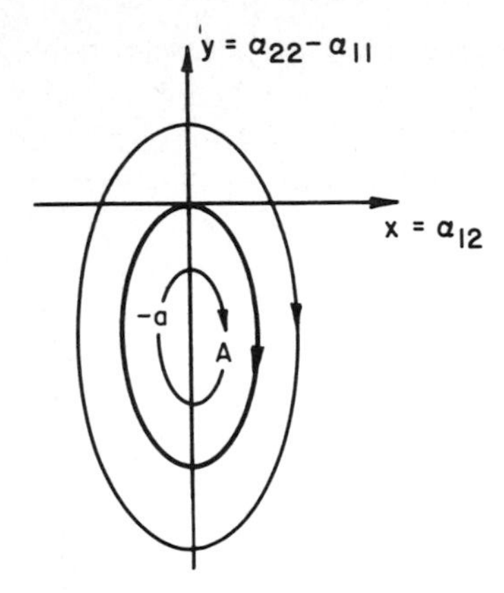

Fig. 3. Prager material.

where δ is the displacement, say, of the point C in Fig. 2 and σ_{12}; σ_{11}, σ_{22} and σ_{33} are the stresses acting on the element. We see that the shear stress remains finite and oscillates as a sine wave about the value k. Similar observations can be made on σ_{11} and σ_{22}.

These surprising results are very similar to those obtained in [6] by computer studies. We note that the stress varies periodically over the course of time with the period π (in time units chosen but not specified here). We compare this with the period 2π associated with the spin $\mathbf{\Omega}$. Clearly these results follow from the overwhelming role played by the spin $\mathbf{\Omega}$ in (20).

We note that the trajectories of the solution of the system of differential equations (22) lie in the (x, y)-plane, Fig. 3. They are ellipses centered at the point $A(0, -a)$ with a 2:1 ratio between major and minor axes. The trajectory corresponding to the initial conditions (22′) is emphasized in this figure.

It is worth noting that one can go to the equilibrium point A of Fig. 3 by making a pure shear test with

$$D = \begin{bmatrix} 1 & 0 & 0 \\ 0 & -1 & 0 \\ 0 & 0 & 0 \end{bmatrix} \quad \text{and} \quad \mathbf{\Omega} = 0.$$

The point A will be reached when the logarithmic strain ϵ in the tension direction reaches the value of 1/2. At A the material can start to undergo a shear flow given by (17) under the *constant* stress components

$$\sigma_{12} = k, \quad \sigma_{11} = a/2, \quad \sigma_{22} = -a/2, \quad \sigma_{33} = 0.$$

Two different aspects of the above results (i.e. (24)) may be cause for worry: (i) the shear does not go to infinity with increasing deformation, whereas it does so in uniaxial stressing; (ii) the shear stress oscillates, giving rise to fears of instability. The question arises (cf. [6] and [7]) as to whether these "undesirable" features can be avoided by a different choice of the function $\mathbf{A}(\mathbf{s}, \boldsymbol{\alpha})$. In [6] the following modification to $\mathbf{A}$ was suggested:

$$\mathbf{A} = a(\mathbf{s} - \boldsymbol{\alpha}) - b(\bar{\alpha})k\boldsymbol{\alpha} \tag{25}$$

where a is a constant, but b is a dimensionless scalar which is a function of $\bar{\alpha}$ defined as

$$\bar{\alpha} = \sqrt{\boldsymbol{\alpha} \cdot \boldsymbol{\alpha}} \tag{26}$$

(25) satisfies the invariance property (10). It can be shown that for this material the requirement (11) takes the form

$$a > 0; \qquad \sqrt{2}a/\bar{\alpha} > b > 0. \tag{27}$$

Calculations based on (13—15) show that in the uniaxial stressing of this material the tangent modulus is given by

$$\frac{d\sigma}{d\epsilon} = \frac{3}{2}\left(a - \frac{1}{\sqrt{2}}\, b(\bar{\alpha})\bar{\alpha}\right) \tag{28}$$

As noted in [6] by choosing

$$b(0) = \text{finite}$$

$$\lim_{\bar{\alpha} \to \infty} \bar{\alpha} b(\bar{\alpha}) = c < \sqrt{2}a \tag{29}$$

one can reproduce a stress—strain curve in uniaxial tension that has an initial slope $3/2\, a$ and an asymptotic slope $3/2\,(a - c)$.

We now study the shear flow of this material. By combining (6) and (25) and using the notation of (21) we find the following system

$$\dot{x} = a - bx + y$$

$$\dot{y} = -4x - by \tag{30}$$

$$\dot{z} = -bz$$

where $b = b(\bar{\alpha})$ and

$$\bar{\alpha}^2 = 2x^2 + \tfrac{1}{2}(y^2 + 3z^2)$$

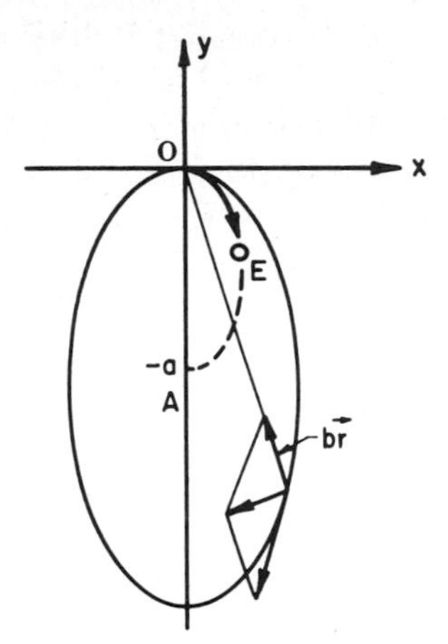

Fig. 4. Nagtegaal and de Jong material.

One can make certain observations on the solutions of the nonlinear system of ordinary differential equation (30) without any numerical calculations. For this purpose consider the solution of (22) with the initial condition $x = y = z = 0$ at $t = 0$. As was observed before, the solution curve in the (x, y, z)-space lies in the (x, y)-plane and on the ellipse

$$4x^2 + (y + a)^2 = a^2.$$

One can then show easily that the solution curve of (30) with the same initial condition must lie within this ellipse. Indeed it is seen from a comparison of (22) and (30) that the points representing the solutions of (30) that start on the above ellipse will move into the interior everywhere except at O (Fig. 4). This in turn implies the above result. (It is worth noting that the equilibrium points of (30) must lie on the elliptical arc OA defined by $4x^2 + y^2 + ay = 0$.)

We see that the modification suggested in [7] cannot give rise to shear stress that increases to infinity. However, numerical integration of (30) shows (cf. [7]) that for certain choices of $b(\bar{\alpha})$ the oscillations in shear stress can be avoided. A standard study of the state of affairs about the equilibria of (30) confirms this result. It is seen, therefore, that the modified form (25) removes one undesirable feature, but as we noted (25) can never give rise to a steadily increasing shear stress.

Finally, we discuss the modification suggested in [7]. The starting point in [7] is the following form for A:

$$\mathbf{A}(\mathbf{s}, \boldsymbol{\alpha}) = a(\mathbf{s} - \boldsymbol{\alpha}) + (\mathbf{b}\boldsymbol{\alpha} - \boldsymbol{\alpha}\mathbf{b}) \tag{31}$$

where a is a positive constant and $\mathbf{b}(\mathbf{s}, \boldsymbol{\alpha})$ is an antisymmetric tensor. In view of (9) this choice leads to the following expression for $\boldsymbol{\alpha}$:

$$\dot{\boldsymbol{\alpha}} = a\mathbf{D} + (\boldsymbol{\Omega} + \lambda\mathbf{b})\boldsymbol{\alpha} - \boldsymbol{\alpha}(\boldsymbol{\Omega} + \lambda\mathbf{b}) \tag{32}$$

We see that the choice of Lee et al. has the potential of lessening the overwhelming influence that $\mathbf{\Omega}$ had in shear flow Prager material. For further progress one must choose a particular form for $\mathbf{b}$. In [7] a definite expression is advocated for $\mathbf{b}$. It seems to us however that many other choices for $\mathbf{b}$ are available. Here is a simple one which is suggested by the work in [7], but quite different from the one used there:

$$\mathbf{b} = c[(\mathbf{s}-\boldsymbol{\alpha})\boldsymbol{\alpha} - \boldsymbol{\alpha}(\mathbf{s}-\boldsymbol{\alpha})] \tag{33}$$

where $c > 0$.
Combining (32) with (33) and using $\mathbf{D} = \lambda(\mathbf{s} - \boldsymbol{\alpha})$ we find

$$\dot{\boldsymbol{\alpha}} = a\mathbf{D} + (\mathbf{\Omega} + c(\mathbf{D}\boldsymbol{\alpha} - \boldsymbol{\alpha}\mathbf{D}))\boldsymbol{\alpha} - \boldsymbol{\alpha}(\mathbf{\Omega} + c(\mathbf{D}\boldsymbol{\alpha} - \boldsymbol{\alpha}\mathbf{D})), \quad a > 0, c > 0 \tag{34}$$

We hasten to note that the above form satisfies the requirements (10) and (11). We also observe that we have in (34) a spin created by $\mathbf{D}$ and $\boldsymbol{\alpha}$.

We consider first the behavior predicted by (34) for uniaxial tension. It is easily seen that this behavior is identical with the one predicted for the Prager material; $\mathbf{D}$ and $\boldsymbol{\alpha}$ have the same principal directions and therefore there is no spin created by $\mathbf{D}$ and $\boldsymbol{\alpha}$.

Next the shear flow of this material is considered. Combining (18) and (34) we find

$$\begin{aligned}
\dot{\alpha}_{12} &= a + (1 + c(\alpha_{22} - \alpha_{11}))(\alpha_{22} - \alpha_{11}) \\
\dot{\alpha}_{11} &= 2(c(\alpha_{22} - \alpha_{11}) + 1)\alpha_{12} \\
\dot{\alpha}_{22} &= -2(c(\alpha_{22} - \alpha_{11}) + 1)\alpha_{12}
\end{aligned} \tag{35}$$

In the notation of (31), (35) becomes

$$\begin{aligned}
\dot{x} &= a + (1 + cy)y \\
\dot{y} &= -4(1 + cy)x \\
\dot{z} &= 0
\end{aligned} \tag{36}$$

$a > 0,\ c < 0;\ x = y = z = 0$ at $t = 0$.

The trajectories of the solutions of (36) are shown* in Figs. 5a–d. It is seen that $z = 0$ throughout. Moreover, it can easily be shown, by

*These trajectories are obtained without much calculation by studying the vector field defined by the R.H.S. of (36). This is what we mean by a geometrical and global study of (36).

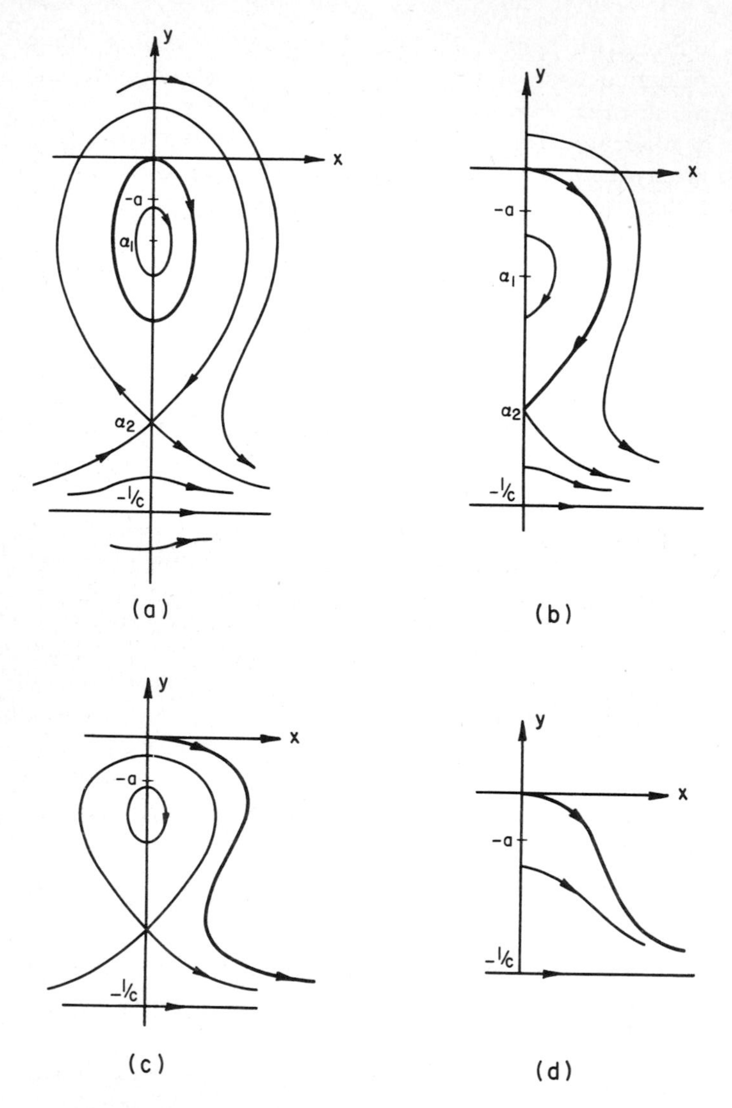

Fig. 5. Lee material: (a) $c < c^* < 1/4a$; (b) $c = c^* < 1/4a$; (c) $c^* < c < c^{**} < 1/4a$; (d) $c > 1/4a$.

noting symmetries with respect to the y-axis that finite trajectories are closed curves. α_1 and α_2 are the equilibrium points and they correspond to the roots of

$$cy^2 + y + a = 0.$$

When c is small, i.e., $c < c^* < 1/4a$ (Fig. 5a) the solution of (3b) with the initial conditions (22′) is a closed orbit. This case is qualitatively similar to the one shown in Fig. 3.

When $c = c^*$ the shear flow solution ends up in the saddle point α_2. In this very special case the shear stress tends to zero with increasing deformation (Fig. 5b).

When $c^* < c < 1/4a$ the shear flow solution tends to infinity: If c is not too large the shear stress reaches a maximum followed by a minimum and then climbs to infinity while y tends to the value of $-1/c$ (Fig. 5c).

Finally when $c > 1/4a$ (Fig. 5d) the right hand side of the first equation of (36) is always positive. Therefore the shear stress increases monotonically to infinity.

We see, without much calculation, that a variety of behaviors are possible in the shear flow of the material defined by (35). It is important to emphasize that when $c > 1/4a$ the shear stress increases to infinity with increasing deformation and the principal directions of $\boldsymbol{\alpha}$ tend towards the bisector directions of the 1—2 coordinate frame.

We end this paper with some notes of caution. The class of representations considered here is highly idealized. We know for instance, that these cannot describe cyclic behavior of many materials adequately. There is a clear need for introducing additional state variables. Nevertheless, these materials can be said to have served us well by teaching us the importance of the joint presence of anisotropy and large rotations.

Acknowledgment

The author is grateful to E.H. Lee and D.M. Parks for long discussions that led to (34). Some of the present results can be found in [8].

References

1 W. Prager, A new method of analyzing stresses and strains in work-hardening solids. J. Appl. Mech., *23* (1956) 493—496.

2 R.T. Shield and H. Ziegler, On Prager's hardening rule, ZAMP, *9* (1958) 260—276.

3 Z. Mroz, On the description of anisotropic work hardening. J. Mech. Phys. Solids, *15* (1967) 163—175.

4 E.H. Lee, Elastic-plastic deformations at finite strains. J. Appl. Mech., *36* (1969) 1—6.

5 E.T. Onat and F. Fardshisheh, On the state variable representation of mechanical behavior of elastic-plastic solids. Proc. Symp. Foundations of Plasticity, Warsaw, 1972, Noordhoff, Leyden, 1973, pp. 89—115.

6 J.C. Nagtegaal and J.E. de Jong, Some aspects of non-isotropic workhardening in finite strain plasticity. To appear in Proc. Res. Workshop on Plasticity, Stanford University, 1981.

7 E.H. Lee, R.L. Mallett and T.B. Wertheimer, Stress Analysis for Kinematic Hardening in Finite-Deformation Plasticity. SUNDAM Report No. 81—11, Division of Applied Mechanics, Stanford University, 1981.

8 E.T. Onat, Representation of inelastic behavior in the presence of anisotropy and of finite deformations, June 1982. To appear in B. Wilshire (Ed.), Creep and Fracture and Engineering Materials, Vol. 2, Pineridge Press, U.K.

Yield Surfaces in the Stress-Temperature Space Generated by Thermal Loading*

A. PHILLIPS and S. MURTHY

Department of Mechanical Engineering, Yale University (U.S.A.)

Abstract

In the stress—temperature space, σ—$\sqrt{3}\tau$—T, pure cyclic thermal loading is used to generate subsequent yield surfaces and their motion for a pure aluminum specimen. The results indicate that the hardening law proposed by the senior author is valid, that there exists a temperature at which the width of the yield curve is zero, and that the concept of yield and loading surfaces is correct.

1. Introduction

In this paper we shall consider a three-dimensional stress—temperature space $\sigma, \sqrt{3}\tau, T$ where σ is the normal stress, τ is the shearing stress and T is the temperature. Thus, in the two-dimensional stress subspace $\sigma, \sqrt{3}\tau$ the plastic region is separated from the elastic region by a yield curve while in the three-dimensional stress—temperature space the plastic region is separated from the elastic region by a yield surface. This yield surface consists of a sequence of yield curves, one for each temperature, all of which are parallel to the stress plane σ—$\sqrt{3}\tau$.

The definition of yield which will be adopted for this paper is the one used consistently by Phillips and his co-workers [1, 2, 3]. It specifies that the yield surface or yield curve delimits the region of purely elastic behavior. The yield point is obtained by means of a well defined operational procedure. This procedure consists of making small excursions of the stress point into the plastic region, by amounts of plastic strain agreed upon in advance (usually 2 μin/in but in some cases up to 5 μin/in) and then by backward extrapolation to the elastic line. For more information concerning this operational procedure we refer to the papers mentioned above. By obtaining the yield point by means of the above operational procedure we ensure that the region enclosed by the yield

*This research was supported by the National Science Foundation.

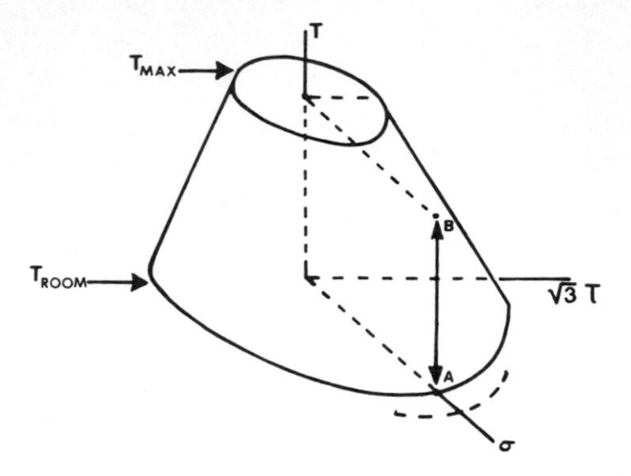

Fig. 1. Typical yield surface in the $(\sigma, \sqrt{3}\tau, T)$ space. *ABA* represents a typical thermal loading cycle. The yield surface is illustrated only in the temperature region between room temperature and maximum testing temperature.

curve (or the yield surface) is purely elastic and that the entire purely elastic region is enclosed by the yield curve or yield surface. Hence, the yield curve or yield surface does not enclose some part of the plastic region as well.

Note that the backward extrapolation used in this paper is radically different conceptually and in its effects on the results from the ones used by Taylor and Quinney [4] and by Mair and Pugh [5]. These authors used penetrations into the plastic region which are of the order of 1000 μin/in or more. In these cases the region enclosed by the resulting yield curve included to a substantial degree a plastic region.

In previous papers [6, 7] the first author and his co-workers determined that for pure aluminum the yield surface has the form shown in Fig. 1. In this figure we see a typical yield surface obtained by isothermal prestressing. We obtained such yield surfaces by performing isothermal tests at different temperatures so that the yield surface is the locus of yield curves obtained at different temperatures. The procedure is to obtain a first yield curve at some temperature T_0, then, while the stress point is well within the region enclosed by this yield curve, change the temperature to another one T_1 and subsequently obtain the yield curve associated with T_1, and so on.

The so-obtained yield surface can be verified by means of a non-isothermal loading. Such non-isothermal loading is illustrated in Fig. 2 and it consists at least in part of variable temperature. While the stress point is located inside the region bounded by the yield surface, the temperature is increased while the stress is either kept constant or changed until at some value of stress and temperature plastic deformation appears. This level of stress and temperature is then compared with the previously, by isothermal testing, obtained yield surface.

We observed [1, 2] that the non-isothermally obtained level of stress and temperature for which plastic deformation begins is located on the

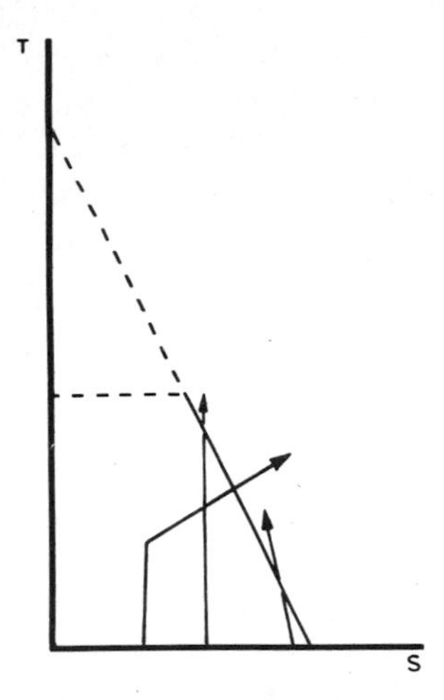

Fig. 2. Typical non-isothermal loading paths.

isothermally obtained yield surface. Hence a stress-temperature path located within the isothermally obtained yield surface does not alter the yield surface, even if the path includes temperature changes.

When a material is stressed beyond the yield surface this surface will change in accordance with the hardening rule valid for that material. The hardening rule which is appropriate for each material must be obtained experimentally and if possible must be deduced from microstructural considerations. However, in the classical theory of plasticity, hardening rules have originally been proposed for reasons of convenience and on some experimental evidence as reliable as it could be obtained at that time. One of these hardening rules was the isotropic hardening rule according to which the yield curve expands isotropically with the loading. Another rule was the kinematic hardening rule according to which the yield curve retains its size and form but with loading it moves about the stress plane like a rigid body. When more accurate experimentation became possible during the past 15 years other hardening rules based on more reliable experiments became possible. Such a hardening rule has been proposed by the senior author [2, 6] and was verified by means of a large number of experiments [3, 6] on aluminum, brass, and copper. These experiments were of isothermal loading type and by means of them the motion and deformation of subsequent yield surfaces was determined. In a recent paper [7] experiments were presented during which the loading was purely thermal and the resulting motion of the yield curve was determined. The thermal loading consisted of a thermal loading cycle originating at a point of the room temperature yield curve where the shearing stress is nearly maximum. The resulting motion of the yield curve was then in the direction of increasing shearing stress.

In the experiments described in this paper the loading will again be a thermal cycle but this thermal cycle will originate at points of the room temperature yield curves where either the normal stress is maximum or minimum, or where the shearing stress is maximum or minimum. In addition, the entire yield surface and its resulting motion will be determined. In [7], on the other hand, only the room temperature yield curve and its motion were determined. In this manner we shall be able to obtain information concerning the hardening rule valid for such a thermal loading cycle.

Another important difference between the experiments reported in [7] and the experiments presented in this paper is that, while in those previous experiments we could determine accurately only the shearing strains, now a modification in the experimental procedure allows us to measure accurately both the shearing strains and the axial strains. In the present experiments we shall find that during a thermal loading cycle the entire yield surface moves in the direction normal to the yield curve at the point where the cycle originates. We shall observe that the motion of the yield surface is without cross-effect at any temperature. The type or magnitude of motion and deformation of the yield surface due to the thermal loading cycle is similar to that which has been observed after isothermal loading. The increase in temperature produces hardening in the neighborhood of the stress point where the thermal loading cycle occurs. It also produces a decrease of the yield stress at all measured temperatures in the opposite direction; that is, of course, an expression of a Bauschinger effect due to thermal cycling.

In a number of papers [8, 9] we introduced the concept of a loading surface which encloses the yield surface. This concept evolved in the senior author's writings since 1965 [10]. To understand the relation of the loading surface motion compared to that of the companion yield surface we shall consider Fig. 3 which depicts a typical stress—strain relation in which there is loading from O to A, then unloading within the elastic range from A to B, which may be extended to C, and then reloading from B to F. We observe that reloading starts producing plastic strain at D which may be below A. From D to E (which is at the same stress level as A) we have small plastic strains but from E to F we have larger plastic strains.

In the stress space, Fig. 4, and on the stress axis σ, Fig. 3, the points corresponding to A, B, C, D, E, and F are the points a, b, c, d, e, and f respectively. The position of the point D will gradually move upwards towards A, and finally coincide with A, the longer the specimen is subjected to the stress corresponding to A, that is a.

Our experimental evidence from [8, 9] shows that if I is the initial yield curve defined by the operational procedure used in the present experiments, then loading to "a" produces two curves, one, II, which follows from I by an *approximately* isotropic hardening process, while another, III, follows from I by a process which exhibits some similarities

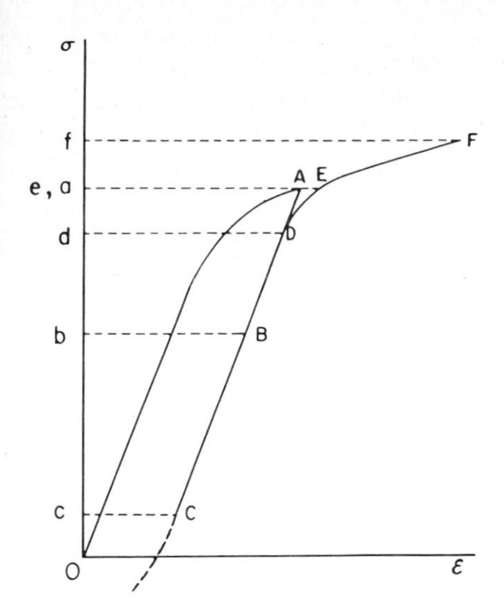

Fig. 3. Typical stress—strain relation with loading OA, unloading AB, and reloading $BDEF$. The corresponding stress space paths will be Oa, ab, $bdef$.

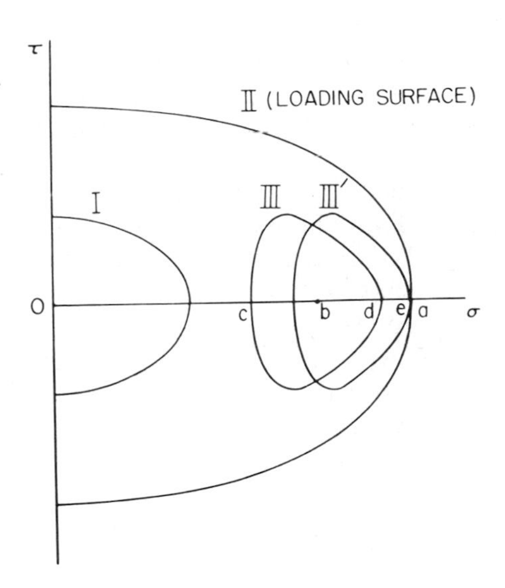

Fig. 4. The loading surface II and the yield surfaces III and III′. When the stress remains stationary at a the yield surface moves slowly from position III to position III′.

with but is substantially different from kinematic hardening. When the point D coincides with the point A, then instead of III we have III′ which is tangential to II at a. The surface II is the *loading surface* which corresponds to the point A or E, and the surfaces III or III′ are the *yield surfaces*, which enclose the purely elastic region as defined by the

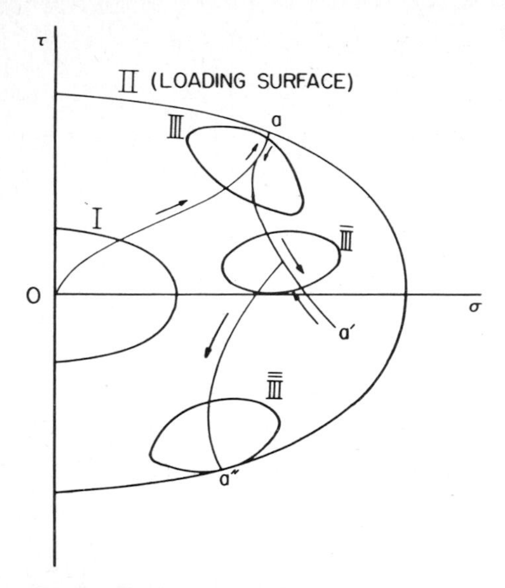

Fig. 5. Motion of the yield surface within the loading surface. The path $Oaa'a''$ generates a motion of the yield surface illustrated by the sequence of positions I, III, $\overline{\text{III}}$, $\overline{\overline{\text{III}}}$. Observe that $\overline{\overline{\text{III}}}$ is tangential to the loading surface.

proportional limit, and correspond to the point D. The region between II and III (or III′) is the region of small plastic strains, that is, the plastic strains developed while the stress–strain curve moves from D to E, Fig. 3. The region outside II is the region of larger plastic strains. From Fig. 4 it can be concluded that staying at A will not eliminate the region between II and III except possibly in one direction. It can also be concluded that staying at A results in the motion of the yield surface with time from III to III′.

Our experiments [8, 9] have shown that this behavior is the same whether loading, unloading, and reloading is a rectilinear one in the σ axis or whether we have a more complex loading–unloading–reloading path. Figure 5 gives an illustration of the behavior to be expected. Loading occurs from O to a and the loading surface expands from I to II as in isotropic hardening. The yield surface III corresponding to "a" follows from I. Suppose we move from inside III to a'; then the yield surface moves from III to $\overline{\text{III}}$. Moving from inside $\overline{\text{III}}$ to a'' changes the yield surface from $\overline{\text{III}}$ to $\overline{\overline{\text{III}}}$. As long as the loading point remains within II or on II there will be no change in the loading surface II. Also if a, a', or a'' are on the loading surface or inside the loading surface but near it, the associated yield surface III, $\overline{\text{III}}$, or $\overline{\overline{\text{III}}}$ will be tangential to the loading surface. When the loading point moves beyond II then both the loading and yield surfaces will change according to their respective laws of hardening. In this paper we shall observe how the relationship between yield surface and loading surface evolves when the yield surface changes because of thermal loading.

From Fig. 1 it follows that as the temperature increases the size of the isothermal yield curve decreases. In a previous paper [11] we concluded that there exists a temperature at which the yield curve degenerates to a straight line (in the $\sigma, \sqrt{3}\tau$, T space), that is the width of the yield curve becomes zero at some value of the temperature. Thus, with increasing temperature, the yield curve instead of being reduced to a single point, it is reduced to a straight line. Only, for the virgin yield surface the above straight line reduces to a single point. We have also seen in the same paper that with increasing prestress the temperature at which the yield curve degenerated to a straight line decreases. In the present paper we shall show the existence of a degenerate yield curve even when the subsequent yield surfaces are generated by thermal loading cycles.

In previous papers [2, 6, 11, 13, 14] the concept of the thermodynamic reference stress was introduced. This is a stress value σ^0_{kl} which in stress space always lies within the yield curve and is expressed by

$$\sigma^0_{kl} = \rho \mathrm{d}\psi / \mathrm{d}\epsilon^{pl}_{kl} \tag{1}$$

where ψ is the specific Helmholtz free energy and ρ is the mass density. For sufficiently high temperature the thermodynamic reference stress will be located on the above-mentioned degenerate yield curve.

The concept of the equilibrium stress—strain lines (or quasistatic stress—strain curves) has been introduced in previous papers [2, 6, 11, 13] and was expanded in [7, 15]. These are the stress—strain lines corresponding to $\dot{\sigma} \cong 0$ for increasing σ and for decreasing σ, for different values of the temperature. They are the upper and lower bounds of the yield curves at a given temperature for different strains, as explained in [11]. In this paper we shall use the concept of the equilibrium stress—strain curves to illustrate the generation of plastic strains in the negative direction.

2. Experimental details

The experiments described in this paper were performed on commercially pure aluminum 1100-0 with tubular specimens described in [1, 3]. Specimen SA-3 was loaded in combined tension and torsion. The specimens were annealed at 650°F for two hours and then slowly furnace cooled. The temperature of the specimen was changed by heat conduction. Heat was supplied by two heating coils, one attached at each end of the specimen around the circumference. A continuously variable transformer was used to regulate the temperature and both the dummy and the active specimen were heavily insulated. The dummy specimen was kept at constant temperature (70°F). Temperature measurements were made by three omega beaded junction 0.010-inch diameter alumel—chromel wire thermocouples which were connected to strain-gage indicators acting as voltmeters to measure temperature. Two specimens were used for preliminary tests and one, SA-3, was used for the main experiments.

Four active 45° rosette BLH-FABR-50-12S13 type strain gages were bounded to the outer surface of the specimen at midlength. The location of the gages were consecutively 90° apart. A single element gage of the same type was mounted transversely above each of the four rosettes. The dummy specimen had similar gages. Digital readout BLH model 800 strain indicators were used.

The testing machine used was of the deadweight type. It is essentially the one used in the tests by Phillips and co-workers [1, 3] except that several modifications appropriate to the present experiments were made. In these experiments only one specimen was used for the determination of the active yield surface, its subsequent yield surfaces and the strains occurring during the thermal loading cycles. To obtain each indication of yield it is necessary to probe into the plastic region to extremely small values, less than 5 μin/in of plastic strain. The exact procedure of obtaining the proportional limit by back extrapolation to the linear elastic line is explained in [1, 3].

Loading rates in tension and torsion were typically 30 and 45 psi per minute, respectively, during both proportional limit probing and isothermal prestressing whenever such prestressing was applied.

The thermal loading was from 70° to 240°F at the rate of 2°F/minute. The temperature—time relation used is shown in Fig. 6. This temperature—time relation is quite different from the ones used in [7]. Preliminary experiments in which at zero stress thermal cycles were applied to the specimen have shown no change in the behavior of the specimen or of the gages after cycling.

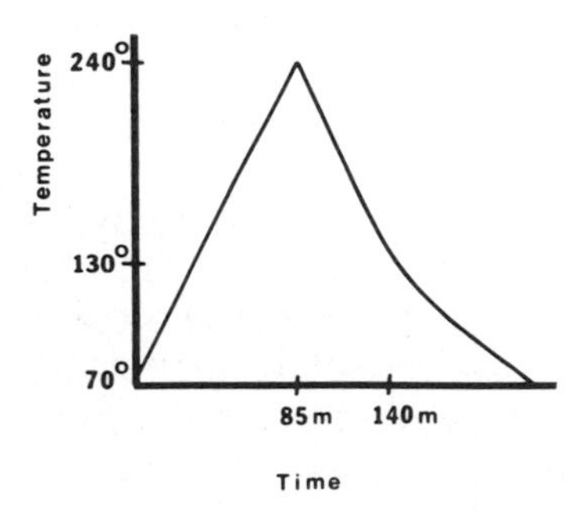

Fig. 6. The temperature—time relation during the thermal loading cycle.

3. Experimental results and discussion

The initial yield surface was first obtained by means of three isothermal yield curves at temperatures $T = 70°$, 108°, and 193°F. These three

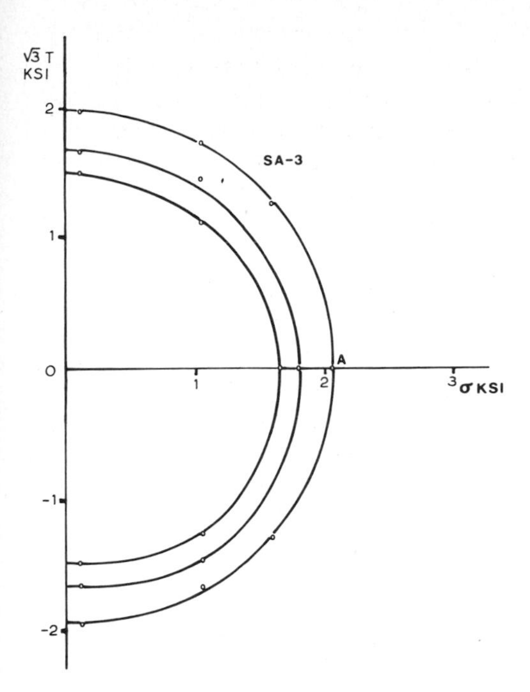

Fig. 7. The initial yield surface. The three curves depict the three isothermals at 70°, 180° and 193°F. The curves lie between the Mises and the Tresca conditions.

curves lie between Mises and Tresca curves but they have a greater similarity with the Mises than with the Tresca curves. They indicate a truncated cone with apex at 605°F. Figure 7 illustrates as an example the three yield curves for specimen SA-3. They are similar to those obtained by the senior author and his co-workers in previous experiments [1, 2, 3, 6].

A thermal loading cycle from $T = 0$ to 240°F at stress point A ($\sigma = 2055$ psi, $\sqrt{3}\tau = 0$) produced permanent strain $\epsilon^{\mathrm{p}} = 30\,\mu$ in/in and the yield surface moved in the direction of the positive σ axis. The subsequent yield surface is shown in Fig. 8 where the location of the point A is also indicated. The subsequent yield surface is indicated by means of three isothermal yield curves at temperatures $T = 70°, 147°, 220°$F. Except for the 70°F yield curve the other yield curves correspond to higher temperatures than the corresponding curves for the initial yield surface. We observe that the subsequent yield surface does not enclose the origin. In addition we observe that the subsequent yield surface has some of the characteristics of subsequent yield surfaces obtained by isothermal prestressing [3]. These characteristics are that the width of the yield surface in the direction of prestressing decreases, and the slope of the yield surface on the side of prestressing is smaller than on the back side. No conclusion could be drawn concerning cross effect since we could not obtain the maximum τ points on the 70° curve of the subsequent yield

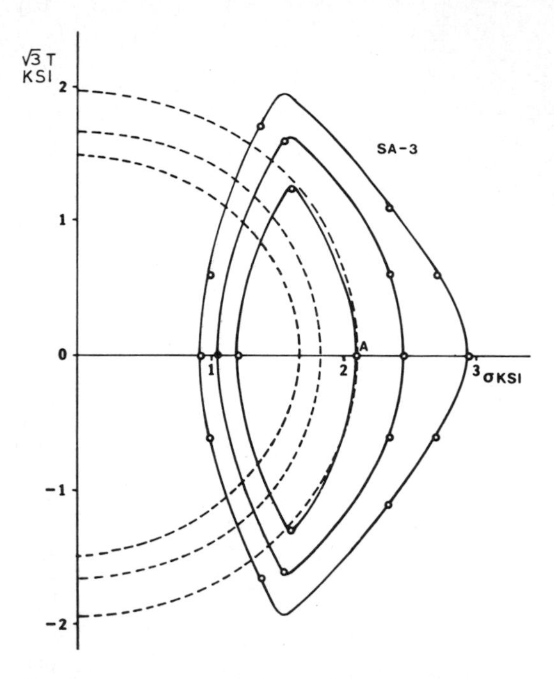

Fig. 8. The first subsequent yield surface (solid lines) compared to the initial yield surface (broken lines). The solid lines are the three isothermals at 70°, 147°, 220°F. Note that the two higher temperatures are different from the corresponding temperatures for the initial yield surface. The first subsequent yield surface was obtained by a thermal loading cycle to 240°F at A.

surface, and the other two curves correspond to higher temperatures, respectively.

Next the specimen was prestressed isothermally at 70°F to the point ($\sigma = 4450$ psi, $\sqrt{3}\tau = 0$) denoted by B in Fig. 9 and then the second subsequent yield surface was determined by means of three isothermal yield curves at $T = 70°$, 147° and 220°F (the same temperatures as for the first subsequent yield surface). The isothermal prestressing at 70°F to B produced permanent strain $\epsilon^p = 9\,\mu$in/in which is a very small amount compared to the previously obtained one. However, this small amount of permanent strain is not indicative of the amount of motion of the yield surface which is quite substantial. In this respect we may consider this alternative small versus large strain increment as a result of the well-known phenomenon of discontinuous yielding of pure aluminum [12]. Again, the width of the yield surface in the direction of prestressing decreases, and the slope of the yield surface on the side of prestressing is smaller than on the back side. No conclusion concerning cross effect could be obtained since the data for the first and second subsequent yield surface did not include the max τ positions.

Next the specimen was subjected to a thermal loading cycle originating at the stress point $\bar{C}$ ($\sigma = 2740$ psi, $\sqrt{3}\tau = 0$), Fig. 9. This point coincides

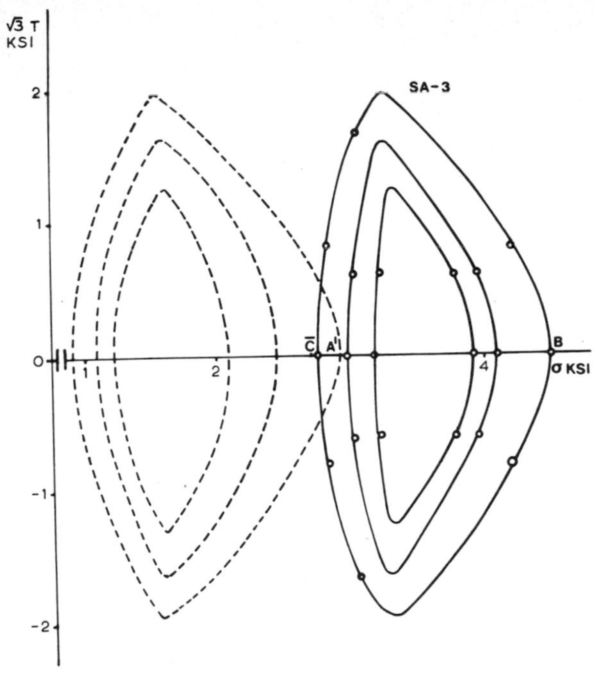

Fig. 9. The second subsequent yield surface (solid lines) compared to the first subsequent yield surface (broken lines). The three isothermals in both cases were obtained at 70°, 147°, and 220°F. The second subsequent yield surface was obtained by isothermal stressing to B along the σ axis.

with the intersection of the σ-axis with the back side of the room temperature curve of the second subsequent yield surface on the σ—$\sqrt{3}\tau$ plane. The amount of plastic strain produced was negligible. This result was as it would be expected since the very substantial slope of the yield surface at $\bar{C}$ allows to a pure thermal loading only a very limited incursion into the plastic region. The result of this incursion into the plastic region was however the movement of the yield surface in the negative σ direction as seen from Fig. 10. Comparing the amount of backward motion of the yield surface due to the thermal loading cycle at $\bar{C}$ with the amount of forward motion of the yield surface due to the thermal loading at A, we conclude that the backward motion is much smaller since the intrusion into the plastic region is also much smaller. In addition, we observe that the width of the yield surface in the direction of prestressing increases, at least for the two higher temperature isothermals. This is as would be expected from previous experiments [2, 4].

Next the specimen was subjected to a thermal loading cycle originating at D ($\sigma = 3080$ psi, $\sqrt{3}\tau = 1925$ psi). This cycle produced a plastic strain ($\epsilon^{pl} = 0$, $\gamma^{pl} = 29\,\mu$in/in) and considerable motion of the yield surface. Indeed, the incursion into the plastic range was considerable. The fourth subsequent yield surface was determined by means of three yield curves at

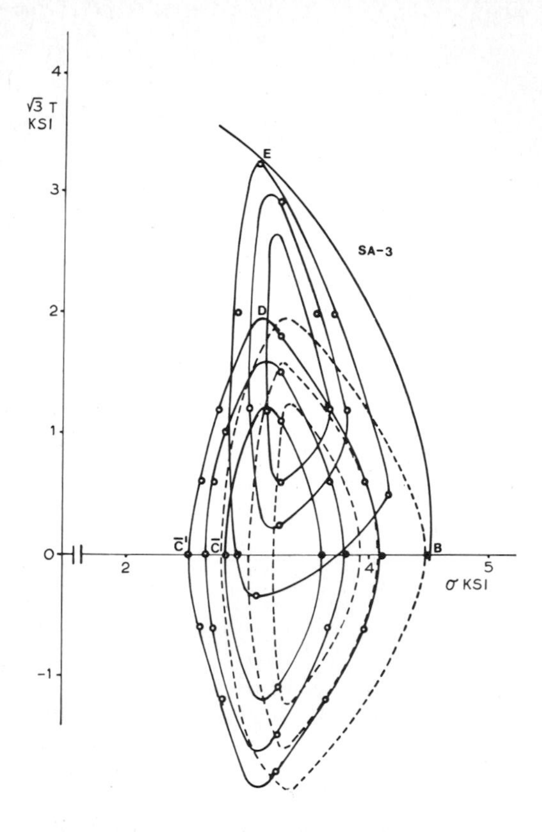

Fig. 10. The third and fourth subsequent yield surfaces (solid lines) compared to the second subsequent yield surface (broken line) and to the loading surface passing through B. The three isothermals in all cases were obtained at 70°, 147°, and 220°F. The third subsequent yield surface was obtained by a thermal loading cycle at $\bar{C}$, while the fourth subsequent yield surface was obtained by a thermal loading cycle at D.

$T = 70°$, $147°$, and $220°$F, Fig. 10. The yield surface in the direction of prestressing decreases in size as shown from all three isothermals. We also observe that the entire yield surface moves in the direction normal to the yield curve at the point where the cycle originates.

Fig. 10 illustrates also the loading surface generated by isothermal loading to B. It is seen that the thermal cycle at D at the third subsequent yield surface moved the yield surface sufficiently in the $\sqrt{3}\tau$ direction to make it tangential to the loading surface at E. In addition, we observe that in this process the fourth subsequent yield surface moved sideways towards increasing values of σ, as if a rotation about E occurred, with the trend to make the yield surface tangential to the loading surface over a wider section of its periphery. This phenomenon has also been observed before for isothermal prestressing to the loading surface [8, 9]. These observations strengthen our confidence towards the validity of the two surfaces concept of yield and loading surfaces introduced in [8, 9, 10].

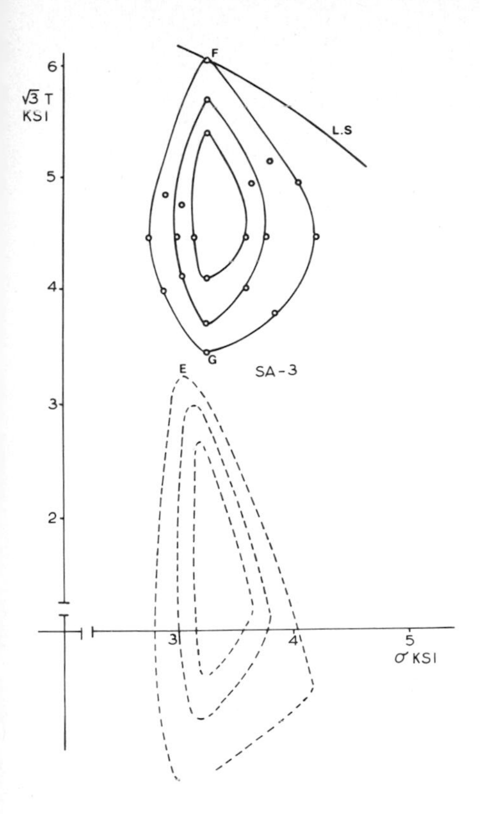

Fig. 11. The fifth subsequent yield surface (solid lines) compared to the fourth subsequent yield surface (broken lines) and to the loading surface. The three isothermals were obtained at 70°, 147° and 220°F. The fifth subsequent yield surface was obtained by isothermal prestressing *EF*.

Next the specimen was prestressed isothermally at 70°F to the point ($\sigma = 3255$ psi, $\sqrt{3}\tau = 6075$ psi) denoted by F in Fig. 11 and then a fifth subsequent yield surface was determined by means of three yield curves at $T = 70°, 147°$, and 220°F. We observe that cross effect was valid since the widths of the three yield curves of the fifth subsequent yield surface were equal to those of the corresponding ones of the fourth yield surface. The size of the fifth subsequent yield surface in the direction of prestressing is considerably smaller than that of the fourth one, as is expected from previous experiments [2, 4]. The fifth subsequent yield surface is nearly tangential to the corresponding loading surface passing through F. The amount of plastic strain produced during the prestressing EF was ($\epsilon^{pl} = 43\ \mu$in/in, $\gamma^{pl} = 183\ \mu$in/in). In addition, while staying at F for the next 42 hours before obtaining the fifth subsequent yield surface an additional creep strain $\epsilon^{cr} = 30\ \mu$in/in, $\gamma^{cr} = 107\ \mu$in/in was developed.

Next we subjected the specimen to a thermal loading cycle at G ($\sigma = 3255$ psi, $\sqrt{3}\tau = 3442$ psi). This cycle generated plastic strain which

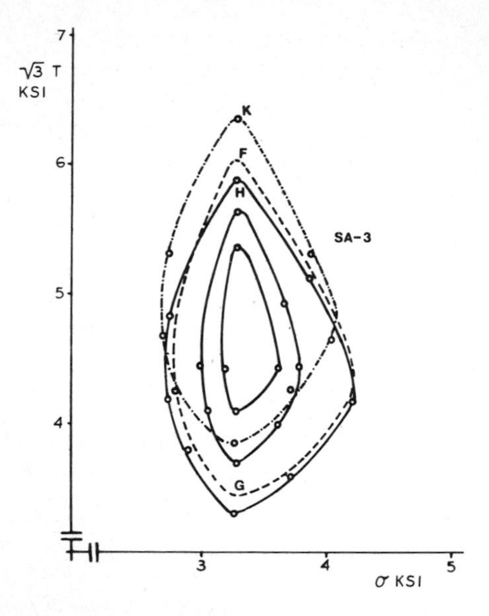

Fig. 12. The sixth subsequent yield surface, (solid curves), and the seventh subsequent room temperature yield curve (dash-dot line), compared to the fifth subsequent room temperature yield curve (broken line). The sixth subsequent yield surface was obtained by a thermal loading cycle at G. The seventh subsequent room temperature yield curve was obtained by a thermal loading cycle at H.

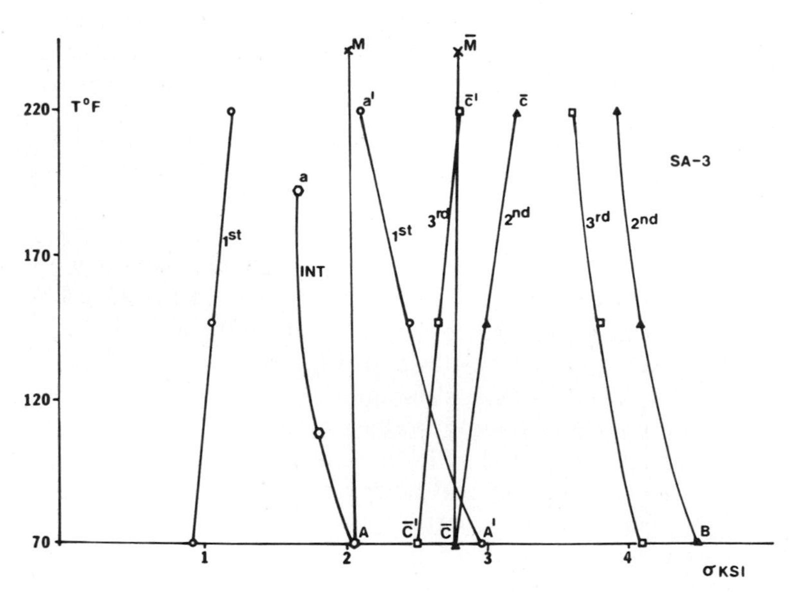

Fig. 13. The intersections of the initial, first, second, and third subsequent yield surfaces with the T—σ plane. The thermal loading cycles are generated at A and at $\bar{C}$.

was very small ($\epsilon^{pl} = 4\,\mu$in/in, $\gamma^{pl} = 8\,\mu$in/in) and the subsequent sixth yield surface is shown in Fig. 12 by means of three yield curves (solid lines). We observe that despite the heavy incursion into the plastic range by the thermal loading cycle at G there is only a very small motion of the 70°F yield curve in the $-\sqrt{3}\tau$ direction and practically no motion of the other two yield curves is observable.

Figure 12 shows also the seventh 70°F yield curve (dash-dot line) due to a thermal loading cycle at H ($\sigma = 3255$ psi, $\sqrt{3}\tau = 5974$ psi). This thermal loading cycle produced substantial plastic strain ($\epsilon^{pl} = 10\,\mu$in/in, $\gamma^{pl} = 50\,\mu$in/in) and a substantial motion of the yield curve. The differences in the consequences of the thermal loading cycle at H versus those at G are understandable if we remark that point G is at the back side of the yield surface while point H is at the front side of the yield surface. From Fig. 12 we also observe again that the thermal loading cycles moved the entire yield surface in the direction normal to the yield curve at the points where the cycles originate.

In Fig. 13 we see the intersections of the initial, first, second and third subsequent yield surfaces with the T—σ plane. We observe that the thermal loading cycles at A and $\bar{C}$ result in the motion of the yield surface in the appropriate directions by approximately the appropriate amounts, that is, the amounts indicated by the depth of penetration into the plastic region. In particular observe how the thermal loading cycle at A moved the initial sloping line Aa to the first subsequent sloping line $A'a'$, while the thermal loading cycle at $\bar{C}$ moved the second subsequent sloping line $\overline{Cc}$ to the third subsequent sloping line $\bar{C}'\bar{c}'$.

In Fig. 14 we see the projections of the third to sixth subsequent yield surfaces on the T—$\sqrt{3}\tau$ plane. Again we observe that the thermal loading cycle at D results in the motion of the yield surface in the appropriate direction by approximately the appropriate amount. We observe that the thermal loading cycle at G did not move the yield surface appreciably. On the other hand, the thermal loading cycle at H moved the yield curve to

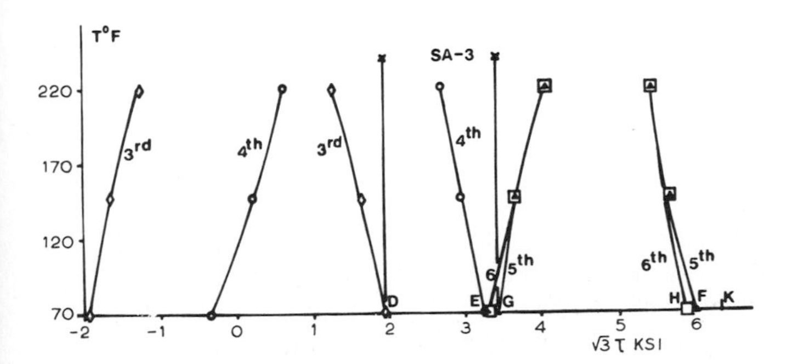

Fig. 14. The projections of the third to the sixth subsequent yield surfaces on the T—$\sqrt{3}\tau$ plane. The thermal loading cycles are generated at D, G, and H.

K which is the position which should be expected from the amount of incursion of the thermal loading path into the plastic range. From Figs. 13 and 14 we observe that the increase in temperature due to thermal loading cycle produces hardening in the neighborhood of the stress point where the cycle occurs, and a decrease of the yield stress in the opposite direction. This is an example of a Bauschinger effect due to thermal load cycling.

In Fig. 15 we repeat the procedure used in [11] to see whether there exists a temperature for which the yield curve degenerates to a ridge line and in such a case to obtain this ridge line as well as the thermodynamic reference stress. Following the procedure explained in [11] for the first subsequent yield surface, Fig. 15a, we draw a series of straight lines AA', BB', CC', DD', EE' parallel to the prestressing direction. The intersections of these lines with the yield curves are plotted in a stress—temperature diagram, Fig. 15b; and we observe that the intersections produce pairs of straight lines (Aa, aA'), (Bb, bB'), ... We see that the intersections a, b, and c lie on a straight line QQ' which represents a high temperature limit of the sequence of yield curves. This phenomenon was predicted in [13] and shown to exist in [11]; here it is shown to exist even when the yield surfaces are produced by thermal cycles. The sequence of yield curves at increasing temperatures suggests the validity of an anisotropic sandhill analogy similar to the classical isotropic sandhill analogy for fully developed ideally plastic torsion. The slope of the sandhill (maximum T gradient) varies with the orientation of the normal to the level curves with the preloading direction. The slope is minimum in the direction of preloading, maximum in the opposite direction. In Fig. 15b, lines Bb, Cc, Dd have approximately the same slope. The slopes of the lines of the front side of the surface bB', cC', dD' are markedly smaller and they are approximately equal. The straight line QQ' to which the yield curves degenerate is the familiar ridge line of the sandhill analogy.

We observed that the size of the elastic region shrinks with increasing temperature and at some temperature the elastic region vanishes. In Fig. 15b the temperature at which the elastic region vanishes is approximately 330°F. This value is very near to what we obtained in [11] for the same material. The thermodynamic reference stress is the projection of the intersection of lines PQ and $P'Q'$ and is located at $\bar{P}$ on the ridge line QQ' as indicated in Figure 15a.

We now introduce schematically the equilibrium stress—strain lines AC and $\overline{AC}$, Fig. 16, which are the upper and lower bounds of the yield curves at a given temperature for different strains, as explained in [11]. Thus, $A\bar{A}$ corresponds to the initial yield curve and $E\bar{E}$, $K\bar{K}$, correspond to subsequent yield curves at different values of plastic strain. The equilibrium stress—strain lines AC and $\overline{AC}$ must change positions with the temperature so that the corresponding initial and subsequent yield surfaces will be generated. As the plastic strain increases we have seen

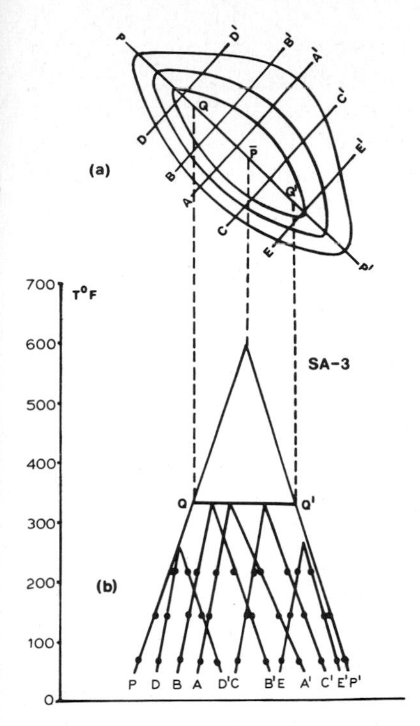

Fig. 15. At an appropriate temperature, the yield curve of the first subsequent yield surface degenerates into a straight ridge line QQ'. The thermodynamic reference stress is located at $\bar{P}$.

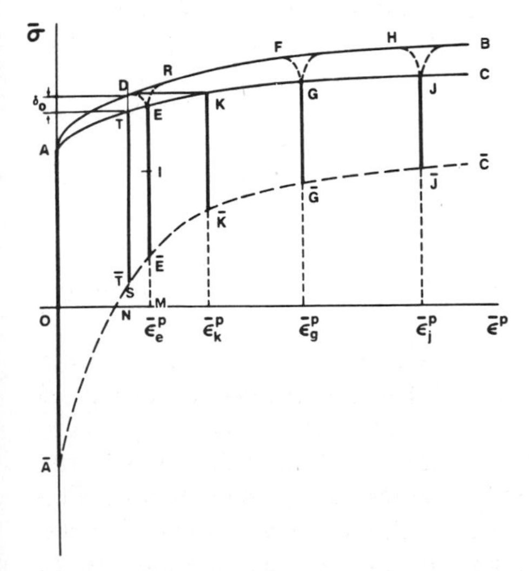

Fig. 16. The equilibrium stress—strain lines AC and $\overline{AC}$.

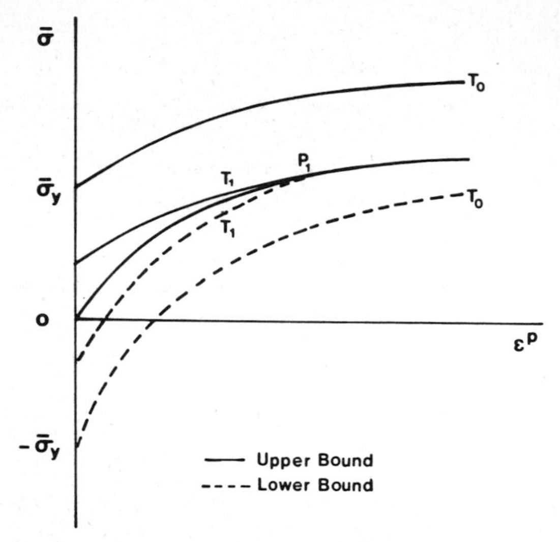

Fig. 17. The locus of the intersections of the equilibrium stress—strain lines AC and $\overline{AC}$ for various values of temperature.

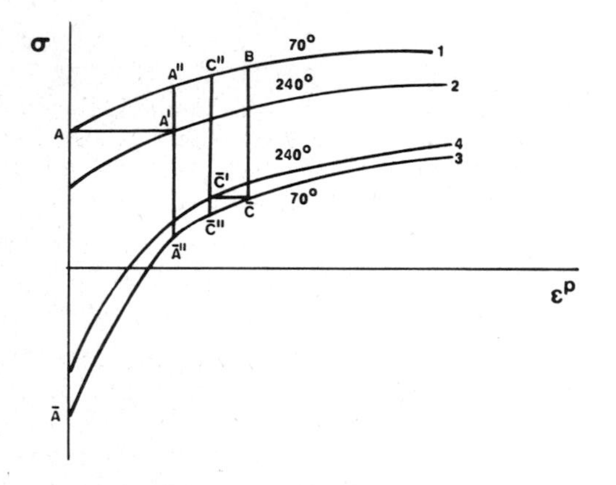

Fig. 18. Illustration of the generation of the plastic strain increment by a thermal loading cycle. Note that $\overline{CC}'$ is in the negative direction.

[11] that the size of the yield curve decreases so that the equilibrium stress—strain lines AC and $\overline{AC}$ must intersect at some value of the plastic strain. Figure 17 shows schematically the locus of the intersections of the equilibrium stress—strain lines AC and $\overline{AC}$ for various values of temperature. In particular, point P_1 represents the intersection of curves AC and $\overline{AC}$ for temperature T_1. The curve OP_1 represents also the location of the thermodynamic reference stress for various amounts of plastic deformation. Figure 17 also shows the sequence of equilibrium stress—strain curves for increasing temperature T.

Figure 18 shows schematically two upper and two lower equilibrium stress—strain lines at 70° and 240°F. We shall now illustrate our experimental findings on the basis of this figure. The initial yield point corresponds to point A, Fig. 8, where the first thermal loading was done. On raising the temperature, the equilibrium stress—strain lines change from curves 1 and 3 to curves 2 and 4. This results in plastic strain and the yield surface moves along AA' until A' is reached. Upon lowering the temperature (return portion of the thermal loading cycle) the equilibrium stress—strain curves change back to curves 1 and 3 and the new yield surface will be $A''\bar{A}''$. This means that the yield surface will move in the direction of the positive σ axis since it moves upward. The back point $\bar{A}$ has moved more than the front point A so that the yield surface no longer encloses the origin.

Next, the isothermal loading will change the yield surface from $A''\bar{A}''$ to $B\bar{C}$ along the equilibrium stress—strain lines 1 and 3. At $\bar{C}$, the thermal loading changes the equilibrium stress—strain curves from 1 and 3 to 2 and 4. Since the stress remains the same and the active branch of the equilibrium stress—strain curve is line 3, the result of this change should be plastic strain in the negative direction and motion of the yield point from $\bar{C}$ to $\bar{C}'$. Upon removal of the temperature the equilibrium stress—strain curves return to 1 and 3. The new room temperature yield curve is $C''\bar{C}''$ which represents a motion of the yield surface towards the origin. This motion and its direction are apparent from Fig. 10. On the other hand, the associated development of plastic strain in the negative direction did not materialize since only negligible plastic strain occurred. But such plastic strain in the negative direction appeared in other tests under similar conditions as indicated in [7].

4. Conclusions

The experiments presented in this paper consisted in obtaining initial and subsequent yield surfaces in the σ—$\sqrt{3}\tau$—T space due primarily to pure cyclic thermal loading but occasionally to isothermal prestressing at room temperature. We observed that during a thermal loading cycle the entire yield surface moves in the direction normal to the yield curve at the point where the cycle originates. The type or magnitude of motion and deformation of the yield surface due to the thermal loading cycle is similar to that which has been observed after isothermal loading. We also observed that the relationship between yield surface and loading surface is the same when the yield surface changes because of thermal loading as when the yield surface changes because of isothermal loading.

We observed that there exists a temperature at which the yield curve degenerates to a straight line (in the $\sigma, \sqrt{3}\tau$, T space). This phenomenon is true whether the yield surface has been generated by a thermal loading cycle or by isothermal prestressing. By means of the concept of the equilibrium stress—strain lines we finally explained the phenomenon of plastic strains directed in the negative direction.

Acknowledgments

The support of this research by the National Science Foundation is gratefully acknowledged. The authors wish to thank Mr. Y. Macheret, graduate student at Yale University, for his help in the timely completion of this paper.

References

1 A. Phillips, C. Liu and J. Justusson, Acta Mechanica, 14 (1972) 119—146.
2 A. Phillips, in: A. Sawczuk (Ed.), Problems of Plasticity, Proc. Int. Symp. Foundations of Plasticity, Vol. II, Warsaw, August 30—September 2, 1972, Noordhoff, Leyden, 1974, pp. 193—234.
3 A. Phillips and J. Tang, Int. J. Solids and Structures, 8 (1972) 463—474.
4 G.I. Taylor and H. Quinney, Phil. Trans. (ser. A), 230 (1931) 323—362.
5 W.M. Mair and H.L.D. Pugh, Journal of Mech. Engr. Science, 6 (1964) 150—163.
6 A. Phillips, The foundations of plasticity. In: Ch. Massonnet, W. Olszak, and A. Phillips, Plasticity in Structural Engineering, CISM Courses and Lectures No. 241, Springer Verlag, Vienna-New York, 1980, pp. 191—272.
7 A. Phillips and W.A. Kawahara, Acta Mechanica (in press), 1983.
8 A. Phillips and H. Moon, Acta Mechanica, 27 (1977) 91—102.
9 A. Phillips and C.W. Lee, Intern. J. Solids and Structures, 15 (1979) 715—729.
10 A. Phillips and R.L. Sierakowski, Acta Mechanica, 1 (1965) 29—35.
11 M.A. Eisenberg, C.W. Lee and A. Phillips, Int. J. Solids and Structures, 13 (1977) 1239—1255.
12 O.W. Dillon, Jr., in: E.H. Lee (Ed.), Proceedings 4th Intern. Congress on Rheology, Part 2, August 26—30, 1963, Interscience, 1965, pp. 377—394.
13 A. Phillips, in: J.L. Zeman and F. Ziegler (Eds.), Topics in Applied Continuum Mechanics, Proc. H. Parkus Symposium, Vienna, March 1—2, 1974, Springer Vienna-New York, 1974, pp. 1—21.
14 A. Phillips and M.A. Eisenberg, Intern. J. of Nonlinear Mechanics, 1 (1966) 247—256.
15 A. Phillips, in: E.H. Lee and R.L. Mallett (Eds.), Plasticity of Metals at Finite Strain: Theory, Computation and Experiment, Proc. of a Research Workshop at Stanford University, June 29—July 1, 1981, Div. of Appl. Mech., Stanford University, 1982, pp. 230—259.

Note on the Mindlin Problem

J. LYELL SANDERS, Jr.

Division of Applied Sciences, Harvard University, Cambridge, Massachusetts 02138 (U.S.A.)

Abstract

A variation on the method of images leads to a method of solution which involves only elementary integrations.

Introductory remarks

The Mindlin problem is to determine the displacements in the elastic half-space $z \geqslant 0$ due to the action of a concentrated force applied at $(0, 0, c)$ when the boundary surface $z = 0$ is free of tractions. The Boussinesq problem is the limiting case $c = 0$. In the Kelvin problem there is no free surface i.e. the force acts at a point in the full infinite elastic space. Modern treatments of these classic fundamental problems generally make use of the Papkovich—Neuber representation for the displacements in terms of four harmonic functions. The mathematics is made easier because so much is known about the properties of these functions.

Mindlin's original papers on the problem [1, 2] appeared before the Papkovich [3], Neuber [4] apparatus became widely known. His third paper [5] presents a simpler method of solution in terms of those developments. At bottom the problem is resolvable in closed form because the boundary conditions are rather simple statements about the boundary values of functions harmonic in a half-space. Mindlin constructs the required functions by means of a form of Green's formula. An even simpler construction is the subject of the present note. A solution by methods applicable in less special circumstances is given in the book [6] by Lur'e.

Solution

The equations needed here can be found in works such as [6—8]. The Navier equations in cartesian tensor notation are

$$G\left(u_{i,jj} + \frac{1}{1-2\nu}\, u_{j,ji}\right) + f_i = 0 \tag{1}$$

where u_i are the displacements, f_i are the body forces, G is the shear modulus, and ν is Poisson's ratio. The Papkovich—Neuber representation of the general solution is

$$2Gu_i = x_j \psi_{j,i} - (3-4\nu)\psi_i + \phi_{,i} \tag{2}$$

where

$$2(1-\nu)\psi_{i,jj} = f_i, \tag{3}$$

$$2(1-\nu)\phi_{,ii} = -x_i f_i. \tag{4}$$

For the Kelvin problem with a force F_i acting at ξ_i the body forces can be taken to be $f_i = F_i \delta(\boldsymbol{x} - \boldsymbol{\xi})$, in terms of Dirac delta functions. The solutions to (3) and (4) are obviously

$$\psi_i = -\frac{F_i}{8\pi(1-\nu)\rho}; \qquad \phi = \frac{\xi_i F_i}{8\pi(1-\nu)\rho} \tag{5}$$

where

$$\rho^2 = (x_i - \xi_i)(x_i - \xi_i). \tag{6}$$

For simplicity the special case of the Mindlin problem corresponding to $F_1 = F_2 = 0, F_3 = 1$, and $\xi_1 = \xi_2 = 0, \xi_3 = c$ will be treated. The other cases yield to the same method. In this case one can safely assume $\psi_1 = \psi_2 = 0$, $\psi_3 = \psi$. The displacements are given by the formulas (with $x_3 = z$)

$$2Gu_\alpha = z\psi_{,\alpha} + \phi_{,\alpha} \qquad (\alpha = 1, 2), \tag{7}$$

$$2Gu_3 = z\psi_{,z} - (3-4\nu)\psi + \phi_{,z} \tag{8}$$

and the stresses are given by

$$\sigma_{\alpha\beta} = (z\psi + \phi)_{,\alpha\beta} - 2\nu\delta_{\alpha\beta}\phi_{,z}, \tag{9}$$

$$\sigma_{3\alpha} = [z\psi_{,z} - (1-2\nu)\psi + \phi_{,z}]_{,\alpha}, \tag{10}$$

$$\sigma_{33} = z\psi_{,zz} - 2(1-\nu)\psi_{,z} + \phi_{,zz}. \tag{11}$$

Assume a solution to the Mindlin problem in the form

$$\psi = \frac{1}{8\pi(1-\nu)}\left(\bar{\psi} - \frac{1}{\rho}\right) \tag{12}$$

$$\phi = \frac{1}{8\pi(1-\nu)}\left(\bar{\phi} + \frac{c}{\rho}\right) \tag{13}$$

in which $\bar{\psi}$ and $\bar{\phi}$ are required to be harmonic and regular everywhere in the half-space $z \geqslant 0$. The terms with ρ account for the concentrated load according to the Kelvin solution. A solution is sought for which the displacements vanish at infinity. Accordingly assume that $\bar{\psi}$, $\bar{\phi}_{,\alpha}$ and $\bar{\phi}_{,z}$ vanish at infinity. These auxiliary hypotheses are justified if a solution can be found for which they are satisfied. From (10) and (11) the boundary conditions on $z = 0$ are

$$\phi_{,z} - (1-2\nu)\psi = 0, \tag{14}$$

$$\phi_{,zz} - 2(1-\nu)\psi_{,z} = 0. \tag{15}$$

The integrated form (14) has been obtained by use of the hypotheses. In terms of $\bar{\phi}$ and $\bar{\psi}$ these read (on $z = 0$)

$$\bar{\phi}_{,z} - (1-2\nu)\bar{\psi} + (1-2\nu)\frac{1}{\rho} + c\left(\frac{1}{\rho}\right)_{,z} = 0, \tag{16}$$

$$\bar{\phi}_{,zz} - 2(1-\nu)\bar{\psi}_{,z} + c\left(\frac{1}{\rho}\right)_{,zz} + 2(1-\nu)\left(\frac{1}{\rho}\right)_{,z} = 0. \tag{17}$$

A form of the method of images can be used to advantage at this stage. Define

$$\bar{\rho}^2 = x_\alpha x_\alpha + (z+c)^2 \tag{18}$$

and note that on $z = 0$

$$\bar{\rho} = \rho, \qquad \left(\frac{1}{\bar{\rho}}\right)_{,z} = -\left(\frac{1}{\rho}\right)_{,z}, \qquad \left(\frac{1}{\bar{\rho}}\right)_{,zz} = \left(\frac{1}{\rho}\right)_{,zz}. \tag{19}$$

Now define two harmonic functions χ_1 and χ_2 by

$$\chi_1 = \bar{\phi}_{,z} - (1-2\nu)\bar{\psi} + (1-2\nu)\frac{1}{\bar{\rho}} - c\left(\frac{1}{\bar{\rho}}\right)_{,z}, \tag{20}$$

$$\chi_2 = \bar{\phi}_{,zz} - 2(1-\nu)\bar{\psi}_{,z} - 2(1-\nu)\left(\frac{1}{\bar{\rho}}\right)_{,z} + c\left(\frac{1}{\bar{\rho}}\right)_{,zz}. \tag{21}$$

By hypothesis and by construction these functions are regular everywhere in $z \geqslant 0$ and vanish at infinity there. The boundary conditions (16) and (17) state that χ_1 and χ_2 vanish on $z = 0$. By a theorem of potential theory χ_1 and χ_2 must vanish everywhere. The equation $\chi_2 = 0$ can be integrated once to yield

$$\bar{\phi}_{,z} - 2(1-\nu)\bar{\psi} = 2(1-\nu)\frac{1}{\bar{\rho}} - c\left(\frac{1}{\bar{\rho}}\right)_{,z}. \tag{22}$$

The equations $\chi_1 = 0$ and (22) can be solved for $\bar{\psi}$ and $\bar{\phi}_{,z}$ to yield

$$\bar{\psi} = 2c\left(\frac{1}{\bar{\rho}}\right)_{,z} - (3-4\nu)\frac{1}{\bar{\rho}}, \tag{23}$$

$$\bar{\phi}_{,z} = (3-4\nu)c\left(\frac{1}{\bar{\rho}}\right)_{,z} - 4(1-\nu)(1-2\nu)\frac{1}{\bar{\rho}}. \tag{24}$$

By an elementary integration and use of the condition on the behavior of the displacements at infinity

$$\bar{\phi} = (3-4\nu)c\frac{1}{\bar{\rho}} - 4(1-\nu)(1-2\nu)\log(z+c+\bar{\rho}). \tag{25}$$

One might observe that terms corresponding to the "image force" are not naturally a part of the solution. The present procedure can be used in case of other boundary conditions such as $u_i = 0$ on $z = 0$. In the parallel load case $F_1 = 1, F_2 = F_3 = 0$ set $\psi_2 = 0$ and make use of

$$x_1\left(\frac{1}{\bar{\rho}}\right)_{,z} = c\left(\frac{1}{\bar{\rho}}\right)_{,1} \qquad \text{on } z = 0. \tag{26}$$

The solution for $\bar{\psi}_1$, $\bar{\psi}_3$, and $\bar{\phi}$ proceeds in an elementary fashion for the traction-free boundary conditions and for the fixed-displacement boundary conditions.

Acknowledgment

Support of the Office of Naval Research under contract N00014-81-K-0668 and the Division of Applied Sciences, Harvard University is gratefully acknowledged.

References

1 R.D. Mindlin, C.R. Acad. Sci. Paris, 201 (1935) 536—537.
2 R.D. Mindlin, Physics, 7 (1936) 195—202.
3 P.F. Papkovich, An expression for a general integral of the equations of the theory of elasticity in terms of harmonic functions. Izvest. Akad. Nauk 10 (1932)
4 H. Neuber, Ein neuer Ansatz zur Lösung räumlicher Probleme der Elastizitätstheorie. Z.A.M.M. 14 (1934).

5 R.D. Mindlin, Force at a point in the interior of a semi-infinite solid. Proc. of the First Midwestern Conference on Solid Mechanics, 1953, pp. 56—59.
6 A.I. Lur'e, Three-Dimensional Problems of the Theory of Elasticity (1955). English translation: Interscience Publishers, New York, 1964.
7 I.S. Sokolnikoff, Mathematical Theory of Elasticity. McGraw-Hill, 1956.
8 M.E. Gurtin, The linear theory of elasticity. Encyclopedia of Physics, Vol. VI a/2, Springer-Verlag, 1972, pp. 1—245.

Rigid-Plastic and Simplified Elastic-Plastic Solutions for Dynamically Side-Loaded Frames*

P.S. SYMONDS and J.L. RAPHANEL**

Division of Engineering, Brown University, Providence, RI 02912 (U.S.A.)

Abstract

The estimation of permanent and maximum (plastic plus elastic) deformations of a portal frame subjected to a pulse of pressure on one column is considered here. Simple techniques are developed which furnish estimates both of the local dishing deformation of the column and of the general side-sway deflection of the frame. Modifications of the simple mode approximation technique [1] are required for this purpose. The first part of the response is treated as wholly elastic. It provides momentum input to a rigid-plastic stage, which we take as starting with a local deformation pattern in the column which leads into the closing mode-form response. Fairly crude estimates of elastic deflection components are added to the rigid-plastic deflections to obtain maximum values. Some comparisons are made with deflection magnitudes computed† by a finite element numerical program, assuming perfectly plastic behavior. The method is applicable to a rate sensitive material, and comparisons for this case are planned for the near future. Emphasis in this paper is on the treatment of the rigid-plastic stage.

1. Introduction

We consider a rectangular portal frame subjected to a short pulse of pressure on one column, of magnitude such that plastic deformation occurs, Fig. 1. As in static "plastic methods" of analysis, the material

*This research was supported in part by grant DAAG29-78-G-0085 from the U.S. Army Research Office, and in part by grant CEE-7926147 from the U.S. National Science Foundation.

**Now at Institut National Polytechnique de Grenoble, France.

†The computations reported on here were made on the Brown University, Division of Engineering, VAX-11/780 computer, the acquisition of which was made possible by grants from the U.S. National Science Foundation (grant ENG7819378), the General Electric Foundation, and the Digital Equipment Corporation.

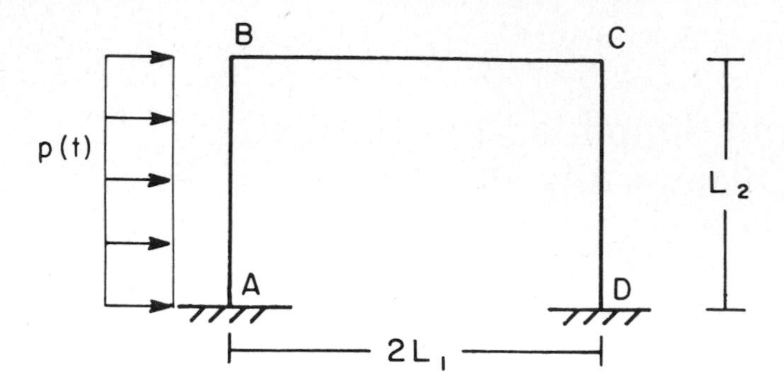

Fig. 1. Frame problem studied.

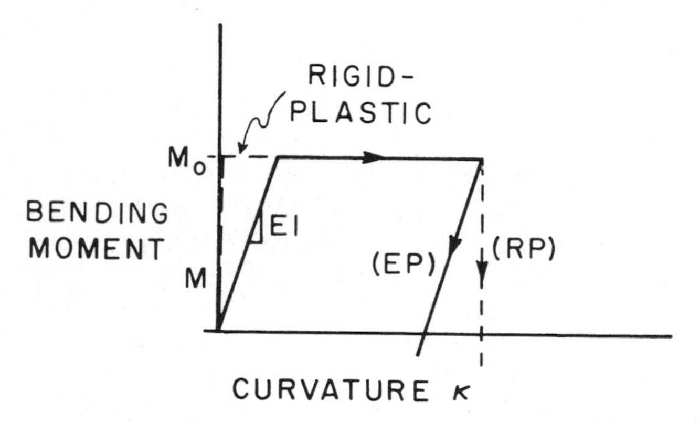

Fig. 2. Idealized moment—curvature relations adopted.

behavior will be taken to be either elastic-perfectly plastic or rigid-perfectly plastic, assuming the bilinear moment-curvature characteristic of Fig. 2. We are mainly interested in understanding the essential structure of the response, and in certain simple approximation techniques.

The side-loaded frame problem involves two kinds of deformation, namely a local "dishing" deformation in the loaded column and general frame "side-sway" deformation. Both are of interest; at the foot of the loaded column the deformations of the two processes are of the same sign, while at the top of this column the first type of deformation causes the right angle at the joint to close and the second process causes it to open. These features have not appeared in problems previously treated by the approximation techniques, and the present work shows modifications to make then applicable.

The approximation techniques of interest include the "mode approximation technique" [1—5] for structures assumed to exhibit rigid-perfectly plastic behavior that are loaded impulsively (with vanishingly short pulse duration). The entire response is taken to occur in a mode-form pattern, in which the shape of the velocity field over the structure remains constant. Solutions of this type, satisfying all the field equations but not necessarily the initial conditions, are possible in rigid-perfectly plastic structures when the loads are constant and geometry change effects are ignored. For the problems defined by these assumptions the complete response to any short pulse loading consists of two phases. In the first "transient" phase the shape of the velocity field is changing. The second and final phase is always in a mode-form pattern.

Thus, the mode approximation technique by-passes the part which is relatively difficult to analyze. A means of determining the initial magnitude of the mode velocity field from the initial velocities caused by impulsive loading was suggested in [1]. In this method the "best" initial velocity minimized the mean square difference Δ_0 between the two initial velocity fields, hence will be referred to as the Δ_0^{min} method. It is also equivalent to a form of momentum conservation [1]. In Section 2 of this paper it will be used in somewhat different ways than previously, since the mode method as previously used is unable to predict local deformations.

Section 3 points out defects of all rigid-plastic analyses, including the lack of a satisfactory general criterion to predict their validity in advance [9]. To avoid the (substantial) difficulties of complete elastic-plastic analysis, a simplified elastic-plastic method was proposed [6—9]. In this, the response is assumed to be either wholly elastic or wholly plastic (i.e. rigid-plastic). No criterion for validity of rigid-plastic response is needed. In the range of loading conditions (pulse magnitude and duration) where elastic effects are important, plastic deformations of "uncontained" type are predicted with fair accuracy. Rigid-plastic solutions are approached in appropriate conditions. The application of this simple approach to the side-loaded frame will be illustrated in Section 4.

2. Rigid-plastic treatments

We first sketch main features of the response assuming rigid-perfectly plastic behavior (Fig. 2, curve marked *RP*), and taking the load as impulsive, i.e. the uniform pressure on the column applied as a delta function pulse with impulse $\hat{I}$, where

$$\hat{I} = L_2 \int_0^t p \, \mathrm{d}t = L_2 i = L_2 \bar{\rho} V_0 \tag{1}$$

where V_0 is the initial velocity of elements of the column, i is the impulse per unit length, $\bar{\rho} = \rho bh$ is the mass per unit length, and L_2, b, H are length, width and thickness of the loaded column, respectively, and ρ is the mass density. To characterize the loading by the initial velocity as in eqn. (1) implies that structural reaction stresses remain finite in magnitude while

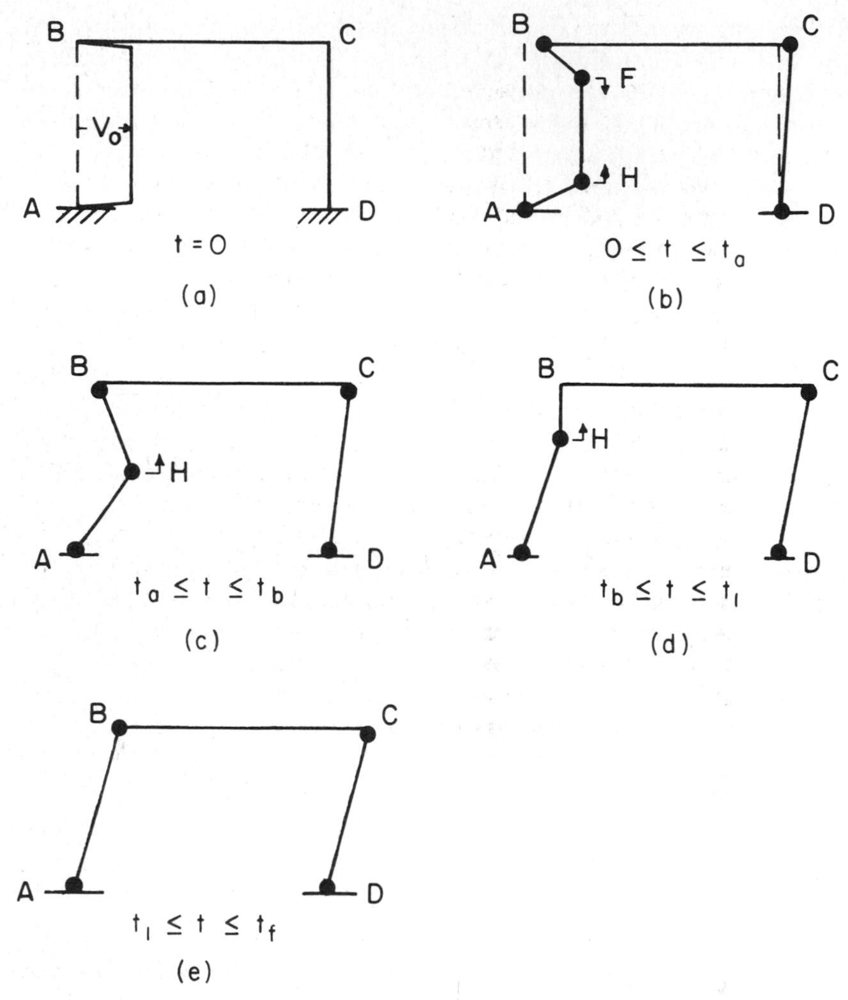

Fig. 3. Velocity patterns of complete rigid-plastic solution.

the pressure magnitude is increased indefinitely. It is assumed here that the load pressure is applied to elements of the column only, so that the transverse (beam) member *BC* has zero initial velocity. At time $t = 0$ there are discontinuities in velocity at the base *A* and at the beam *B*, but if shearing deformation is neglected these disappear for $t > 0$. The sequence of velocity patterns is sketched in Fig. 3. Plastic hinges are indicated by black circles. The interior hinges in column *AB* move in the directions shown. In Fig. 4 the final shape is sketched.

In this problem the transient phase lasts until time t_1 when the final 4-hinge side-sway mode is reached, Fig. 3(e). During the transient phase there are three distinct intervals in which travelling hinges occur, Fig. 3 (b–d). Analyzing the response in this phase requires deriving the equations

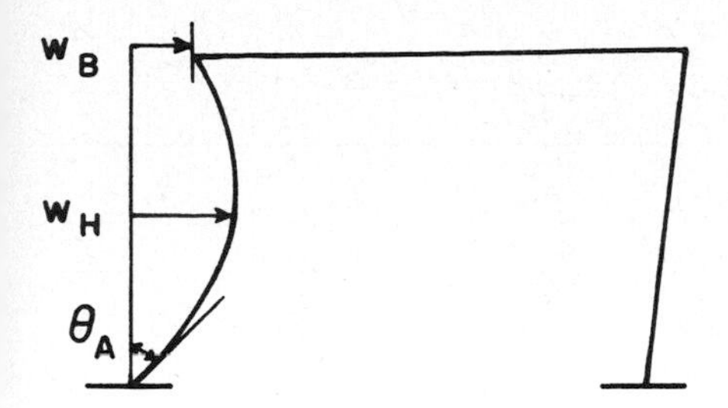

Fig. 4. Final deflection pattern of complete rigid-plastic solution.

of motion in each of these intervals in terms of the propagation speeds of the hinges and the acceleration magnitudes at discrete points. These nonlinear differential equations must be solved numerically to obtain velocities and displacements. The equations may be derived from free body diagrams or by Tamuzh's principle [10—12]. Discontinuities in acceleration occur at moving hinges, and discontinuous changes of the acceleration and stress field occur at the end of each interval when a plastic hinge disappears or changes sign (as at *B*, see Figs. 3(c, d)). Setting up and solving the successive systems of equations is straightforward but tedious. Velocity and deflection magnitudes are listed in Table 1.

TABLE 1
"Exact" rigid-plastic

	Start	t_a	t_b	t_1	t_f
$(M_0/L_2\,\hat{I})t$	0	0.0197	0.0535	0.0648	0.152
$\dot{w}_H/V_0$	1	1	0.171	0.080	0
$\dot{w}_B/V_0$	0	0.104	0.171	0.160	0
$(\bar{\rho}M_0/\hat{I}^2)w_H$	0	0.0197	0.0392	0.0404	0.0438
$(\bar{\rho}M_0/\hat{I}^2)w_B$	0	0.0015	0.0057	0.0076	0.0146

We show an alternative simpler solution which provides estimates of certain deformation quantities. The travelling hinges of the "exact" rigid-plastic solution sketched above are eliminated by adopting a model of the structure in which plastic hinges may occur only at discrete sections (nodal points) chosen in advance. The equations of motion are then linear in each interval, but of course change when the pattern of plastic hinges changes. The solution is simpler than the exact one, although if a large number of nodes is assumed the process of setting up and solving the equations in successive time intervals remains tedious. We adopt a simple model with nodes at the midpoints *H*, *K* of the two columns as well as at *A*, *B*, *C*, *D*.

The successive velocity patterns are shown in Fig. 5 and the final deflections indicated in Fig. 6. Table 2 gives velocity and deflection

TABLE 2
6-Hinge rigid-plastic model

	Start	t'_a	t'_b	t'_c	t_1	t_f
$(M_0/L_2\hat{I})t$	0	0.0007	0.0037	0.0535	0.0609	0.152
$\dot{w}_H/V_0$	1.483	1.465	1.392	0.172	0.084	0
$\dot{w}_B/V_0$	0.070	0.070	0.071	0.172	0.168	0
$(\bar{\rho}M_0/\hat{I}^2)w_H$	0	0.0011	0.0054	0.0443	0.0453	0.0491
$(\bar{\rho}M_0/\hat{I}^2)w_B$	0	0.0001	0.0003	0.0063	0.0076	0.0153

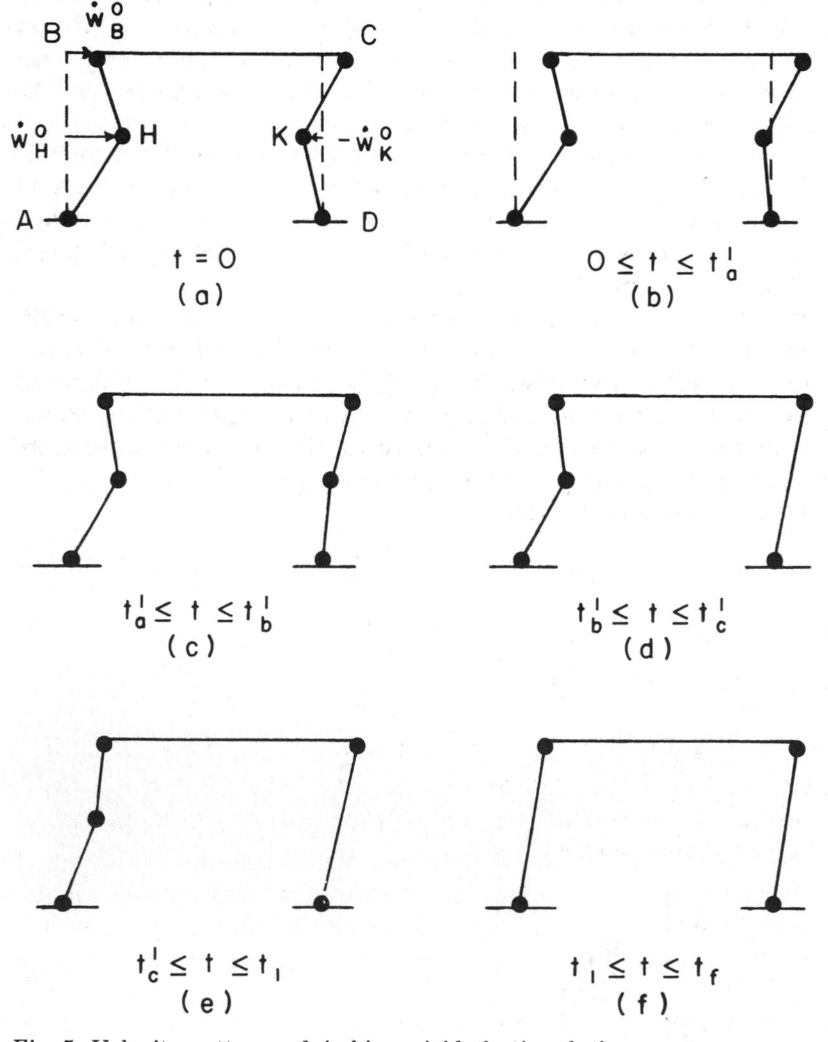

Fig. 5. Velocity patterns of six-hinge rigid-plastic solution.

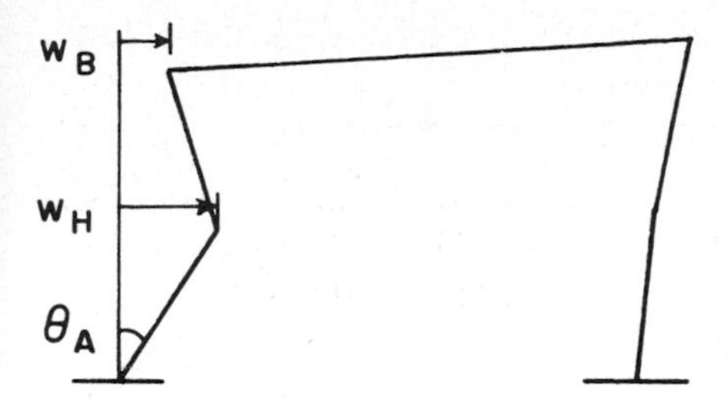

Fig. 6. Final deflection pattern of six-hinge rigid-plastic solution.

magnitudes. The choice of a model with node at K means that the transient phase has four intervals; the velocity at K is first negative and then positive, while in intervals (c) and (d) (see Fig. 5) the hinge at B is first closing and then opening. In each case the midpoint hinge disappears when the velocity at the midpoint becomes half the beam velocity. The model is acknowledged to be artificial; although a negative velocity at K occurs in an elastic-plastic model, no plastic hinge would occur at K in such a model. For the rigid-plastic analysis a 5 node model as in Fig. 5(d) would be simpler and equally realistic. However a model with 6 symmetric nodes is preferable for the elastic-plastic analysis discussed later, and is therefore shown here.

The determination of velocity and displacement magnitudes is straightforward from the constant stress and acceleration fields for each hinge pattern, once the initial velocities ($\dot{w}_H^0$, $\dot{w}_B^0$, $\dot{w}_K^0$) are found for pattern (a) of Fig. 5. It is worth emphasizing two ways of determining these. A dynamical method involves writing equations of motion for a pressure $p(t)$ applied to column AB, assuming hinge pattern (a).

In matrix form these equations are

$$\frac{1}{12}\bar{\rho}L_2\begin{bmatrix}4 & 1 & 0\\ 1 & 4+24\beta & 1\\ 0 & 1 & 4\end{bmatrix}\begin{bmatrix}\ddot{w}_H\\ \ddot{w}_B\\ \ddot{w}_K\end{bmatrix} = \frac{1}{4}pL_2\begin{bmatrix}2\\ 1\\ 0\end{bmatrix} + \frac{M_0}{L_2}\begin{bmatrix}-8\\ 0\\ 8\end{bmatrix} \tag{2}$$

where $\bar{\rho} = \rho bh$ and $\beta = \bar{\rho}_1 L_1/\bar{\rho}L_2$ for mass per unit length $\bar{\rho}_1$ of the beam members and $\bar{\rho}$ of the columns. In our sample $\bar{\rho}_1 = \bar{\rho}$ and $\beta = 0.75$. The mass matrix is recognized as "consistent" for linear interpolation functions. Solving, the accelerations are

$$\begin{bmatrix}\ddot{w}_H\\ \ddot{w}_B\\ \ddot{w}_K\end{bmatrix} = \frac{3}{172}\frac{p}{\bar{\rho}}\begin{bmatrix}85\\ 4\\ -1\end{bmatrix} + \frac{M_0}{\bar{\rho}L_2^2}\begin{bmatrix}-24\\ 0\\ 24\end{bmatrix} \tag{3}$$

Integrating with respect to time, and taking the pressure as a delta function $p = i\delta(t)$, for specific impulse i, we have

$$\int_0^t p\,dt = i = \bar{\rho} V_0 \tag{4}$$

where V_0 is the uniform initial velocity of column AB corresponding to uniform impulse i per unit length. The initial velocities are

$$\{\dot{w}_H^0, \dot{w}_B^0, \dot{w}_K^0\} = \frac{3}{172}\{85, 4, -1\}\, V_0 \tag{5}$$

The alternative method is that of minimizing the function Δ_0 of the difference (in the mean square sense) between the given initial velocity field (V_0 over column AB, zero elsewhere) and the velocity field of pattern Fig. 5(a) with initial magnitudes $\dot{w}_H^0$, $\dot{w}_B^0$, $\dot{w}_K^0$:

$$\Delta_0 = \frac{1}{2}\int_0^{L_2/2} \bar{\rho}(V_0 - \xi\dot{w}_H^0)^2\, ds + \frac{1}{2}\int_0^{L_2/2} \bar{\rho}[V_0 - (1-\xi)\dot{w}_H^0 - \xi\dot{w}_B^0]^2\, ds$$

$$+ \bar{\rho}L_1\dot{w}_B^{02} + \frac{1}{2}\int_0^{L_2/2} \bar{\rho}[\xi\dot{w}_B^0 + (1-\xi)\dot{w}_K^0]^2 ds + \frac{1}{2}\int_0^{L_2/2} \bar{\rho}(\xi\dot{w}_K^0)^2 ds \tag{6}$$

where $\xi = 2s/L_2$. Minimization with respect to $\dot{w}_H^0$, $\dot{w}_B^0$ and $\dot{w}_K^0$ by differentiating with respect to these quantities and solving the resulting equations furnishes the values of eqn. (5). This is an illustration of the equivalence of the Δ_0^{min} method to a scalar momentum conservation equation, as pointed out in [1]. It is re-emphasized here because at first sight the Δ_0^{min} device appears concerned only with kinematics rather than dynamics; and although the momentum interpretation was stated in [1], emphasis was put on minimization of Δ_0 as a measure of "error" of an approximating velocity field.

The above are full rigid-plastic solutions in the sense that the initial conditions are satisfied as closely as permitted by the model. The initial velocities of the discrete hinge model Fig. 5(a) differ greatly from those of the continuous structure, but would tend toward the latter as the number of elements in column AB is increased. There are doubts and difficulties about rigid-plastic analyses, but we postpone discussion of these and consider next simple approximation techniques for rigid-plastic response. First we apply the "mode approximation technique" as proposed in [1]. In this method the velocity field is taken in the four-hinge side-sway pattern of Figs. 3(e) and 5(f); this is the stable mode-form solution for the impulsively loaded frame. We write this solution as

$$\dot{w} = \dot{w}_B(t)\,\phi(s) \tag{7}$$

where $\phi(s)$ expresses the shape and $\dot{w}_B(t)$ is the velocity of the transverse member *BC*. If s is measured from the base of each column,

$$\text{in } AB\text{: } \phi = s/L_2\,;\ \text{in } BC\text{: } \phi = 1;\ \text{in } CD\text{: } \phi = s/L_2\,. \tag{8}$$

The equation of motion for pressure $p(t)$ is then

$$\bar{\rho}L_2^2(4+12\beta)\ddot{w}_B = 3pL_2^2 - 24M_0\,. \tag{9}$$

Thus for impulsive loading $p = \bar{\rho}V_0\delta(t)$, for $t > 0$

$$\ddot{w}_B = -\frac{6}{1+3\beta}\frac{M_0}{\bar{\rho}L_2^2} = -\frac{24}{13}\frac{M_0}{\bar{\rho}L_2^2} \tag{10a}$$

(for $\beta = L_1/L_2 = 0.75$). The initial velocity $\dot{w}_B^0$ is

$$\dot{w}_B^0 = 3V_0/(4+12\beta) = 3V_0/13. \tag{10b}$$

Again we note that the last result is obtainable by the Δ_0^{min} device, minimizing with respect to $\dot{w}_B^0$ the function

$$\Delta_0 = \frac{1}{2}\int \bar{\rho}\,[\dot{w}^0(s) - \dot{w}_B^0\,\phi(s)]^2\,ds \tag{11}$$

taking $\phi(s)$ given by eqn. (8) and $\dot{w}^0(s)$ as shown in Fig. 3(a). The final deflection at beam level (for impulsive load) is

$$w_B^f = -\dot{w}_B^{02}/2\ddot{w}_B = 0.0144\,\bar{\rho}L_2^2\,V_0^2/M_0 \tag{12a}$$

The structure comes to rest in time t_f given by

$$t_f = -\dot{w}_B^0/\ddot{w}_B = 0.125\,\bar{\rho}L_2^2\,V_0/M_0 \tag{12b}$$

Comparing with Table 1, these trivially simple calculations are seen to provide useful estimates of the rigid-plastic deflection at beam level and of the duration time. However, the "local" deformation in the column is not treated; no reasonable estimate is provided for the large displacement at midpoint of the loaded column, from which rotations at joints *A* and *B* can be found.

Such estimates may be obtained with least additional work by taking the velocity field first to be in the pattern of Fig. 7(b), i.e. the third pattern Fig. 5(d) of the six-hinge model. For this shape of field the equations of dynamics for uniform pressure $p(t)$ on column *AB* and $\beta = L_1/L_2 = 0.75$ are

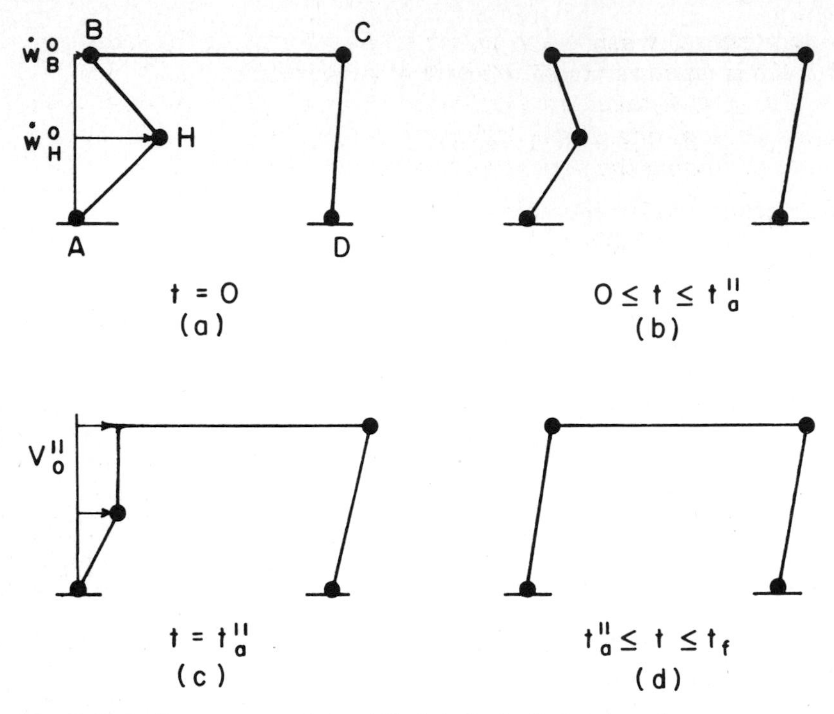

Fig. 7. Velocity patterns of simplified rigid-plastic treatment.

$$\frac{\bar{\rho}L_2}{12}\begin{bmatrix}4 & 1\\ 1 & 24\end{bmatrix}\begin{bmatrix}\ddot{w}_H\\ \ddot{w}_B\end{bmatrix}=\frac{pL_2}{4}\begin{bmatrix}2\\ 1\end{bmatrix}+\frac{M_0}{L_2}\begin{bmatrix}-8\\ 2\end{bmatrix} \tag{13a}$$

Their solution is

$$\begin{bmatrix}\ddot{w}_H\\ \ddot{w}_B\end{bmatrix}=\frac{3}{95}\begin{bmatrix}47\\ 2\end{bmatrix}\frac{p}{\bar{\rho}}+\frac{24}{95}\begin{bmatrix}-97\\ 8\end{bmatrix}\frac{M_0}{\bar{\rho}L_2^2} \tag{13b}$$

For impulsive loading, $p = \bar{\rho}V_0\,\delta(t)$, these integrate to

$$\begin{bmatrix}\dot{w}_H\\ \dot{w}_B\end{bmatrix}=\frac{3}{95}\begin{bmatrix}47\\ 2\end{bmatrix}V_0+\frac{24}{95}\begin{bmatrix}-97\\ 8\end{bmatrix}\frac{M_0 t}{\bar{\rho}L_2^2} \tag{13c}$$

The initial velocities $\{\dot{w}_H^0\,\dot{w}_B^0\}$ are given by the first vector on the righthand side of (13c). (They are also obtained by the Δ_0^{min} device, with simple modifications of Δ_0 as given by eqn. (6)). Since eqns. (13) apply to the velocity pattern Fig. 5(d) with closing plastic hinge at B they cease to be valid after time t_a'' when $\dot{w}_H(t_a'') = \dot{w}_B(t_a'') = V_0''$. For an approximate treatment we disregard as insignificant the succeeding interval in which $\frac{1}{2}\dot{w}_B < \dot{w}_H < \dot{w}_B$, and take the velocity field at t_a'' (Fig. 7(c)) as input to the mode field of Fig. 7(d). The Δ_0^{min} device again provides a convenient means of finding the appropriate mode velocity; it gives

$$\dot{w}_B^0 = \frac{19 + 48\beta}{16 + 48\beta} V_0'' = 1.058\ V_0'' \tag{14a}$$

By eqn. (13c) the time t_a'' at which the two velocities are equalized and their values are given by

$$M_0 t_a'' = 0.0536\ L_2 \hat{I};\ V_0'' = 0.1714\ V_0 \tag{14b}$$

TABLE 3A
Approximate rigid-plastic

	Start	t_a''	t_f
$(M_0/L_2\hat{I})t$	0	0.0536	0.1518
$\dot{w}_H/V_0$	1.484	0.1714	0
$\dot{w}_B/V_0$	0.0632	0.1714	0
$(\bar{\rho}M_0/\hat{I}^2)w_H$	0	0.0444	0.0488
$(\bar{\rho}M_0/\hat{I}^2)w_B$	0	0.0063	0.0152

TABLE 3B

Mode technique
$M_0 t_f/L_2\hat{I} = 0.125$
$\dot{w}_H = \dot{w}_B/2$
$\dot{w}_B^0 = 0.231\ V_0$
$\bar{\rho}M_0 w_H^f = 0.0072\ \hat{I}^2$
$\bar{\rho}M_0 w_B^f = 0.0144\ \hat{I}^2$

In Table 3A are collected the times, velocities, and final displacements (Fig. 8) obtained by applying eqns. (12) and (13) with the above results. Shown in Table 3B are values from the simplest "mode technique".

We have given results only for impulsive loading specified by impulse $\hat{I}$. This has the limitation that, for any chosen frame model or technique, the patterns of velocity and the final deflected shape are the same for all impulses; all deflection magnitudes are proportional to $\hat{I}^2$. Real loads of course are always finite, and the velocity pattern depends on the load level (Fig. 9) while final deflections w_H^f, w_B^f have different ratios depending

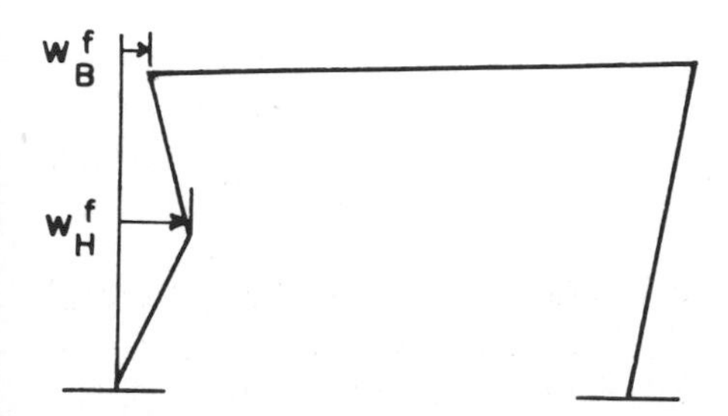

Fig. 8. Final deflection pattern of simplified rigid-plastic treatment.

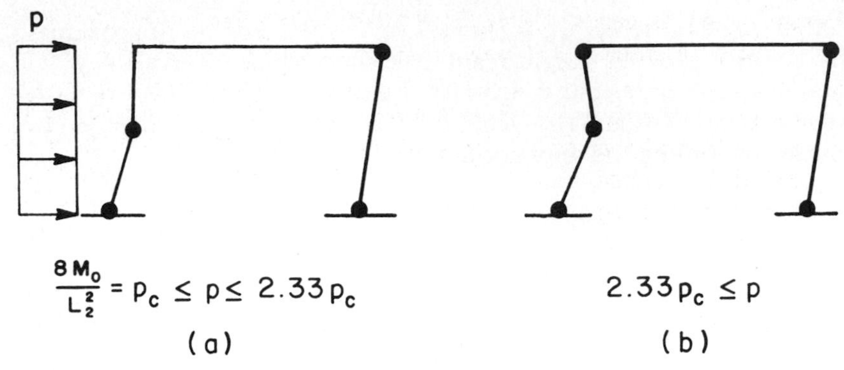

Fig. 9. Velocity patterns for step loading by pressure p in two ranges, for simplified rigid-plastic model.

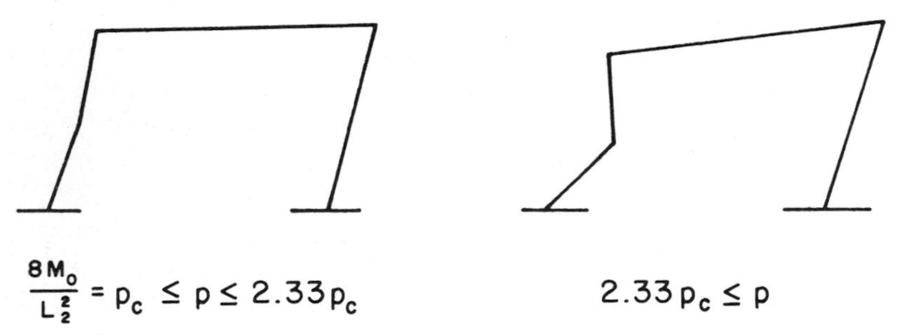

Fig. 10. Final deflection patterns for step loading by pressure p in two ranges, for simplified rigid-plastic model.

on both the maximum load and the pulse duration. Figure 10 illustrates this. If the load level is not much above the static plastic collapse magnitude, the velocity and displacement patterns will resemble the simple side-sway (four-hinge) mode. For a pressure pulse with any specified $p(t)$ the four- or five-hinge models of Fig. 9 and the simplified treatment sketched above would often be adequate as far as rigid-plastic analysis is concerned. The dynamical equations (13) would be used until the velocities are equalized as in Fig. 7(c). The remaining motion would be taken in the side-sway mode Fig. 7(d) with initial velocity given by eqn. (13c) in terms of V_0'', and deflections and response time by integrating eqn. (9). It is rare that load pulses are known accurately in detail. Youngdahl's [13] definition of "effective load" pulse involving only the total impulse and centroid time of the force pulse provides a convenient and realistic representation.

3. Defects of rigid-plastic treatments

The basic defect of rigid-plastic analyses has been mentioned: the lack of a general criterion for validity. For impulsive loads an energy ratio criterion [14] is easily applied, although the necessary minimum value of the ratio remains uncertain. It is of little use for general finite pulses [9, 15].

The response of the frame, whatever the type of pulse, must in an initial period be wholly elastic; finite stresses and strains at the yield condition are only reached in a finite time, in any real structure. Thus the assumption of "rigid-perfectly plastic behavior" is clearly false, as far as this period is concerned, and the predictions of an "exact" rigid-plastic analysis for it must be regarded as dubious. As already mentioned, no plastic hinge at K in the six-hinge model of Fig. 5 would actually occur. The various moving hinges in the "exact" solution of Fig. 3 are probably equally artificial.

The simple mode approximation technique for impulsive loading not only by-passes the early (unrealistic) response period, but by the Δ_0^{min} method transmits the specified initial momentum in a way that closely (or exactly) matches that of the transient period of the rigid-plastic solution. In its original form it provides no information on deformations in the transient stage, but the modification outlined above does include both the local deformations of the transient and general frame deflections of the final mode-form stage. The suggested further modifications allow finite load pulses to be treated similarly. The simplicity of these methods is a plus, but as rigid-plastic treatments their validity is uncertain, and methods are needed for the cases where rigid-plastic analysis is inappropriate or questionable.

Many computer programs are available (using finite differences or finite elements) for elastic-plastic dynamic analysis. Despite the availability of such programs their use in problems no more complex than the present one is likely to be expensive and time consuming. For example, a particular small scale frame of the type considered here requires computations for at least 50 milliseconds of response time in order to estimate maximum and permanent deflections. Using a finite element program with a conventional implicit integration scheme, it was found* after experimentation that a time step not greater than 2.5 microseconds was necessary for accuracy for an impulse causing moderate plastic deformations. Thus 20,000 time steps were needed even in this small problem.

We have done further studies with the program ABAQUS**. This is a general finite element program whose special advantage for our purpose is automatic control of the time step. Several complete response runs have been made of a number of pulse loaded structures including the present frame problem. These are treated as virtual experiments. They

---

*By a group in the Civil Engineering Department, University of Cape Town, with whom the writer worked in July and August 1982.

**ABAQUS was kindly made available for our research use by Hibbitt, Karlsson, and Sorensen, Inc., Providence, R.I.

allow much greater resolution of detail than do real experiments, but of course have their own limitations. Details will be published elsewhere. The elastic-plastic response to a suddenly applied load is initially complex. However it becomes "smoother" with time. In some cases the operation of a convergence theorem [2] is apparent, and it can be seen that the whole response is made up of a transient period, an interval in which the response is close to a rigid-plastic mode form, and finally a residual elastic vibration [16]. Here the "mode-form solution" is an exact solution of the elastic-plastic structure which consists of the mode-form solution of the rigid-plastic structure (with time-independent stress field) plus the elastic strains computed from that stress field.

4. Simplified elastic-plastic method

Here we briefly describe the application of the simple elastic-plastic approach mentioned in the Introduction to the side-loaded frame problem. A more complete description, considering also rate sensitive plastic behavior, will be published separately [17].

The response is assumed to occur in three stages, an initial wholly elastic stage being followed by a rigid-plastic stage which brings the structure to rest. This stage in turn is followed by one of elastic vibrations that are eventually damped out by the usual damping processes, leaving the permanent deflections of the structure. The simplification comes from the separation of elastic from plastic response, and from the relatively simple nature of the rigid-plastic stage. In the previous applications [6—8] this was taken in the simplest possible form, namely, that of a one-degree-of-freedom mode-form motion (the final stage of the complete rigid-plastic solution). In the present case of a side-loaded frame, it is necessary to consider the non-modal behavior prior to the final mode response in order to estimate the local dishing deformation of the loaded column. Thus the general shape of the initial velocity field of the rigid-plastic stage is that shown in Fig. 7a. This field is defined by velocities $\dot{w}_H^{p0}$, $\dot{w}_B^{p0}$ at H and B, and from any prescribed values of these the final plastic deformations w_H^f, w_B^f are readily calculated.

The rigid-plastic response described above is preceded by an elastic response which is determined by the specified impulsive loading (or, more generally, by the specified load pulse). The elastic response is terminated at a time t_e, and its velocity field at this instant furnishes the initial velocities $\dot{w}_H^{p0}$, $\dot{w}_B^{p0}$ of the rigid-plastic motion which follows. This is appropriately done by applying the Δ_0^{min} method to the two fields. We may write

$$\Delta_0 = \frac{1}{2} \int \bar{\rho}(\dot{w}^e - \dot{w}^{p0})^2 \, ds \qquad (15)$$

where $\dot{w}^{e}(s)$ is the terminal elastic velocity field and $\dot{w}^{p0}(s)$ the initial field of the rigid-plastic solution, s being the distance measured along the center-line of the frame, and the integration covering the frame. The field $\dot{w}^{p0}(s)$ is a function of $\dot{w}_H^{p0}$ and $\dot{w}_B^{p0}$, and the "best" values of the latter are those that minimize $\Delta_0(\dot{w}_H^{p0}, \dot{w}_B^{p0})$.

The condition for terminating the elastic stage must be stated. The initial velocity field of the rigid-plastic stage has plastic hinges at A, H, B, C, D (Fig. 7a, b), with moments $\pm M_0$ at these sections. We take the terminating condition to be

$$F[M_\alpha(t_e)] = [(\tfrac{1}{5})(M_A^2 + M_H^2 + M_B^2 + M_C^2 + M_D^2)]^{1/2} = M_0 \tag{16}$$

Here $M_A(t_e)$ etc. are moments computed at the five hinge sections from the elastic solution at time t_e, which is determined by eqn. (16). This condition is arbitrary and provisional. Other forms of $F(M_\alpha)$ have been tried; results appear insensitive to this. When the applied impulse $\hat{I}$ is large the time t_e is small, and a rigid-plastic solution is approached. Below a limiting value $\hat{I}_L$, the condition cannot be satisfied. For such impulses the plastic deformations according to the present approximation are taken to be zero. The limiting value $\hat{I}_L$ may be regarded as a threshold for the occurrence of "uncontained" plastic deformations. In general some plastic flow will occur for impulse magnitudes below $\hat{I}_L$, but only for $\hat{I} \geqslant \hat{I}_L$ will the type of rigid-plastic response considered here take place, involving both local deformation in column AB and general frame deformation resulting from the mode-form (side-sway) motion that terminates the rigid-plastic response stage.

The shape of the final velocity field of the elastic stage will be different for different impulse magnitudes (considering impulsive loading), and hence the final permanent deflection shape will depend on this magnitude. When finite duration and pulse magnitudes are considered, the final deflected shape would be expected to depend more strongly on these variables. In general, applying the estimation technique should begin with a study of the rigid-plastic response to a step loading with load magnitudes first at the static collapse value and then above this value by amounts appropriate to the particular problem. Mode-form dynamic solutions appropriate to various step load magnitudes are easily found [4], and from the estimated effective load value [13] of a particular problem, the shape can be estimated of the initial velocity field of the rigid-plastic stage that is appropriate to use in the present method.

The procedure outlined above for determining the permanent displacements is described in more detail in [17]. Results for final displacements w_B^f and w_H^f are shown in Figs. 11(a) and 11(b), respectively, plotted against $\sqrt{R_1} = \hat{I}/\hat{I}_0$, where $\hat{I}_0 = \frac{1}{2}\sqrt{3}\sqrt{\rho/E}\sigma_0 bHL_2$; note that $R_1 = 3.5R$, where $R = \frac{1}{2}\bar{\rho}L_2V_0^2/(7M_0^2L_2/4EI)$ = ratio of initial kinetic energy to capacity elastic strain energy.

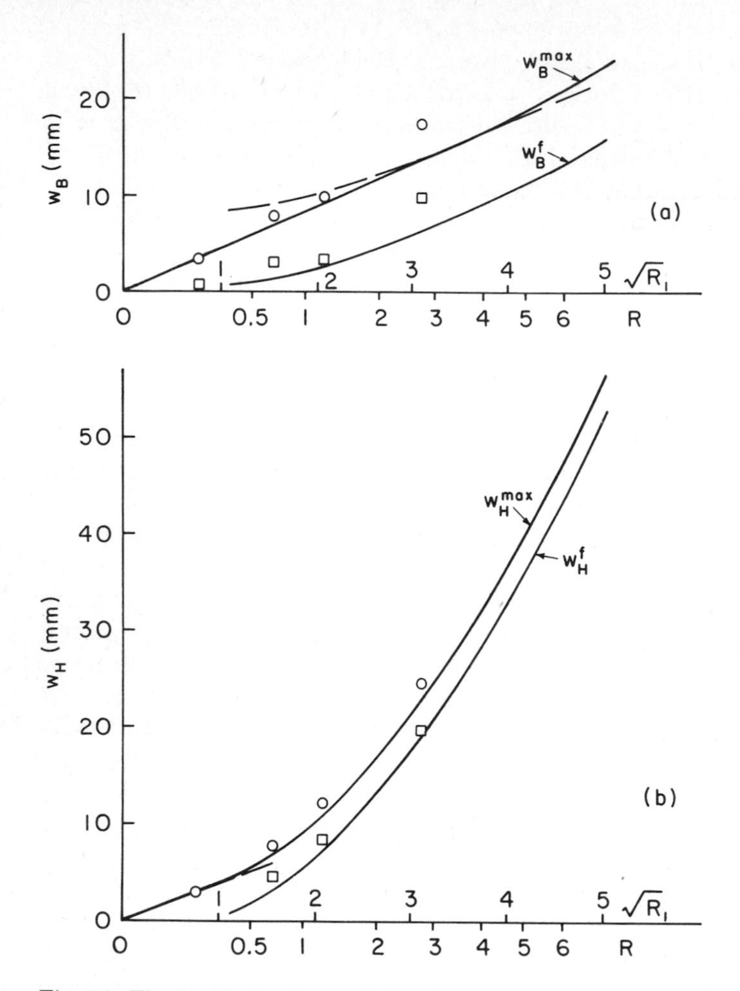

Fig. 11. Final and maximum deflections at corner B (a) and at midpoint H of loaded column (b). Predictions of present method are shown by curves. Values computed by ABAQUS finite element program are shown by squares for final deflections and circles for maximum displacements.

The maximum deflections at these points, including elastic as well as plastic components, can also be estimated. No attempt is made to find the total deflections that actually occur during the response, which is a complex and rapidly changing mixture of elastic and plastic effects. The simple procedure outlined below and illustrated in Figs. 11 and 12 is intended to provide a rough measure of the maximum deflection magnitudes.

We assume that maximum deflections will be attained in the latter part of the second stage of response, namely, in the mode-form phase of the rigid-plastic stage. In this motion the moment field $M^{rp}(s)$ is as sketched in Fig. 12. This defines a field of elastic curvatures $M^{rp}(s)/EI$.

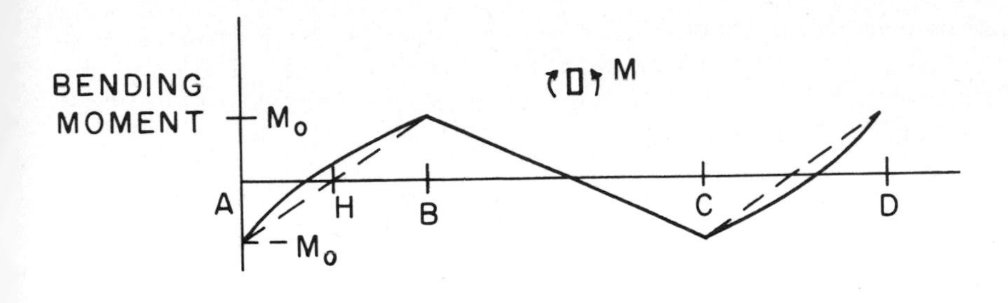

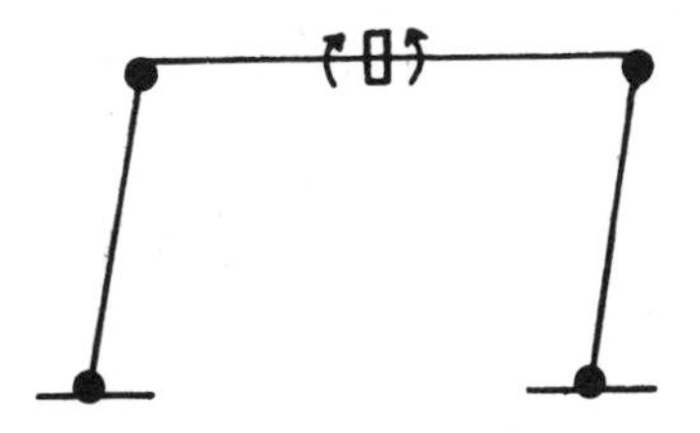

Fig. 12. Bending moment diagram for closing rigid-plastic pattern (four-hinge side-sway mode).

These can be regarded as strains due to relaxing the flexural rigidity from infinity (in the rigid-plastic model) to the actual value EI. The corresponding displacements are the elastic components to be added to those of the rigid-plastic motion. If we neglect the relatively small effects of the accelerations (assume linear moment variations), the elastic displacements at H and B are approximately

$$w_H^e = \frac{M_0 L_2^2}{12EI}, \qquad w_B^e = \frac{M_0 L_2^2}{6EI} \tag{17a, b}$$

The dashed curves in Figs. 11 (a) and (b) are obtained by adding these constant values to the curves marked w_H^f or w_B^f. These obviously cannot provide even rough estimates when the applied impulse is too small. In the elastic range, the maximum displacement is directly proportional to the initial velocity, and in Figs. 11 (a) and (b) would be shown by straight lines from the origin. When plastic deformation occurs, the slope of the maximum displacement versus $\sqrt{R_1}$ curve increases. A simple way of estimating the maximum displacement for small impulses is to draw a straight line from the origin which is tangential to the dashed curve. This would over-estimate purely elastic deflections but should provide fair estimates of maximum deflections in the range of mixed elastic-plastic response. For impulse magnitudes larger than that of the contact point C, the elastic increments given by eqn. (17) would be appropriate, and the dashed curve furnishes the maximum deflections.

Final displacements for the ABAQUS calculations are shown by square symbols, and maximum displacements by circles in Figs. 11 (a) and (b). The simple methods outlined do appear to provide fair estimates of both final and maximum magnitudes.

References

1 J.B. Martin and P.S. Symonds, Mode approximation for impulsively loaded rigid-plastic structures. J. Eng. Mech. Div., Proc. ASCE, 92 (EM5) (1966) 43—66.
2 J.B. Martin and L.S.S. Lee, Approximate solutions for impulsively loaded elastic-plastic beams. J. Appl. Mech., 35 (1968) 803—809.
3 P.S. Symonds and C.T. Chon, On dynamic plastic mode form solutions. J. Mech. Phys. Solids, 26 (1978) 21—35.
4 J.B. Martin, The determination of mode shapes for dynamically loaded rigid-plastic structures. To be published in Meccanica (1981/1982).
5 P.S. Symonds, The optimal mode in the mode approximation technique. Mech. Res. Comm., 7 (1) (1980) 1—6.
6 P.S. Symonds and J.L. Raphanel, Large deflections of impulsively loaded plane frames — extensions of the mode approximation technique. In: J. Harding (Ed.), Institute of Physics Conf. Ser. No. 47 London and Bristol, 1979, pp. 277—287.
7 P.S. Symonds, Elastic, finite deflection and strain rate effects in a mode approximation technique for plastic deformation of pulse loaded structures. J. Mech. Eng. Sci., 22 (4) (1980) 189—197.
8 P.S. Symonds, Finite elastic and plastic deformations of pulse loaded structures by an extended mode technique. Int. J. Mech. Sci., 22 (1980) 597—605.
9 P.S. Symonds, Elastic-plastic deflections due to pulse loading. In: Gary Hart (Ed.), Proc. Second Specialty Conference on Dynamic Response of Structures: Experimentation, Observation, Prediction, and Control New York, N.Y., ASCE, 1980 pp. 887—901.
10 J.L. Raphanel, Plastic Deformations of a Portal Frame Under Short Pulse Loading in One Column by an Approximate Method Including Effects of Elasticity and Plastic Rate Sensitivity. Ph.D. Thesis to Brown University, (Division of Engineering) June, 1982.
11 V.P. Tamuzh, On a minimum principle in the dynamics of rigid-plastic bodies. Prik. Mat. Mekh., 26 (1962) 715—722.
12 J.B. Martin, Extremum principles for a class of dynamic rigid-plastic problems. Int. J. Solids Structures,. 8 (1972) 1185—1204.
13 C.K. Youngdahl, Correlation parameters for eliminating the effect of pulse shape on dynamic plastic deformation. J. Appl. Mech., 37 (1970) 744—752.
14 E.H. Lee and P.S. Symonds, Large plastic deformation of beams under transverse impact. J. Appl. Mech., 19 (3) (1952) 308—314.
15 P.S. Symonds, Survey of Methods of Analysis for Plastic Deformation of Structures Under Dynamic Loading. Report BU/NSRDC/1-67, from Brown University to Office of Naval Research, Naval Ship Research and Development Center, Contract Nonr 3248(01)(X), June, 1967.
16 P.S. Symonds and W.T. Fleming, Jr., Parkes Revisited. Report in preparation.
17 J.L. Raphanel and P.S. Symonds, Estimation of Large Deformations of Portal Frames Subjected to Asymmetric Impulsive Loading. Report in preparation.

Fracture Mechanics for Delamination Problems in Composite Materials

S.S. WANG

Department of Theoretical and Applied Mechanics, University of Illinois at Urbana-Champaign, Urbana, IL 61801 (U.S.A.)

Abstract

A fracture mechanics approach to the well-known delamination problem in composite materials is presented. Based on theory of anisotropic laminate elasticity and interlaminar fracture mechanics concepts, the composite delamination problem is formulated and solved. The exact order of the delamination crack-tip stress singularity is determined. Asymptotic stress and displacement fields for an interlaminar crack are obtained. Fracture mechanics parameters such as mixed-mode stress intensity factors, K_{I}, K_{II}, K_{III}, and the energy release rate, G, for composite delamination problems are defined. To illustrate the fundamental nature of the delamination crack behavior, numerical solutions for delaminated graphite-epoxy composites under uniform axial extension are presented. Effects of fiber orientation, ply thickness, and delamination length on the interlaminar fracture are examined.

1. Introduction

Delamination, sometimes also called interlaminar cracking, is one of the most frequently encountered types of damage in advanced composite materials. The problem is not only of great theoretical interest but also of significant technical importance. The presence and growth of delamination cracks in composite laminates may lead to severe reliability and safety problems, such as reduction of structural stiffness, exposure of the interior to adverse environment, and disintegration of the material, which may cause the final fracture. Thus understanding the basic mechanics of delamination is of critical importance in characterization of flaw criticality and assessment of structural integrity of advanced composite materials and structures.

The delamination problem is generally very complex in nature and difficult to solve, because it involves not only geometric and material discontinuities, but also inherently coupled, mode I, II and III fracture in an anisotropic, layered material system. This situation is especially true for angle-ply composite laminates, in which delamination is inherently three-dimensional in nature. The composite delamination is basically an interlaminar fracture problem, involving debonding or separation between two highly anisotropic, fiber-reinforced laminae. The problem of an interfacial crack between two dissimilar isotropic materials has received much attention recently (see [1—12], for example). But advancements in the study of delamination between dissimilar anisotropic fiber composites have been very limited, to the author's knowledge.

In this paper, a study on the fundamental nature of the delamination problem based on recently developed interlaminar fracture mechanics theory for composite materials [13—15] is presented. Governing partial differential equations derived from theory of anisotropic elasticity and general solutions for the composite mechanics problem are briefly outlined. Stress singularities of a delamination crack in composite laminate fracture are determined. Fracture mechanics parameters such as mixed-mode stress intensity factors K_{I}, K_{II} and K_{III} as well as the energy release rate G are introduced. To illustrate the fundamental mechanics of delamination, numerical examples on the important and well-known edge-delaminated composite problem subjected to uniform axial extension are presented. Effects of geometric, lamination, and crack variables on the interlaminar fracture behavior are studied also.

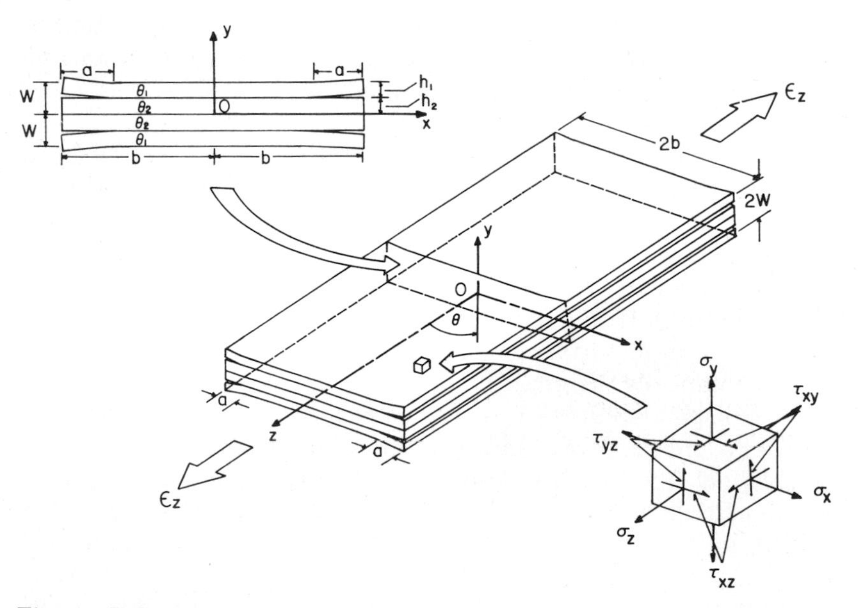

Fig. 1. Delamination cracks in a $[\theta_1/\theta_2/\theta_2/\theta_1]$ composite laminate under uniform axial extension $\epsilon_z = e$.

2. Assumptions

Consider a composite material (Fig. 1) composed of unidirectional fiber-reinforced plies of uniform thicknesses, $h_1, h_2, \ldots, h_n$. For simplicity and without loss of generality, we restrict our attention to the cases of symmetric composite laminates with $[\theta_1/\theta_2/\ldots/\theta_2/\theta_1]$ fiber orientations and ply thicknesses, $h_1 = h_n$, $h_2 = h_{n-1}, \ldots$ The composite has a width $2b$ and is subjected to a uniform axial extension, $\epsilon_z = e$, along the z axis. The composite laminate is sufficiently long that, in the region far away from the ends, end effects are negligible by virtue of Saint Venant's principle. Consequently, stresses in the composite are independent of z. Delamination occurs in the form of interlaminar cracks between dissimilar plies with fiber orientations θ_k and θ_{k+1}. Perfect bonding is assumed in the composite everywhere except in the region of delamination.

3. Basic equations and general solutions

A theory for the mechanics of delamination fracture in composite materials has been developed recently by the author, and some of the preliminary information has been reported in [13, 14]. Due to space limitations, only a brief outline of the theory is given here; detailed mathematical derivations, solution accuracy and convergence, and interlaminar fracture mechanics interpretations are given in [13—15]. It is only stated that formulation of the delamination problem is based on the theory of anisotropic laminate elasticity and that well-known Lekhnitskii's stress functions [16] are used to establish the governing partial differential equations. An eigenfunction expansion method is employed for determining the stress singularity at the delamination crack tip. The boundary collocation technique is then used to evaluate the complete solution for finite-dimensional composite laminates having delamination cracks under mechanical loading.

3.1 GOVERNING PARTIAL DIFFERENTIAL EQUATIONS

For each individual fiber-reinforced lamina with rectilinear anisotropy of a general form, the constitutive equations may be expressed by the generalized Hooke's law in contracted notation as

$$\epsilon_i = S_{ij}\sigma_j \qquad (i, j = 1, 2, \ldots, 6), \tag{1}$$

where ϵ_i and σ_j are strain and stress components, and S_{ij} is the compliance tensor, respectively. The engineering strains, ϵ_i, are defined in a Cartesian coordinate system by $\epsilon_1 = \epsilon_x$, $\epsilon_2 = \epsilon_y, \ldots, \epsilon_5 = 2\epsilon_{xz}$, $\epsilon_6 = 2\epsilon_{xy}$, and the stresses σ_i are defined in an analogous manner.

Introducing Lekhnitskii's stress potentials $F(x, y)$ and $\Psi(x, y)$ [16], and following the procedure given in [13, 16], one can obtain a pair of coupled, governing partial differential equations for each composite lamina as

$$\begin{cases} L_4 F + L_3 \Psi = 0, & \text{(2a)} \\ L_3 F + L_2 \Psi = 0, & \text{(2b)} \end{cases}$$

where L_2, L_3 and L_4 are linear differential operators of the second, third and fourth order, respectively, defined by

$$L_2 = \tilde{S}_{44}\,\partial^2/\partial x^2 - 2\tilde{S}_{45}\,\partial^2/\partial x \partial y + \tilde{S}_{55}\,\partial^2/\partial y^2, \tag{3a}$$

$$L_3 = -\tilde{S}_{24}\,\partial^3/\partial x^3 + (\tilde{S}_{25} + \tilde{S}_{46})\,\partial^3/\partial x^2 \partial y - (\tilde{S}_{14} + \tilde{S}_{56})\,\partial^3/\partial x \partial y^2 + \tilde{S}_{15}\,\partial^3/\partial y^3, \tag{3b}$$

$$L_4 = \tilde{S}_{22}\,\partial^4/\partial x^4 - 2\tilde{S}_{26}\,\partial^4/\partial x^3 \partial y + (2\tilde{S}_{12} + \tilde{S}_{66})\,\partial^4/\partial x^2 \partial y^2 - 2\tilde{S}_{16}\,\partial^4/\partial x \partial y^3 + \tilde{S}_{11}\,\partial^4/\partial y^4, \tag{3c}$$

in which $\tilde{S}_{ij}$ is the reduced compliance tensor defined as

$$\tilde{S}_{ij} = S_{ij} - S_{3i} S_{3j}/S_{33} \qquad (i, j = 1, 2, 4, 5, 6). \tag{3d}$$

3.2 GENERAL SOLUTIONS FOR STRESSES AND DISPLACEMENTS

The governing equations, eqns. 2a and 2b, are coupled, linear partial differential equations with constant coefficients related to elastic constants of each fiber-reinforced lamina. Wang et al. [13] have shown that, by introducing proper forms for the Lekhnitskii complex stress functions, the general solution for the problem can be expressed as

$$\sigma_x = \sum_{k=1}^{3} [C_k \mu_k^2 Z_k^\delta + C_{k+3} \bar{\mu}_k^2 \bar{Z}_k^\delta] + \sigma_{x0}, \tag{4a}$$

$$\sigma_y = \sum_{k=1}^{3} [C_k Z_k^\delta + C_{k+3} \bar{Z}_k^\delta] + \sigma_{y0}, \tag{4b}$$

$$\tau_{yz} = -\sum_{k=1}^{3} [C_k \eta_k Z_k^\delta + C_{k+3} \bar{\eta}_k \bar{Z}_k^\delta] + \tau_{yz0}, \tag{4c}$$

$$\tau_{xz} = \sum_{k=1}^{3} [C_k \mu_k \eta_k Z_k^\delta + C_{k+3} \bar{\mu}_k \bar{\eta}_k \bar{Z}_k^\delta] + \tau_{xz0}, \tag{4d}$$

$$\tau_{xy} = -\sum_{k=1}^{3} [C_k \mu_k Z_k^\delta + C_{k+3} \bar{\mu}_k \bar{Z}_k^\delta] + \tau_{xy0}, \tag{4e}$$

and

$$u = \sum_{k=1}^{3} [C_k p_k Z_k^{\delta+1} + C_{k+3}\bar{p}_k \bar{Z}_k^{\delta+1}]/(\delta+1) + u_0, \tag{5a}$$

$$v = \sum_{k=1}^{3} [C_k q_k Z_k^{\delta+1} + C_{k+3}\bar{q}_k \bar{Z}_k^{\delta+1}]/(\delta+1) + v_0, \tag{5b}$$

$$w = \sum_{k=1}^{3} [C_k t_k Z_k^{\delta+1} + C_{k+3}\bar{t}_k \bar{Z}_k^{\delta+1}]/(\delta+1) + w_0, \tag{5c}$$

where $Z_k = x + \mu_k y$, and μ_k and η_k are roots of the algebraic characteristic equation given in [13, 16]. The p_k, q_k and t_k in eqns. 5a—c are related to lamina elastic constants and can also be found in Refs. [13] and [16]; σ_{i0} and u_{j0} are known quantities determined from the remote loading conditions for each individual case studied. The expression for σ_z can then be obtained as

$$\sigma_z = (e - S_{3j}\sigma_j)/S_{33} \qquad (j = 1, 2, 4, 5, 6). \tag{5d}$$

4. Delamination crack-tip stress singularity

Consider a delamination between the kth and $(k+1)$th plies in a composite laminate. Assuming that interlaminar crack surfaces are free from traction, one can immediately introduce the following stress boundary conditions:

$$\sigma_y^{(i)} = \tau_{yz}^{(i)} = \tau_{xy}^{(i)} = 0 \qquad (i = k, k+1), \text{ along } \phi = \pm\pi. \tag{6a}$$

The continuity conditions for displacements and interlaminar stresses along the ply interface require that

$$\{\sigma_y^{(k)}, \tau_{yz}^{(k)}, \tau_{xy}^{(k)}\} = \{\sigma_y^{(k+1)}, \tau_{yz}^{(k+1)}, \tau_{xy}^{(k+1)}\}, \tag{6b}$$

$$\{u^{(k)}, v^{(k)}, w^{(k)}\} = \{u^{(k+1)}, v^{(k+1)}, w^{(k+1)}\}. \tag{6c}$$

Substituting eqns. 4a—e and 5a—c into the above constraint conditions leads to twelve algebraic equations in $C_m^{(k)}$ and $C_m^{(k+1)}$. The existence of nontrival solutions for $C_m^{(k)}$ and $C_m^{(k+1)}$ requires that the determinant of coefficients of the twelve algebraic equations vanish, i.e.,

$$\|\Delta(\delta)\|_{12\times 12} = 0. \tag{6d}$$

This gives a standard eigenvalue problem in which δ may be solved numerically from the transcendental equation, eqn. 6d, involving material constants, S_{ij}, μ_k, η_k of plies adjacent to the delamination. The general form of the solution for δ can be written as

$$\delta_n = n \qquad \text{or} \qquad \delta_n = (n - \tfrac{1}{2}) \pm i\gamma \quad (n = 0, 1, 2, \ldots), \tag{7}$$

where γ is a real constant related to anisotropic elastic constants of fiber-reinforced laminae adjacent to the interlaminar crack only.

Due to the requirement of positive definiteness of strain energy in the elastic composite, the δ_n bounded by

$$-1 < \mathrm{Re}[\delta_n] < 0 \tag{8}$$

characterize the order of delamination crack-tip stress singularity. For the delaminated $[\theta/-\theta/-\theta/\theta]$ graphite-epoxy composites with various fiber orientations, the eigenvalues δ_n, which satisfy the aforementioned constraint condition, are given in Table 1. The stress singularities for an interlaminar crack between two highly anisotropic laminae are observed to contain a pair of complex conjugates, $\delta_{1,2} = -0.5 \pm i\gamma$, and a constant, $\delta_3 = -0.5$. This situation is unique and different from that of an interface crack between two dissimilar isotropic or orthotropic media [1—6] in the sense that δ_1, δ_2 and δ_3 occur simultaneously in the present delamination problem. In the degenerated cases such as $\pm\theta = 0°$ and $90°$, the composite laminates become unidirectional. The delamination is located in an orthotropic material; thus, the classical inverse square-root singularity for crack-tip stresses is recovered fully.

TABLE 1

Dominant stress singularities for delamination in $[\theta/-\theta/-\theta/\theta]$ graphite—epoxy composites*

θ	$\delta_{1,2}$	δ_3
0°	−0.5	−0.5
15°	−0.5 ± 0.00642 i	−0.5
30°	−0.5 ± 0.02399 i	−0.5
45°	−0.5 ± 0.03434 i	−0.5
60°	−0.5 ± 0.02942 i	−0.5
75°	−0.5 ± 0.01579 i	−0.5
90°	−0.5	−0.5

*The composite ply properties given in [13, 14] are used in the computation here

From eqns. 4, 5 and 7, it is clearly seen that the asymptotic stress field at a delamination crack tip possesses the well-known oscillatory singularity and the associated displacement field also exhibits oscillatory characteristics, leading to the controversial crack-surface overlapping or interpenetration, which is physically inadmissible. Similar phenomena have also been observed in studying an interface crack between dissimilar isotropic materials [1—9]. As first pointed out by England [5] and Malyshev et al. [7] in examining the crack behavior in dissimilar isotropic solids, the crack surface overlapping is found to be confined to an extremely small region, and the interpenetration may not be of practical significance in the context of linear elastic fracture mechanics for an interface crack in a tensile field. Refined models to account for interface crack closure in dissimilar isotropic media have been reported in [10—12]. However, recent studies by Wang et al. [14, 15] have shown that, in composite laminates with certain combinations of laminar elastic properties, ply orientations, and loading conditions, the delamination crack surface closure may become extremely large. Thus, under these circumstances, the current model and approach need to be modified to take into account the interlaminar crack closure and crack surface contact stresses.

5. Stress intensity factors for delamination in composites

The eigenfunctions in eqns. 4 and 5 are determined by imposing appropriate material and geometric symmetry conditions and remote traction boundary conditions. Hence, complete stresses and displacements, σ_i and u_i, in the composite laminate can be fully established. Neglecting the higher-order terms, the asymptotic solution structure of the crack-tip stress field can be shown to have the following form:

$$\sigma_i = \sum_{j=1}^{3} \sum_{k=1}^{3} [f_{ijk} Z_k^{\delta_j} + g_{ijk} \bar{Z}_k^{\delta_j}] \qquad (i = 1, 2, 3, \ldots, 6), \tag{9a}$$

where Z and $\bar{Z}$ have their origin located at the delamination crack tip; f_{ijk} and g_{ijk} are related to material constants, laminate geometry, and boundary conditions, and δ_j are the eigenvalues bounded by eqn. 8. For the convenience of further developments, the asymptotic stress field around the crack tip can be written as

$$\sigma_i = \sum_{j=1}^{3} \sigma_{ij}(x, y; \delta_j) + \mathrm{O}(\text{non-singular, higher-order terms}), \tag{9b}$$

where σ_{ij} is the jth component of the singular stress σ_i corresponding to the eigenvalue δ_j which meets the constraint condition eqn. 8.

In view of the stress solution structure given by eqns. 9a and 9b, it is possible to define in the context of mechanics of fracture the delamination stress intensity fractors for the composite in a manner analogous to that given in Refs. [4—6] by considering the interlaminar stresses, σ_2, σ_4 and σ_6 (i.e., σ_y, τ_{yz}, and τ_{xy}), along the ply interface ahead of a delamination crack as

$$K_{\text{I}} = \lim_{x \to 0^+} \sum_{j=1}^{3} \sqrt{2\pi}\, x^{-\delta_j} \sigma_{2j}(x, 0; \delta_j), \tag{10a}$$

$$K_{\text{II}} = \lim_{x \to 0^+} \sum_{j=1}^{3} \sqrt{2\pi}\, x^{-\delta_j} \sigma_{6j}(x, 0; \delta_j), \tag{10b}$$

$$K_{\text{III}} = \lim_{x \to 0^+} \sum_{j=1}^{3} \sqrt{2\pi}\, x^{-\delta_j} \sigma_{4j}(x, 0; \delta_j), \tag{10c}$$

in which the origin of the Cartesian coordinates has been transferred to the crack tip.

It is noted that the K_{I}, K_{II} and K_{III} defined in eqns. 10a—c for a delamination crack between dissimilar anisotropic composites are different from those for a crack in a homogeneous solid. Thus, the stress intensity factors may not carry the conventional physical interpretation as in the cohesive (or homogeneous) fracture. While the overall stress intensity factor [3—7] may be used to depict the maximum amplitude of the crack-tip stress and to correlate crack extension, an accurate description of the asymptotic stress field at the composite delamination crack tip requires the detailed knowledge of individual stress intensity factors. Because of the complexities involved in the problem, numerical methods are generally required to obtain this information.

6. Strain energy release rate for delamination

While the stress intensity factors K_1, K_{II} and K_{III} describe the details of an interlaminar crack-tip field, the strain energy release rate G is also of significant interest, since this is a quantity physically measurable in experiments and mathematically well defined. The fracture energy release rate, G, in a delaminated composite may be evaluated by using Irwin's virtual crack extension concept [17] as

$$\begin{aligned} G &= G_{\text{I}} + G_{\text{II}} + G_{\text{III}} \\ &= \lim_{\delta\beta \to 0} \frac{1}{2\delta\beta} \int_0^{\delta\beta} \{\sigma_y(r, 0)[v^{(k)}(\delta\beta - r, \pi) - v^{(k+1)}(\delta\beta - r, -\pi)] \\ &\quad + \tau_{xy}(r, 0)[u^{(k)}(\delta\beta - r, \pi) - u^{(k+1)}(\delta\beta - r, -\pi)] \\ &\quad + \tau_{yz}(r, 0)[w^{(k)}(\delta\beta - r, \pi) - w^{(k+1)}(\delta\beta - r, -\pi)]\}\mathrm{d}r, \end{aligned} \tag{11}$$

where polar coordinates (r, ϕ) are used for the convenience of computation, and $\delta\beta$ is the length of virtual crack extension. The interlaminar stresses, σ_y, τ_{xy}, and τ_{yz}, in eqn. 11 can be obtained from the crack-tip stress field equations such as eqns. 4a—e. The corresponding displacements are also those of the asymptotic field equations obtained in the previous section, i.e., eqns. 5a--c.

It is noted here that, even though the near-field stresses possess an oscillatory singularity and $K_i(i = \text{I, II, III})$ may not have the usual significance attached as in the cohesive (homogeneous) crack case, the energy release rates G_i and G are well defined quantities in the problem, and should be of practical importance in studying the composite delamination. The G and its components G_{I}, G_{II} and G_{III} can be evaluated theoretically and experimentally to provide basic measures for the delamination problem.

7. Numerical examples and discussion

For the purpose of illustrating the fundamental nature of the delamination crack behavior in composite materials, graphite-epoxy laminates with symmetric $[\theta/-\theta/-\theta/\theta]$ fiber orientations containing delamination cracks along the θ and $-\theta$ ply interface are studied here. The composite laminate is subjected to a uniform axial extension and has a geometry

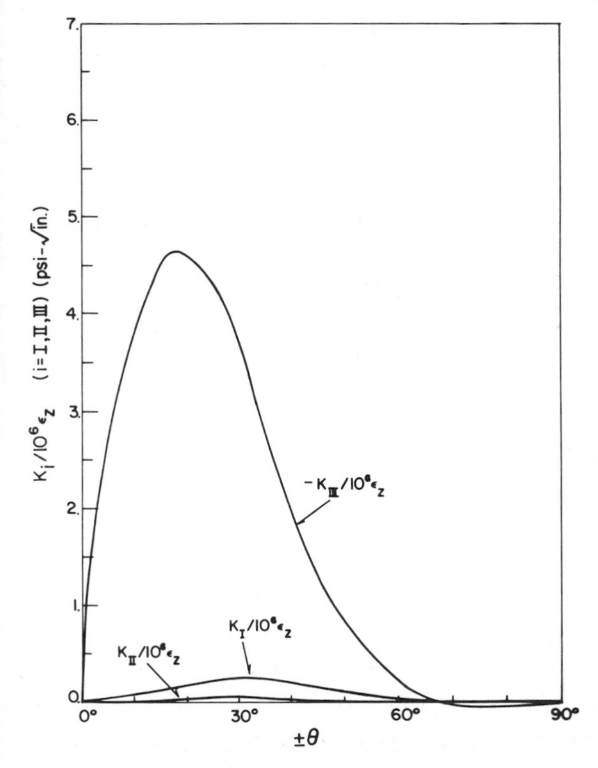

Fig. 2. Delamination stress intensity factors, K_i ($i = \text{I, II, III}$), in angle-ply $[\theta/-\theta/-\theta/\theta]$ graphite-epoxy composite under uniform axial extension ϵ_z ($h_1 = h_2 = 1$ in., $b = 8$ in., $a = 1$ in.).

shown in Fig. 1 with a width-to-thickness ratio $2b/2W$ and uniform ply thickness h_i. Delamination cracks of length a are assumed to emanate from the edges of the composite. Ply properties typical of high-modulus unidirectional graphite-epoxy composite given in [13, 14] are used in the numerical computation.

7.1 INFLUENCE OF FIBER ORIENTATION ON DELAMINATION CRACK BEHAVIOR

Consider the $[\theta/-\theta/-\theta/\theta]$ graphite-epoxy composites with various fiber orientations subjected to uniform axial strain, $\epsilon_z = e$. For illustration, we choose the composite with an aspect ratio $2b/(2h)$ equal to 8, ply thickness $h_1 = h_2 = 1$ in., and delamination length $a = 1$ in. The mixed-mode K_I, K_{II} and K_{III} are determined and given in Fig. 2 for various θ's. It is observed that even though the composite laminate is under the simplest loading condition, the delamination crack-tip response is very complicated indeed due to the complex interlaminar stress transfer, the nonhomogeneity of the composite, the anisotropy of ply properties, and the unusual delamination configuration with respect to the loading direction. The tearing-mode stress intensity factor K_{III} associated with the interlaminar shear τ_{yz} is about one or two orders of magnitude higher than K_I and K_{II} in general in the laminates studied. The opening-mode stress intensity K_I is also very appreciable due to the significant interlaminar normal stress σ_y. The simultaneous presence of K_I, K_{II} and K_{III} in a composite delamination problem is unique to fiber-reinforced laminates, not observable in fracture problems of bonded dissimilar isotropic media in general. The total energy release rates, G, for the $[\theta/-\theta/-\theta/\theta]$ graphite-epoxy composite laminates with the aforementioned delamination are shown in Fig. 3. The results reveal that the energy rate, i.e., the driving force for delamination fracture, is highest for a delaminated $[\pm\theta]_s$ composite with $\theta \approx 18^\circ$, and has rather small values for $\theta \geqslant 45^\circ$, indicating that the composite with $\theta \geqslant 45^\circ$ may be more prone to intralaminar cracking than delamination. Obviously, from the present results, the delamination behavior is observed to be inherently three-dimensional in nature. Thus for composites with more general lamination, crack geometry, and loading conditions, fully three-dimensional fracture analyses are essential for obtaining complete information.

7.2 EFFECT OF PLY THICKNESS ON DELAMINATION

The effect of laminate geometric variables on the delamination behavior is best illustrated by examining the change of K_I, K_{II} and K_{III} with ply thickness in a $[45^\circ/-45^\circ/-45^\circ/45^\circ]$ graphite-epoxy composite with $h_1 + h_2 = W = 2$ in. Given the delamination crack length ($a = 1$ in.), laminate width ($2b = 4$ in.), and the loading condition as before, the

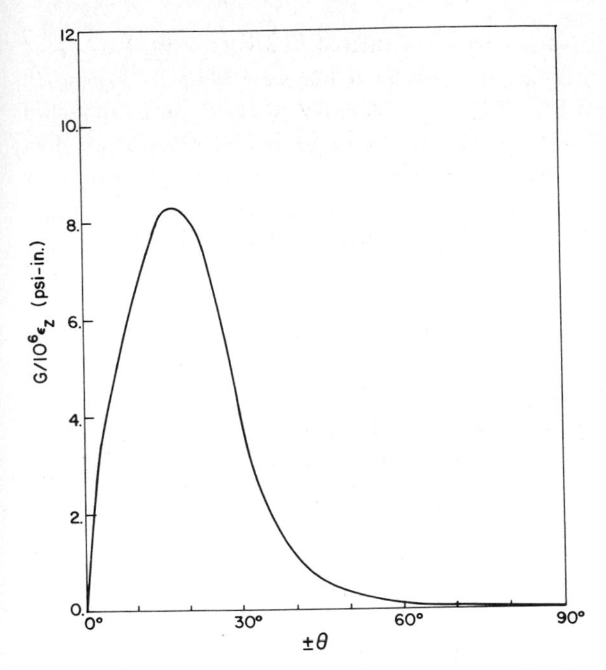

Fig. 3. Energy release rate G for edge delamination in angle-ply $[\theta/-\theta/-\theta/\theta]$ graphite-epoxy composite under uniform axial extension ϵ_z ($h_1 = h_2 = 1$ in., $b = 8$ in., $a = 1$ in.).

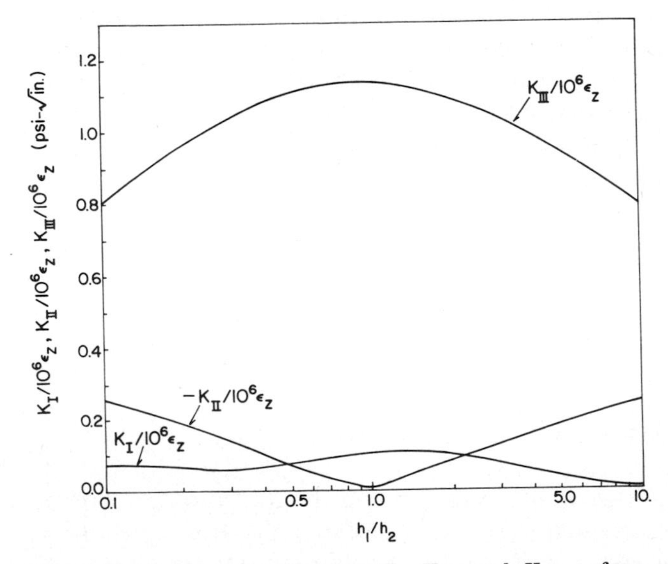

Fig. 4. Stress intensity factors, K_{I}, K_{II} and K_{III}, of an edge delamination crack in $[45^\circ/-45^\circ/-45^\circ/45^\circ]$ graphite-epoxy composite with various ply-thickness ratios h_1/h_2.

delamination stress intensity factors for various h_1/h_2 ratios are obtained in Fig. 4. The crack-tip tearing and opening stresses τ_{yz} and σ_y (or K_{III} and K_I) have maximum intensifications, as the ply thicknesses h_i become identical, i.e., $h_1/h_2 \approx 1$. The K_{II}, however, reaches a minimum due to the reduction in τ_{xy}. It should be noted here that the K_I, K_{II} and K_{III} depend on material constants of all plies as well as the overall laminate geometry. Therefore, values of K_i are related not only to local crack properties but also to global lamination and geometric variables as well as the loading condition.

The delamination strain energy release rate is also a function of the ply thickness. In the $[45°/-45°/-45°/45°]$ graphite-epoxy composite with a geometry given before, the change of G with h_1/h_2 is shown in Fig. 5, where maximum driving force occurs at $h_1 = h_2$, indicating the criticality of the relative ply thickness for delamination fracture in composites.

7.3 INTERLAMINAR CRACK LENGTH vs. DELAMINATION ENERGY RELEASE RATE

To illustrate the basic nature of delamination extension in laminated composites, strain energy release rates in the $[45°/-45°/-45°/45°]$ graphite-epoxy with various interlaminar crack lengths are presented. Effects of laminate width on delamination crack extension is also investigated. The change of total strain energy release rate G with delamination length a is given in Fig. 6 to illustrate the fundamental characteristics of delamination fracture. For the composite laminates with various $2b/2h$'s, the G is observed to change with delamination length in a unique manner. The energy release rate (or crack-extension driving force) reaches a maximum or a relative constant at a length a^* approximately equal to one or two ply thicknesses in the composites studied, depending upon the $(2b/2h)$ ratio. As the delamination exceeds this characteristic dimension a^*, the G decreases monotonically.

From a fracture mechanics perspective, several important features regarding the basic nature of composite delamination are revealed. Assuming that the material resistance to delamination growth remains constant (i.e., the failure criterion, G_c = constant, is used), one may find that a critical dimension of delamination length associated with the maximum G for each composite laminate may exist (e.g., $a^* \simeq 2h$ for the case $b/h = 8$); the word "critical" refers to the one that experiences a stable crack extension at the lowest load. It also indicates that any inherent interlaminar flaw a_0 in the composite, which is less than a^* will experience initially unstable growth as the load or G reaches the critical level, and is anticipated to be arrested at a later stage. Any initial delamination greater than a^* will experience a stable growth under a monotonically rising load. Thus there may exist an inherently built-in crack arrest mechanism for the delamination. The phenomena indicated in the G—a diagram have been noted by several researchers conducting experimental and analytical studies on the delamination fracture. The a^* may be an important quantity in the life prediction for delaminated composite materials and structures subjected to static and cyclic loading.

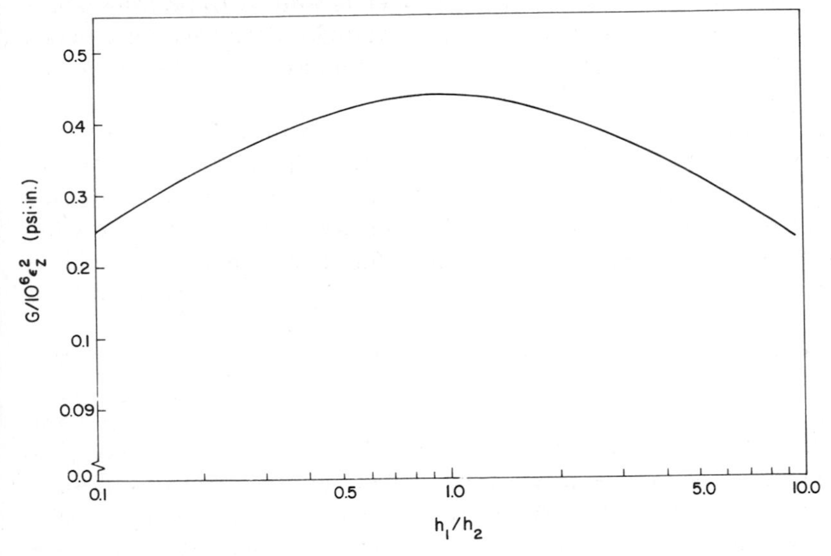

Fig. 5. Effects of ply thickness h_1/h_2 on strain energy release rate G fore edge delamination in [45°/ — 45°/ — 45°/45°] graphite-epoxy composite ($a = 1$ in., $2b = 4$ in.).

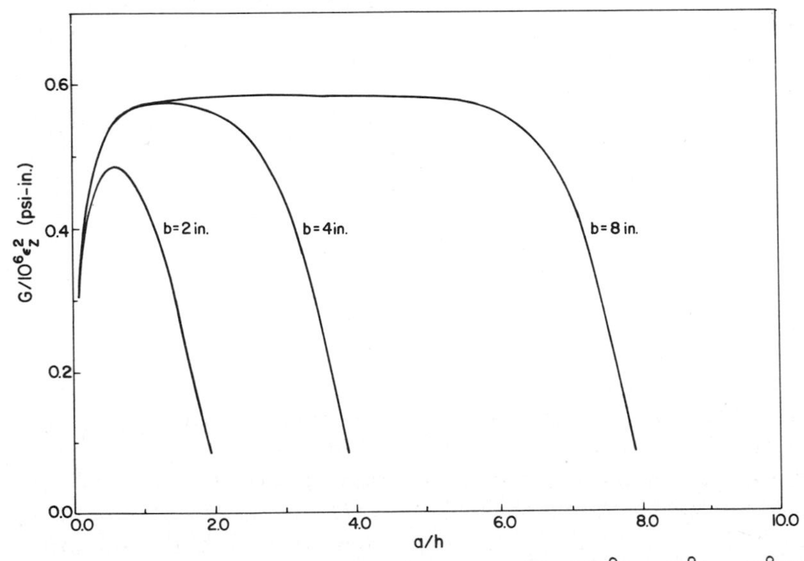

Fig. 6. Delamination strain energy release rate G in [45°/ — 45°/ — 45°/45°] graphite-epoxy composites with various delamination crack lengths a and laminate widths b.

8. Concluding remarks

Based on the information obtained in the paper, the following remarks may be suggested:

(1) Stress singularities for a delamination crack between highly anisotropic laminae are obtainable using an eigenfunction analysis. The order of delamination crack-tip stress singularity is different from that of an interface crack between dissimilar isotropic or orthotropic media by the simultaneous presence of three distinct characteristic eigenvalues, $-0.5 + i\gamma$, $-0.5 - i\gamma$, and -0.5.

(2) The fracture mechanics concept may be extended to delamination problems in anisotropic composite laminates by properly defining the interlaminar crack-tip stress intensity factors and strain energy release rates. For delaminated angle-ply composite laminates, K_{I}, K_{II} and K_{III} always occur simultaneously with K_{III} being one or two orders of magnitude higher than the other two.

(3) The crack extension driving force or strain energy release rate for delamination in composite laminates can be determined by using Irwin's virtual crack extension concept. Delamination stability in composite laminates may be assessed for any inherent interlaminar flaw relative to the critical delamination size a^* obtained in the G—a diagram.

(4) Fiber orientation, ply thickness, stacking sequence and related lamination variables are found to influence the delamination crack behavior (in terms of K_{I}, K_{II}, K_{III} and G) significantly.

9. Acknowledgments

The work described in this paper was supported in part by the Office of Naval Research and the National Aeronautics and Space Administration — Lewis Research Center (NASA-LRC), Cleveland, Ohio under Grant NSG 3044. The author is grateful to Dr. Y. Rajapakse, Dr. C.C. Chamis of NASA-LRC, Dr. G.P. Sendeckyj of the Air Force Flight Dynamics Laboratory, and Professor H.T. Corten of the University of Illinois at Urbana-Champaign for valuable discussion and encouragement during the course of this study.

References

1 M.L. Williams, The stresses around a fault or crack in dissimilar media. Bull. Seis. Soc. Am., 49 (1959) 199—204.

2 A.R. Zak and M.L. Williams, Crack point stress singularities at a bi-material interface. J. Appl. Mech., 30 (1963) 142—143.

3 G.C. Sih and J.R. Rice, The bending of plates of dissimilar materials with cracks. J. Appl. Mech., 31 (1964) 477—482.

4 F. Erdogan, Stress distributions in bonded dissimilar materials with cracks. J. Appl. Mech., 32 (1965) 403—410.

5 A.H. England, A crack between dissimilar media. J. Appl. Mech., 32 (1965) 400—402.
6 J.R. Rice, Plane problems of cracks in dissimilar media. J. Appl. Mech., 32 (1965) 418—423.
7 B.M. Malyshev and R.L. Salganik, The strength of adhesive joints using theory of fracture. Int. J. Fracture Mech., 1 (1965) 114—128.
8 F. Erdogan and G. Gupta, The stress analysis of multilayered composites with a flaw. Int. J. Solids Struct., 7 (1971) 39—61.
9 F. Erdogan and G. Gupta, Layered composite with an interfacial flaw. Int. J. Solids Struct., 7 (1971) 1089—1107.
10 M. Comninou, The interface cracks. J. Appl. Mech., 44 (1977) 631—636.
11 J.D. Achenbach, L.M. Keer and S.H. Chen, Loss of adhesion at the tip of an interface crack. J. Elasticity, 9 (1979) 397—424.
12 A.F. Mak, L.M. Keer, S.H. Chen and J.L. Lewis, A no-slip interface crack. J. Appl. Mech., 47 (1980) 347—350.
13 S.S. Wang, Edge delamination of fiber-reinforced composite laminates. Proc. of the 21st AIAA/ASME/ASCE/AHS Structures, Structural Dynamics and Materials Conference, April 6--8, 1981, Atlanta, GA, 1981, pp. 473—484.
14 S.S. Wang and I. Choi, Mechanics of delamination in composite materials. Int. J. Fract., in press, 1983.
15 S.S. Wang and I. Choi, The interface crack between dissimilar anisotropic, composite materials. J. Appl. Mech., 50 (1983) 169—178.
16 S.G. Lekhnitskii, Theory of Elasticity of an Anisotropic Body. Holden-Day, San Francisco, 1963.
17 G.R. Irwin, Analysis of stresses and strains near the end of a crack traversing a plate. J. Appl. Mech., 24 (1957) 361—364.